Once, Only Once, and in the Right Place

Residence Rules in the Decennial Census

Panel on Residence Rules in the Decennial Census

Daniel L. Cork and Paul R. Voss, *Editors*

Committee on National Statistics

Division of Behavioral and Social Sciences and Education

NATIONAL RESEARCH COUNCIL
OF THE NATIONAL ACADEMIES

THE NATIONAL ACADEMIES PRESS
Washington, D.C.
www.nap.edu

THE NATIONAL ACADEMIES PRESS 500 Fifth Street, NW Washington, DC 20001

NOTICE: The project that is the subject of this report was approved by the Governing Board of the National Research Council, whose members are drawn from the councils of the National Academy of Sciences, the National Academy of Engineering, and the Institute of Medicine. The members of the committee responsible for the report were chosen for their special competences and with regard for appropriate balance.

The project that is the subject of this report was supported by contract no. YA132304CN0005 between the National Academy of Sciences and the U.S. Census Bureau. Support of the work of the Committee on National Statistics is provided by a consortium of federal agencies through a grant from the National Science Foundation (Number SBR-0112521). Any opinions, findings, conclusions, or recommendations expressed in this publication are those of the author(s) and do not necessarily reflect the views of the organizations or agencies that provided support for the project.

International Standard Book Number 10: 0-309-10299-5
International Standard Book Number 13: 978-0-309-10299-5

Library of Congress Control Number 2006935988

Additional copies of this report are available from the National Academies Press, 500 Fifth Street, NW, Washington, D.C. 20001; (202) 334-3096; Internet, http://www.nap.edu.

Printed in the United States of America.

Suggested citation: National Research Council (2006). *Once, Only Once, and in the Right Place: Residence Rules in the Decennial Census.* Panel on Residence Rules in the Decennial Census. Daniel L. Cork and Paul R. Voss, eds. Committee on National Statistics, Division of Behavioral and Social Sciences and Education. Washington, DC: The National Academies Press.

THE NATIONAL ACADEMIES

Advisers to the Nation on Science, Engineering, and Medicine

The **National Academy of Sciences** is a private, nonprofit, self-perpetuating society of distinguished scholars engaged in scientific and engineering research, dedicated to the furtherance of science and technology and to their use for the general welfare. Upon the authority of the charter granted to it by the Congress in 1863, the Academy has a mandate that requires it to advise the federal government on scientific and technical matters. Dr. Ralph J. Cicerone is president of the National Academy of Sciences.

The **National Academy of Engineering** was established in 1964, under the charter of the National Academy of Sciences, as a parallel organization of outstanding engineers. It is autonomous in its administration and in the selection of its members, sharing with the National Academy of Sciences the responsibility for advising the federal government. The National Academy of Engineering also sponsors engineering programs aimed at meeting national needs, encourages education and research, and recognizes the superior achievements of engineers. Dr. Wm. A. Wulf is president of the National Academy of Engineering.

The **Institute of Medicine** was established in 1970 by the National Academy of Sciences to secure the services of eminent members of appropriate professions in the examination of policy matters pertaining to the health of the public. The Institute acts under the responsibility given to the National Academy of Sciences by its congressional charter to be an adviser to the federal government and, upon its own initiative, to identify issues of medical care, research, and education. Dr. Harvey V. Fineberg is president of the Institute of Medicine.

The **National Research Council** was organized by the National Academy of Sciences in 1916 to associate the broad community of science and technology with the Academy's purposes of furthering knowledge and advising the federal government. Functioning in accordance with general policies determined by the Academy, the Council has become the principal operating agency of both the National Academy of Sciences and the National Academy of Engineering in providing services to the government, the public, and the scientific and engineering communities. The Council is administered jointly by both Academies and the Institute of Medicine. Dr. Ralph J. Cicerone and Dr. Wm. A. Wulf are chair and vice chair, respectively, of the National Research Council.

www.national-academies.org

PANEL ON RESIDENCE RULES IN THE DECENNIAL CENSUS

PAUL R. VOSS *(Chair)*, Department of Rural Sociology, University of Wisconsin–Madison (emeritus)
JORGE CHAPA, Center on Democracy in a Multiracial Society, University of Illinois at Urbana-Champaign
DON A. DILLMAN, Social and Economic Sciences Research Center and Departments of Sociology and Community and Rural Sociology, Washington State University
KATHRYN EDIN, Department of Sociology, University of Pennsylvania
COLM A. O'MUIRCHEARTAIGH, National Opinion Research Center and Harris School of Public Policy, University of Chicago
JUDITH A. SELTZER, Department of Sociology, University of California, Los Angeles
C. MATTHEW SNIPP, Department of Sociology, Stanford University
ROGER TOURANGEAU, Joint Program in Survey Methodology, University of Maryland, and Survey Research Center, University of Michigan

DANIEL L. CORK, *Study Director*
MICHAEL L. COHEN, *Senior Program Officer*
AGNES E. GASKIN, *Senior Program Assistant*
BARBARA A. BAILAR, *Consultant*
MEYER ZITTER, *Consultant*

Acknowledgments

THE PANEL ON RESIDENCE RULES in the Decennial Census of the Committee on National Statistics (CNSTAT) is pleased to submit this final report and wishes to thank the many people who have contributed to our work over the panel's lifetime.

We thank the staff of the U.S. Census Bureau, under the leadership of director C. Louis Kincannon, deputy director Hermann Habermann, and associate director for decennial census Preston Jay Waite, for their accessibility and cooperation in providing information and materials to the panel and for several valuable interactions with the panel. Philip Gbur and Frank Vitrano acted superbly as lead liaisons between the Census Bureau and the panel, and Vitrano was a particular pleasure to work with as the lead technical contact between the panel and the Bureau. Ed Byerly, head of the Census Bureau's internal residence rules working group, merits recognition for guiding panel members and other participants through lengthy, comprehensive "walk-through" sessions at two of the panel's five public meetings. In plenary sessions and in smaller working group activities, the panel also benefited from its interaction with other talented members of the Census Bureau staff, including Robert Fay, Eleanor Gerber, Nancy Gordon, Deborah Griffin, Karen Humes, Elizabeth Krejsa, John Long, Sue Love, Elizabeth Martin, Louisa Miller, Laurel Schwede, Dave Sheppard, Annetta Clark Smith, and Maria Urrutia.

Our Panel on Residence Rules on the Decennial Census was one of three simultaneous CNSTAT panels studying different topics related to the upcoming 2010 census and the emergence of the American Community Survey as a data collection vehicle. As our work has progressed, we have found multiple points of overlap with the other two panels—the Panel on the Functionality and Usability of Data from the American Community Survey and the Panel on Coverage Measurement and Correlation Bias in the 2010 Census. We have benefited from our interaction with our colleagues on these panels, and we particularly thank their respective chairs, Graham Kalton and Robert Bell, for their cooperation with activities of our panel.

To assist in its work, the panel commissioned two papers for presentation at its meetings and to inform our deliberations. Terri Ann Lowenthal, an independent consultant and a former congressional staff member with expertise in the census, outlined the congressional and regulatory perspectives on census residence issues and reviewed legislative and judicial precedents. We thank her for her contribution, as well as for her ongoing work of informing the broader census stakeholder community of legislative developments in her series of "News Alerts" from the Census Project (http://www.censusproject.org). In the second paper, futurist Joseph F. Coates reviewed broad societal trends that may complicate the definition and interpretation of residence in the next 25 years. His paper provoked a stimulating discussion at the panel's December 2004 meeting, and we appreciate his work.

This report has been reviewed in draft form by individuals chosen for their diverse perspectives and technical expertise, in accordance with procedures approved by the Report Review Committee of the National Research Council (NRC). The purpose of this independent review is to provide candid and critical comments that will assist the institution in making the published report as sound as possible and to ensure that the report meets institutional standards for objectivity, evidence, and responsiveness to the study charge. The review comments and draft manuscript remain confidential to protect the integrity of the deliberative process.

We thank the following individuals for their participation in the review of this report: Margo Anderson, History and Urban Studies, University of Wisconsin–Milwaukee; Beth Osborne Daponte, Institution for Social and Policy Studies, Yale University; Vincent Fu, Department of Sociology, University of Utah; Kimberly Goyette, Department of Sociology, Temple University; Martha Jones, Division of Workers' Compensation Research Unit, Department of Industrial Relations, State of California; Steven Ruggles, Minnesota Population Center, University of Minnesota; Nora Cate Schaeffer, Department of Sociology, University of Wisconsin–Madison; and John H. Thompson, Office of the Executive Vice President, National Opinion Research Center, Chicago, Illinois.

Although the reviewers listed above provided many constructive comments and suggestions, they were not asked to endorse the conclusions or recommendations nor did they see the final draft of the report before its release. The review of the report was overseen by Kenneth Wachter, Department of Demography, University of California, Berkeley, and Stephen E. Fienberg, Department of Statistics, Carnegie Mellon University. Appointed by the NRC, they were responsible for making certain that an independent examination of the report was carried out in accordance with institutional procedures and that all review comments were carefully considered. Responsibility for the final content of this report rests entirely with the authoring panel and the institution.

In addition to the Census Bureau staff, we wish to thank the other expert speakers who contributed to our plenary meetings: Patricia Allard, Brennan Center for Justice, New York University School of Law; Robert Goldenkoff, U.S. Government Accountability Office; David McMillen, National Archives (formerly with the U.S. House Committee on Government Reform); Jim Moore, U.S. House Committee on Government Reform; and Peter Wagner, Prison Policy Initiative.

Our meeting drew attendants from several other federal agencies and interested groups. We can not list them all, but we do wish to thank those whose active contributions helped further the work of the panel: Allen Beck, Bureau of Justice Statistics; John Drabek, Office of the Assistant Secretary for Planning and Evaluation, U.S. Department of Health and Human Services; Robert Parker, U.S. Government Accountability Office (retired); D.E.B. Potter, National Center for Health Statistics; Susan Schechter, U.S. Office of Management and Budget; Ed Spar, Council of Professional Associations on Federal Statistics; and Katherine Wallman, U.S. Office of Management and Budget.

The panel appreciates the efforts of the reports office of the Division of Behavioral and Social Sciences and Education. Eugenia Grohman provided careful editing of the manuscript, Kirsten Sampson-Snyder patiently shepherded the report through scheduling and review processes, and Yvonne Wise managed the production of the finished volume.

Logistical support for the panel was provided with great skill and cheerfulness by Agnes Gaskin, senior program assistant. Research assistance was provided by Marisa Gerstein prior to her resumption of graduate studies in early 2005. The panel also benefited greatly from the long experience and wise counsel of CNSTAT consultants Barbara Bailar and Meyer Zitter. The panel is particularly indebted to the regular and active participation in its meetings of Constance Citro, director of CNSTAT. We simply could not have wished for a more experienced and talented group of committee staff as we worked our way through the history of residence rules in the census and particular problems regarding the concept of residence and its implementation in the 2000 census, and as we explored alternative ways to better ensure that future censuses will count each person living in the country once, and only once, and in the correct place.

I speak for the entire panel in expressing our profound gratitude to the panel's study director, Daniel Cork. His uncanny ability to somehow bring to the screen exactly the relevant paragraph from some obscure report or a needed statistic from some data set regularly delighted panel members and routinely kept us on task. He carefully guided the panel during the process of coming to consensus regarding a final set of recommendations, and he drafted the text of our report during a time when his efforts were also very much in heavy demand by another CNSTAT panel.

Finally, I thank my fellow panel members for their generous contributions of time and expert knowledge. We worked extraordinarily well together, somehow always maintaining a wonderful sense of spirited camaraderie despite occasional disagreements over matters of emphasis or substance.

Paul R. Voss, *Chair*
Panel on Residence Rules in the Decennial Census

Contents

List of Figures

List of Tables

List of Boxes

Executive Summary

THE FUNDAMENTAL GOAL of the U.S. decennial census is to count each person living in the country once, only once, and in the correct place. Since its inception, the census has followed a variant of a *de jure* standard for defining residence, seeking to count people at a single "usual residence." The census does contain some elements for which the alternative *de facto* standard—counting people at their current residence or where they are found at census time—is used, including operations to count the homeless. However, what was true for the first U.S. census in 1790 remains so for the 2010 and future censuses: residence can be extremely difficult to define and measure. Though most census respondents can readily identify a single usual residence, some people may have ties to two or more residences, and others may lack ties to any fixed residence. The basic concept of "residence" has evolved over time and can vary greatly across segments of the population, as can related concepts like "house," "home," and "family."

The U.S. Census Bureau used a set of 31 formal residence rules for the 2000 census. As in previous censuses, respondents generally saw only the limited extract of these rules that formed the instructions to the first question on the census questionnaire. Rather than subject respondents to the complete and complex list of rules, the Bureau tried to find a mix of instructions and cues to lead respondents to conceptualize residence in the same way as the Bureau.

As it began reworking the residence rules for the 2010 census, the Census Bureau requested that the National Academies' Committee on National Statistics (CNSTAT) convene this Panel on Residence Rules in the Decennial Census, charged to:

> examine census residence rule issues and make recommendations for research and testing to develop the most important residence rules for the

2010 census. Recommendations will address potential ways to modify census residence rules to facilitate more accurate counting of the population or identify the reasons why the rules should stay the same.

Though the concept inherits from a long tradition of practice dating to the 1790 census act, active census law and regulation do not define the residence standard for the decennial census (*de jure* or *de facto*), nor do they define what constitutes "usual residence." Accordingly, census rules and procedures for defining residence are prime candidates for periodic review, such as we provide in this report.

A PRINCIPLE-BASED APPROACH TO RESIDENCE

Census residence rules must satisfy several functions at once—among them, a reference for enumerators and interested respondents, a guide to the construction of census questions and instructions, and a template for the design of census operations. We find that, as developed and used in the 2000 census, the residence rules for the decennial census were too complicated and difficult to communicate. The set of 31 formal residence rules was not organized for ease in comprehension, and instead seemed to be a loose amalgamation of previously encountered problematic residence situations. The sheer number and redundancy of the rules detract from their effectiveness in training temporary census enumerators.

A basic flaw of the residence rules for the 2000 census is their lack of a conceptual base: they were essentially a set of exceptions to a concept of usual residence, and not an explication of one. Inferences as to the underlying logic of the residence rules required careful scrutiny. As a result, the residence rules for the 2010 and future censuses should be substantially rewritten (relative to those used in 2000), and the Census Bureau should make a concerted effort in 2010 to improve the communication of residence rules. Core concepts should be expressed as a small number of concise residence principles. These residence principles should then be used to develop other products, such as any instructions or cues to respondents on the census questionnaire, training materials for enumerators, census processing and editing routines, and a "frequently asked questions" list for enumerator and respondent reference and posting on the Internet.

As a candidate set, the panel recommends the following suggested statement of residence principles: The fundamental purpose of the census is to count all persons whose usual residence is in the United States and its territories on Census Day.

1. All persons living in the United States, including non-U.S. citizens, should be counted at their usual residence. Usual residence is the place where they live or sleep more than any other place.

2. Determination of usual residence should be made at the level of the individual person, and not by virtue of family relationship or type of residence.
3. If a person has strong ties to more than one residence, the Census Bureau should collect that information on the census form and subsequently attempt to resolve what constitutes the "usual residence."
4. If a usual place of residence cannot be determined, persons should be counted where they are on Census Day.

As indicated above, the core set of residence principles should be used to develop specific products, such as an operations guide, the instructions and questions on the census form, and enumerator training materials. An operations guide should be something akin to—but clearer and better organized than—the formal rules of the 2000 census, describing how the residence principles apply to living situations in which residence is complex or ambiguous to define. Such a mapping of the principles to specific living situations would be a valuable resource, not only for Census Bureau staff who provide questionnaire assistance to respondents, but also for interested respondents. This guide could usefully be structured as a "frequently asked questions" list.

A focus on residence principles suggests a specific implication for census operations. Under the normal census calendar, the mass mailout of census questionnaires begins in mid-March; this early mailout is essential, given the operational complexity of the census and statutorily fixed deadlines for the release of selected census results. However, the problem lies in encouraging rapid return of census questionnaires, possibly *before* the Census Day reference date. The 1990 census flatly instructed respondents to reply by April 1 while simultaneously asking them to report their household composition as of that date; in 2000, language on the questionnaire in two pre-Census Day mailings asked for prompt response. However small the effect may be, it is a logical contradiction to actively encourage people to use a reference date but to report their situation before that date arrives; the Census Bureau should refrain from overtly directing that respondents commit this basic violation of residence principles. To be consistent with the principle of the basic residence question on the census form—where did you live on April 1?—the Census Bureau should encourage prompt response but make clear that the form should be completed and returned on Census Day or as soon thereafter as possible.

CHANGING THE STRATEGY FOR COLLECTING RESIDENCE INFORMATION

Survey research, particularly work on cognitive response to self-administered survey instruments, shows clearly that their form significantly affects the responses. Responses to self-administered census forms depend upon the visual layout and design of questionnaires as well as the actual wording of questions and residence cues. Evidence suggests that people often ignore instructions on questionnaires. In addition, they may disregard instructions with which they disagree, even if they do read them. They may find the instructions confusing or contradictory with their own views, or may feel that they do not need direction. The Census Bureau should continue and strengthen its research on the combined effects of visual layout features and specific wording situations in the development and testing of questionnaires and their effectiveness. That said, the Bureau should vigorously pursue approaches to collecting residence data that do not depend critically on the wording of a limited set of instructions.

Questions, Not Instructions

The 2000 and other recent censuses put the onus of interpreting and applying residence concepts on respondents, using limited instructional text and lists of specific groups of people to include or exclude to try to lead respondents to follow the Census Bureau's unwritten concept of residence. The 2000 census form was more visually attractive and easier to follow than its instruction-heavy 1990 counterpart, but it still forced respondents to adhere to instructions that they may not have bothered to read or could not easily understand, absorb a complex concept on the basis of limited information, and accept interpretations of residence they may not share.

Instead, we recommend an approach based on asking guided questions—and multiple questions, as necessary—to elicit residence data. This approach would shift the burden of deciding what constitutes "usual residence" from respondents to the Census Bureau: the census form would allow collection of sufficient information from respondents to allow the Bureau to determine residence during processing and editing for situations that are not straightforward. In the 2010 census, the Census Bureau should conduct a major experiment to test a form that asks a sufficient number of residence questions to determine the residence situation of each person, rather than requiring respondents to follow complicated residence instructions in formulating their answers. The results of this test, and associated research, should guide decision on full implementation of the approach in 2020.

A question-based form might take a "worksheet" approach, spreading the burden of answering over a small number of questions by asking for counts of specific types of people; for example, people who are staying temporarily on Census Day with no other place where they usually reside. We believe that adoption of a question-based rather than instruction-based form could be particularly advantageous given the prospect of multiple response modes to the census—the standard mailed census forms, nonresponse follow-up interviews using handheld computers, and possibly responses by telephone or the Internet. A question-based approach may be more robust against differences in answers due to the mode by which the questionnaire is administered. Regardless of the final structure of residence questions chosen for 2010, research must be done on response effects created by mode of administration—mail, phone, Internet, and interview with handheld computers.

The Short Form Is Too Short: "Any Residence Elsewhere" and Other Questions

In 2000, respecting the demand to keep burden on respondents as low as possible as well as following direction from congressional oversight authority, the Bureau made sure that every question on the census short and long forms was matched to specific legal and regulatory uses. But paring the questionnaire too far can prevent the census from achieving its core mission of gathering accurate resident counts. Collection of enough information to determine "usual residence" requires the addition of some additional questions.

The principal addition that is needed is a question that asks whether each person has any other residence. Foreign censuses have found ways to collect auxiliary address information with an economy of space, and the Bureau's own valuable work in matching the complete set of 2000 census results against itself (using probability models based on name and date of birth) suggests that the computational power needed to process and retain auxiliary address information is at hand. Consequently, information on "any residence elsewhere" (ARE) should be collected from census respondents. This information should include the specific street address of the other residence location. A follow-up question should ask whether the respondent considers this ARE location to be their usual residence, the place where they live or sleep more than any other place.

Though we believe that ARE data collection is something that could be implemented for the 2010 census, it may be prudent to include it as a major experiment instead. A major test of census residence concepts, conducted in conjunction with the 2010 census, should be the basis for postcensal development leading to the 2020 census. This test should include both a question-based approach to collecting resident count information and a provision for ARE reporting by all census respondents, including those living in group quarters (nonhousehold) situations. The information should be gathered and processed, field verified on at least a sample basis, and reported on in census evaluations, in order to direct research over the next decade and fuller implementation in 2020.

Finally, no recent census has allowed respondents the ability to directly indicate that they believe that address information on their census questionnaire is inaccurate. Respondents have been unable to indicate, for example, that they have received the form at a seasonal home or that the Postal Service delivered the form to the wrong unit or apartment. The census questionnaire should allow respondents to correct the address printed on the form if it is wrong (e.g., address is listed incorrectly or questionnaire is delivered to wrong unit or apartment number). In addition, respondent-corrected address information should be one source of information to update the Master Address File.

Related Census Operations: Master Address File

The importance of the Master Address File (MAF)—the address list used by the Census Bureau to mail out questionnaires and assign nonresponse follow-up work—to the census process is difficult to overstate. An accurate MAF is crucial to the quality of the census, as well as the Bureau's other major survey programs. Together with the Topologically Integrated Geographic Encoding and Referencing (TIGER) system database, the MAF provides the key linkage between personal census questionnaire responses and specific geographic units. Inaccuracy in MAF and TIGER detracts from the quality of the decennial census, producing errors of inclusion and omission.

We endorse the recommendations of previous CNSTAT panels that the Census Bureau's efforts to continually update the MAF are vitally important and need careful planning. The Bureau must find ways to solicit and use input from local and tribal authorities in updating and correcting the MAF on a regular basis, with particular attention to obtaining information on unusual housing stock (such as multiple housing units inside family homes, leased hotel or motel quarters).

GROUP QUARTERS

Several of the most prominent situations where usual residence is not easily defined involve people living in group quarters or nonhousehold situations. These include college students, prisoners, and patients and residents in health care facilities. Consistent with the findings of predecessor CNSTAT panels, we find that, as implemented in the 2000 and recent censuses, group quarters enumeration is unacceptably bad. Failure to reconcile the group quarters roster with the MAF contributed to a host of census errors. Group quarters frames were constructed without sufficient standardization and awareness of diversity in housing unit and group quarters stock, and data from the 2000 census long-form sample were particularly marred by extremely high levels of item nonresponse. The latter failure—the quality of long-form-sample data—will be obviated in the 2010 census by the advent of the American Community Survey (ACS) and, consequently, the use of only a short form. However, the challenge of collecting even the basic census items from group quarters' populations remains.

It is particularly important that the Bureau's planned integration of the MAF with its roster of group quarters/nonhousehold locations—which were completely separate in 2000—be executed effectively. Further, it is essential that the Bureau establish programs for continuous update and refinement of group quarters address listings. In 2000, the segment of the Local Update of Census Addresses (LUCA) program dedicated to the review of "special places" (larger entities like colleges and military bases that contain group quarters) was particularly rushed and not given ample consideration. We strongly urge that participants in 2010 census local geographic partnership programs should be allowed to review address listings for group quarters in their jurisdictions, not just the household population listings. In addition, the Census Bureau should consider an improved special place LUCA program under which colleges and universities, medical facilities, and other group quarters locations may review the Bureau's address listings for their facilities.

New Approaches

The Census Bureau has begun serious work to redefine group quarters, to provide more meaningful categories, and to be more consistent with terms used by practitioners. Although some interim definitions were apparently finalized too late to be tested effectively in 2004, these revised definitions should be tried out in the 2006 census test and the 2008 dress rehearsal. We encourage the redefinition efforts but also suggest a broader focus: we conclude that there is sufficient diversity in what the Census Bureau has treated as the "group quarters" population that the term "group quarters" no longer makes conceptual sense. Its compartmentalization as a separate list and a separate operation—trying to force this entire segment of the population to respond to the census using a single form—is fundamentally flawed. Other dimensions may be more meaningful and easier to implement, such as a distinction between institutionalized and noninstitutionalized populations or a distinction based on length of stay (short-term versus long-term facilities).

As recommended above, ARE information should be collected for all group quarters/nonhousehold residents, just as we advocate its collection in the main household census form. In 2000, the Individual Census Reports administered to all group quarters residents asked whether the person had a "usual home elsewhere" (UHE) and asked for address information. However, the formal residence rules only made residents of a few group quarters types eligible to be counted at their UHE location (if that address was found to be valid). The Census Bureau's failure to analyze the rich UHE data collected from other group quarters is highly regrettable; ARE information should be collected from all group quarters residents in 2010, and those data should be analyzed and evaluated extensively.

We favor a system in which group quarters/nonhousehold residents are approached and enumerated in the same manner as the general household population. Direct enumeration (questionnaires distributed to and filled out by respondents, or administered by enumerator interview) is preferable to other means of data collection. However, the 2000 census experience underscored a key practical reality: even with its vigorous and generally successful partnership operations with community organizations, only about one-half of census records for the group quarters/nonhousehold population were obtained from direct enumeration. Instead, many were culled from facility and administrative records. Direct questionnaire distribution and enumerator access to all parts of the group quarters/nonhousehold universe cannot be assumed; many group quarters administrative staff will be either unwilling or unable to permit such direct access. In these cases, the Bureau needs to consider optimal use of enumerator time and facility resources. The Census Bureau should produce a small number of alternative census forms that collect a common core of information for different types of residence settings, such as those that are known to have long lengths of stay rather than short-term stays. The Census Bureau should also develop a spreadsheet-type ledger form that reflects the reality that some "responses" will have to be obtained from facility administrative records or a central "gatekeeper."

People in Prisons

A particular issue involving the group quarters/nonhousehold population that has drawn considerable attention in the buildup to the 2010 census is whether prisoners should be counted at the prison location or at some other place. A provision in the Census Bureau's 2006 appropriations required it to provide Congress with a report on the feasibility of counting prisoners at a "permanent home of record."

Major growth in the prison population, accompanied by expansion in the number of correctional facilities maintained by the federal government and the states, has prompted challenges to the Census Bureau's "usual residence" standard regarding the counting of the incarcerated population. The prison population includes disproportionate numbers of racial minorities and persons from large urban areas; that this population is counted in the largely rural areas where prisons tend to be located, and that they are included in redistricting calculations despite being barred from voting in most cases, raises legitimate concerns of equity and fairness in the census.

Under the panel's recommended principles for determining residence, federal and state prisoners would be counted at the prison location because that location is the place where the prisoner lives and sleeps more than any other place; this is consistent with current Census Bureau practice. However, two corollaries are in order:

- Our principles hold that determination of usual residence should be made at the level of the individual; this would mean that persons in prison need not have their residency fixed solely by virtue of their location in a structure identified as a prison. Using the panel's recommended question-based approach and revised nonhousehold enumeration operations, the census could obtain individual-level information on time spent in prison and expected date of release. The Bureau would then have the information needed to implement an individual-level assignment of residence under whatever set of rules is in place prior to the actual enumeration.

  In 2000, less than 20 percent of the population in correctional facilities was enumerated through self-response to a questionnaire or through enumerator interview. Even with enhanced efforts to facilitate direct interviews in prisons, it is only realistic to assume that a major share of the prison population in 2010 will have to be counted using administrative and prison records. Hence, any prospect for counting prisoners at locations other than the prison depends vitally on the completeness, consistency, and accessibility of records maintained by individual prisons or by state and federal departments of corrections. The quality of these data resources is not well known. Absent this knowledge, it is difficult even to identify the alternative to counting at the prison: Should prisoners be counted at their last preincarceration address (if one can be specified), or do records only permit prisoners to be allotted to the county or city (much less the tract or block) from which they were sentenced? The Census Bureau should participate in a comprehensive review of the consistency of content and availability of prison records. The accuracy of prisoner-reported prior addresses is uncertain, and should be assessed as a census experiment. Though an "enduring ties" argument is frequently invoked to argue for changes to the Bureau's prisoner counting policy, the strength of those ties merits empirical assessment. For example: Has the property (to which a prisoner is connected) changed ownership? Do respondents at the address have any contact with or relation to the prisoner?

- The evidence of political inequities in redistricting that can arise due to the counting of prisoners at the prison location is compelling. Short of counting prisoners at some location other than the prison—for which there is currently insufficient information as well as the lack of any principled way to do so—a partial remedy might be to provide tract- or block-level counts of prisoner populations as part of the Bureau's data products for redistricting. State redistricting bodies would then have the capacity to decide whether to include or exclude prisoners from proposed districts. The states' interest in having such a separate prisoner count should be assessed by the Bureau as part of its work with state officials to determine the layout of the standard redistricting data file in 2010.

In the future, the Bureau may decide on an exception being made or a principle being added to count prisoners at a location other than the prison. However, the information necessary for such a decision does not now exist. A research and testing program, including experimentation as part of the 2010 census, should be initiated by the Census Bureau to evaluate the feasibility and cost of assigning incarcerated and institutionalized individuals, who have another address, to the other location. Such a program would include collecting and analyzing individual-level residence information (either self-reported ARE locations or the location coded in corrections department files).

EXPERIMENTATION AND TESTING FOR THE FUTURE

Our review of residence-related research at the Census Bureau, including the design of mid-decade census tests in 2005 and 2006, suggests critical deficiencies in the Bureau's overall research agenda. One such deficiency is that the Census Bureau often relies on small numbers (20 or less) of cognitive interviews or very large field tests (tens or hundreds of thousands of households, in omnibus census operational tests) to reach conclusions about the effectiveness of changes in census enumeration procedures. As a consequence many important questions about the effectiveness of residence rules do not get addressed effectively.

We strongly suggest that the Bureau make fuller use of moderate-scale surveys and tests, combining field tests of a few hundred households with cognitive interviewing. The Census Bureau should undertake analytical research on specific problems in order to better evaluate the effectiveness of residence and other questions on the census forms. These studies should be designed to focus on particular populations of interest. Candidates for such research include:

- why babies are often omitted from the census form (targeted at households with newborns);
- whether census respondents find a pure *de facto* residence rule easier to follow and interpret than a *de jure* rule (generally, and with specific reference to large households);
- whether additional residence and location probes on questionnaires—increasing the length of the survey—impairs response or other operational activities (e.g., page scanning);
- the difficulty and advantages of including a reference date or time frame;

- multilingual and linguistically isolated households; and
- whether the Census Bureau standard of "live or sleep most of the time" is consistent with respondent notions of "usual residence."

On a similar note, data similar to those collected by the 1993 Living Situation Survey should be conducted on a regular basis. A convenient form for a more regular study could be inclusion of a supplement to the ACS or a stand-alone survey.

An advantage of the use of moderate-scale experiments is that they can focus on specific problems; the major omnibus tests and experiments performed by the Bureau often combine multiple topics and treatments so that effects of specific changes are difficult or impossible to determine. When designing experimental tests, the Census Bureau should always include a control form—either the questionnaire items used in the preceding census or the exact items used in immediately previous census tests—so that individual modifications can be more effectively assessed.

Another deficiency in the Bureau's research program involves unanalyzed data from the 2000 (and previous) censuses. Through its own resources as well as contacts with outside researchers, the Census Bureau has data on diverse residence situations that could be used to inform residence-related decisions. Although the research done to date does provide some information on the nature of omissions and duplicates in the 2000 census, the analyses are not sufficient to fully sort out important effects, and the data that have been collected need further analysis. One specific example of this deficiency concerns census reporting in large households. The decennial census form only allows for detailed data collection on six persons (names can be listed for an additional six people); this may contribute to an undercount of children and babies if census respondents tend to list their household members in descending order of age. Possible trends in age reporting should be able to be detected using census operational data: extant data from the 2000 census on large households of seven or more members should be reanalyzed for better understanding of the nature of the households and to inform better practices to collect data for large households. Generally, the Census Bureau should conduct and facilitate further research using its detailed census and survey results; as needed, the Bureau should consider ways to facilitate this work through contracts with outside researchers.

The ACS that will replace the long-form sample in the 2010 census should be a major focus of residence-related research by the Census Bureau because its residence standard differs from the decennial census. Specifically, the ACS uses a "two-month rule" or "current residence" concept that is akin to a *de facto* standard, while the decennial census "usual residence" standard is a *de jure* type. The nature and extent of interpretation problems that may arise due to these discrepant standards is vital to evaluating both. The Census Bureau should plan to ask a question on the usual residence of each household member in the ACS questionnaire, in order to evaluate the extent of incongruity of residence standards between the long-form replacement survey and the decennial census. The usual residence question should first be tested using the survey's experimental "methods panel"; the resulting data should be fully evaluated and analyzed to refine final versions of the question.

The Census Bureau is considering unduplication methodologies for the 2010 census, building from innovations in the 2000 census and its success in probabilistically matching census records based on name and date of birth. In particular, the Bureau is exploring real-time unduplication during census processing and is developing an expanded coverage follow-up operation to provide data to help identify potential duplicates. Focus on duplication is important; however, Census Bureau research on living situations that do not easily fit census residence rules should strive to gather data on the sources of omissions in the census, as well as sources of duplication. In addition, a comprehensive assessment of the components of gross coverage error (both undercount and overcount) should be added as a regular part of the census evaluation program.

The mechanics of censustaking have changed greatly since marshals were first sent out on horseback in 1790; as times have changed, the "usual residence" concept has endured even though its exact interpretation has shifted. The most recent paradigm shift in defining residence in the census came with the adoption of mail-based enumeration for most of the census population in 1970; that shift included drawing a linkage between census residence and a specific mailing address. Looking to the future, over the long term, the Census Bureau research program needs to consider broader shifts that lie ahead—the impact of the Internet and e-mail and the diminished importance of traditional mailing addresses (and paper mail) in people's lives, more transitory living arrangements, the changing need for census data as private and public databases grow in completeness.

There is a serious need for additional quantitative information on the magnitude of emerging social trends for groups, as well as a need for further qualitative assessment and better definitions of concepts. Important hypotheses can emerge from qualitative techniques such as ethnographic research, but these need to be tested quantitatively. People's attachment to households and group quarters has changed significantly over several decades and is likely to continue to change in ways that cannot now be predicted with confidence. The Census Bureau should establish a standing research office whose task it is to continually monitor changes in factors influencing people's attachments to locations where they are counted, and the connectedness of changes among them, using such information to generate appropriate research and recommendations for changes in how people can be more accurately enumerated in the decennial census.

Part I

Residence and the Census

– 1 –

Introduction

THE 23RD U.S. DECENNIAL CENSUS in 2010 promises to be markedly different from its predecessors in several important respects, if current plans hold. Data from the American Community Survey (ACS), a continuous monthly survey of households, are intended to replace data traditionally gathered by the census "long-form" sample. Through this change, the U.S. Census Bureau hopes not only to provide more immediate and accurate data on the detailed economic and demographic topics formerly covered on the census long form, but also to streamline and focus the decennial census on the basic "short-form" questions that provide the population counts used to reapportion legislatures and allocate federal funds. It is also hoped that technology—in particular, the use of handheld computers by temporary enumerators for nonresponse follow-up—will improve the quality of the census data.

In spirit and effect, though, the modern decennial census draws fundamentally from the legislation enacted by Congress to conduct the very first decennial census of the United States in 1790. That legislation dictated that residents should be counted at their "usual place of abode" (1 Stat. 101, §5). Every subsequent decennial census has adopted the same basic goal of counting every resident of the United States once, and only once, at what is determined to be his or her usual place of residence.

The verbs "enumerate" and "count" are used interchangeably in everyday discourse, but they are fundamentally different in the context of population censuses and surveys. In a census, enumeration is the actual collection of data from a person: a person is "enumerated" at the location where he or she is found, whether through a mailed questionnaire, personal contact, or other means. This location may or may not be the same as the location where a person is "counted"—recorded or tabulated in the census returns. Indeed, the difference between "enumerate" and "count," between collection and recording, defines two distinct concepts of residence that have historically shaped censuses and surveys. In purest form, a *de facto* approach to defining residence directly equates the two locations: people are counted exactly where they are found or enumerated at the time of the census. Alternatively, a pure *de jure* approach counts people at a place of legal residence, which may not be the same as the place of enumeration.

Both the *de jure* and *de facto* approaches are ideal models that are difficult or impossible to execute in their purest forms. For instance, unless a nation maintains a central population register or national identification program (and all residents comply with the program), different views of what constitutes legal residence complicate a purely *de jure* count. Likewise, a pure *de facto* snapshot of the population of a country would require so massive a deployment of resources—covering every house, hotel, and vehicle at the designated time—as to be technically infeasible. Instead, hybrid approaches that may be close to (but not strictly) one of the ideal models are commonplace. Accordingly, when a census or survey is identified as *de jure* or *de facto*, it typically has a residence standard that is closer to one of these models than the other, and the label is used as a convenient shorthand. The U.S. census—based on the concept of "usual residence"—is commonly referred to as a *de jure* census even though it is not purely so. It contains elements of a *de facto* system, such as programs for counting people in temporary or transient residences. It is also not a pure *de jure* system because the definition of "usual" residence is not written into legal code or regulation; rather, it is open to varying interpretations.

"Usual residence" is both a very simple and a bewilderingly complex concept, due to the inherent ambiguity of defining "usual" for some living situations. Although, for most respondents, identifying a usual residence is a straightforward exercise, for many others it is not. Moreover, neither is it always clear that respondents' notions of their usual residences are the same as the Census Bureau's concept. Even the marshals on horseback dispatched to conduct the 1790 census had to confront the potential residence ambiguities posed by people splitting time at more than one residence or whose travels put them away from home for lengthy periods. Modern censuses have to find ways to count a population that is increasingly mobile and diverse, with complex living arrangements: for example, children in the joint custody of divorced parents, "snowbirds" and other second-home owners who spend large parts of the year in different places, and long-term recreational vehicle users and others who may be true nomads who do not have any place where they stay "most of the time." Complex living situations also challenge core definitions and approaches to taking a census: how do terms like "household" apply in migrant worker communities and densely populated immigrant enclaves of major cities? Should prisoners be counted in exactly the same manner, or asked the same questions, as residents of college dormitories or nursing homes?

As a guide to how different living situations should be reconciled with the "usual residence" standard, the Census Bureau maintains a set of residence rules for the decennial census. By 2000, the Census Bureau's listing included 31 formal residence rules. The actual compilation of residence rules is rarely if ever viewed by the general public. Instead, census respondents typically see only the instructions and questions on the census form, designed to distill the basic residence concepts and lead respondents to provide answers consistent with the "usual residence" standard. Though the full set of rules is rarely seen by census respondents, a clear concept of the meaning of residence, coupled with an effective mapping of that concept to the actual conduct of the census, is crucial to the accuracy of a census.

1–A THE PANEL AND ITS CHARGE

In 2004 the U.S. Census Bureau requested that the Committee on National Statistics (CNSTAT) of the National Research Council convene a Panel on Residence Rules in the Decennial Census, with the following charge:

> This study will examine census residence rule issues and make recommendations for research and testing to develop the most important residence rules for the 2010 census. Recommendations will address potential ways to modify census residence rules to facilitate more accurate counting of the population or identify the reasons why the rules should stay the same.
>
> The panel would consider residence rules in terms of how they contribute to or inhibit an accurate count of the population. Its deliberations may include the appropriate geographic location for enumerating each person but would not include the issue of who should be enumerated in the census—for example, whether civilian citizens who live abroad or undocumented immigrants should be included.[1]

[1]The latter issue, as to whether illegal immigrants—or, for that matter, any non-U.S. citizen—should be included in the census count, remains a contentious one. The issue was the subject of major legal challenges in the 1980 and 1990 censuses (*Federation for American Immigration Reform v. Klutznick* and *Ridge v. Verity*, respectively). A segment of the 1986 Council of Professional Associations on Federal Statistics workshop on residence rules (described in Section 1–B) provoked a lengthy discussion on this question. Most recently, H.J.Res. 53 introduced in the 109th Congress proposes that the Constitution be amended so that census totals used to apportion the House of Representatives "shall be determined by counting the number of persons in each State who are citizens of the United States." As of August 2006, the bill had not been acted upon by the House Judiciary Committee; the House Subcommittee on Federalism and the Census (of the Government Reform committee) held hearings on the bill in December 2005. (See Massey and Capoferro [2004] for a recent overview of the limitations of current data sets in examining the size and trends in undocumented migration.)

The proposal for the panel study further identifies basic questions of interest:

1. What kinds of population groups are most likely to be double counted or omitted due to confusion about or inappropriate application of residence rules?
2. What relevant socioeconomic trends are occurring that could explain why one or more residence rules may no longer be relevant or may be difficult to enforce in the census context?
3. What kinds of solutions to residence rule problems should be explored?
4. If the panel proposes possible changes to the rules, what kind of test design is necessary to determine if the proposed changes have the anticipated impact?

Our panel is one of three CNSTAT panels simultaneously studying different aspects of the decennial census; the others are the Panel on the Functionality and Usability of Data from the American Community Survey and the Panel on Coverage Evaluation and Correlation Bias in the 2010 Census. Though the three panels differ in their core topics, there are areas of substantive overlap between them—for instance, the difference in the underlying residence concept between the ACS and the decennial census and the design of residence probe questions for census follow-up and coverage measurement operations. Accordingly, members of our panel have participated in activities of our sister panels when applicable, and their members have joined some of our discussions.

1–B PREVIOUS EFFORTS

As our panel examines the residence rules and concepts for the 2010 and future censuses, we obviously build on decades of work and experience by the U.S. Census Bureau. Significantly, we also draw from the experience of previous conferences and studies related to census residence.

In December 1986 the Council of Professional Associations on Federal Statistics (COPAFS) convened a Workshop on Residence Rules, sponsored by the Census Bureau. That conference featured several overview papers on the residence concept, as well as detailed presentations on specific longstanding residence concerns such as Americans living overseas, prisoners, and migrant workers. The proceedings of the workshop are printed in Council of Professional Associations on Federal Statistics (1987), which includes a summary of recommendations (CEC Associates, 1987). The panel has benefited greatly from the comprehensive coverage of the 1986 conference; throughout this report, we refer to this event as "the COPAFS residence rules workshop."

The 1986 workshop provided particularly strong discussion of the counting of military personnel and of American civilians living abroad and a vigorous discussion of the inclusion of illegal immigrants in the census. Given the boundaries of our charge, we do not examine these topics with the same level of detail, but direct interested readers to the conference proceedings for fuller discussion.[2]

Other previous CNSTAT study panels have also discussed issues of census residence as part of their work. The Panel to Evaluate Alternative Census Methods (National Research Council, 1994) included a subgroup on response and coverage issues, which commented on residence rules and methods for enumerating hard-to-count populations. In its interim report, that panel argued strongly for continued ethnographic studies and improved methods for collecting information from traditionally hard-to-enumerate populations, of the sort which typically present difficulties for the "usual residence" concept (National Research Council, 1993). Our panel's work is also informed by the work of two immediate predecessor panels dedicated to the decennial census, the Panel to Review the 2000 Census (National Research Council, 2004c), which analyzed duplication in the 2000 census in great detail, and the Panel on Research on Future Census Methods (National Research Council, 2004b), which analyzed the developing plans for the 2010 census.

The existing record may lead some to wonder why this study is being conducted, at this time. The basic arguments in response to this question are twofold. First, the topic of residence rules in the decennial census sounds, deceptively, like a very narrow focus. In fact, the topic is quite massive. Developing residence rules and effectively conveying residence concepts to census respondents requires careful attention to changes in the demography of the United States (no small task, in itself) along with the latest experience and practice in the psychology of survey data collection. Given the breadth of subject areas, residence rules in the census is a ripe topic for periodic reexamination and evaluation based on the latest research. Second, and more directly, the findings of coverage evaluations of the 2000 census—in particular, high levels of duplication and an estimated net *overcount* (as discussed below)—are a major impetus for this work. The 2000 census experience heightens the need for attention to both undercoverage and duplication in the census, two components of gross census error.[3] Problems with residence rules can contribute to both undercoverage and duplication; as part of a fuller examination of census error, then, it is important to examine whether residence concepts are being conveyed most effectively.

[2]Several of the papers were later printed in the March/April 1988 issue of *Society*. Some of the discussion related to the evolution of rules for counting Americans living overseas is summarized in Appendix C.

[3]"Error" is used in the statistical sense in this discussion, meaning a difference between an estimate and the (unknown) true value of a quantity.

1–C PLANS FOR THE 2010 CENSUS

As we begin discussion of the role of residence rules in the 2010 census, it is useful first to briefly consider the proposed general shape for the 2010 census and the testing milestones that will lead to the decennial count. (Additional detail regarding the plan, and analysis of it, is provided in National Research Council [2004b].)

We have already mentioned one major innovation planned for the 2010 census: the replacement of the decennial census long form with the ongoing ACS. Though it is not a direct part of this panel's focus, we discuss the ACS in more detail in Section 8–C. With the arrival of the ACS, the Census Bureau is in the position of having two flagship products—the decennial census and the ACS—follow two complex residence standards that are very different conceptually.

A second major component of the current planning for 2010 is a major upgrade of the Census Bureau's geographic resources, the Master Address File (MAF) and the Topologically Integrated Geographic Encoding and Referencing (TIGER) system database. The Census Bureau uses the MAF to mail questionnaires and to follow up with nonrespondents. For 2010, the Bureau is planning to integrate the MAF so that the roster of group quarters locations—places such as college dormitories, prisons, and nursing homes—is no longer maintained separately from the roster of housing units. TIGER is the Census Bureau's digital map for the entire nation and is thus critical for geocoding addresses in order to tabulate census returns at the correct locations. The major share of activities in the Census Bureau's MAF/TIGER Enhancements Program is aimed at realigning TIGER features using local geographic information system sources and other sources, such as satellite and aerial photography.

The Census Bureau has planned for a mix of mail-only and field census tests (the latter involving deployment of enumerators to follow up with nonresponding households) prior to 2010. The first of these operational trials, the 2003 National Census Test, was mail-only and focused on revised layout of the questions on race and ethnicity. The test was also intended to assess the use of alternative response modes in answering the questionnaire—mail, telephone/interactive voice response, and Internet. In 2004, the Census Bureau conducted a census test in four areas: a part of the borough of Queens in New York City and Colquitt, Thomas, and Tift counties in Georgia. This mail and field test was principally intended to examine the feasibility of handheld computing devices for follow-up data collection. The test also included a dry run with new group quarters definitions and revised race and ethnicity questions (the latter based on input from the 2003 test).

In 2005, the Bureau conducted another mail-only National Census Test. As we will discuss in detail in Section 6–D, this test examined a number of revised residence questions and coverage "probe" questions. In October 2005, the Census Bureau posted notice of a further mailout experiment, which was to occur in early 2006 (*Federal Register*, October 5, 2005, pp. 58180–58182). Plans call for 24,000 housing units to be mailed an experimental form that restructures the basic residence question on the census form, using a design from the 2005 test; the experimental forms add items that prompt respondents to make sure their forms are complete. The experimental form will also prominently display a "due date" to try to determine the effects on the rate and speed of response.

The last test in the sequence is planned to be a mail-and-field test in 2006, on the Cheyenne River Reservation and Off-Reservation Trust Land in South Dakota, and in part of Travis County (Austin), Texas. This test will feature additional testing of handheld computers; for our purposes, the test is important because it is supposed to be an operational test of final residence rules concepts.

The culmination of the testing cycle is a dress rehearsal in 2008; in January 2006, the Census Bureau announced that this rehearsal will take place in San Joaquin County, California, and a nine-county region around Fayetteville, North Carolina. Like the other mail-and-field tests, it will be conducted in a small number of selected test sites. The Census Bureau's intent is to resolve major experimental components in the earlier tests so that the 2008 dress rehearsal is a true rehearsal. By comparison, the 1998 dress rehearsal for the 2000 census was itself a major test of competing census models, pitting "traditional" census techniques against variants that would include probability sampling of nonresponding households rather than comprehensive follow-up.

The Census Bureau's test schedules—and the difficulty of making changes in census procedure in a short time frame—limit the ability of this panel to effect major changes in census operations for the next census. The decennial census is a highly complex and resource-intensive operation, requiring substantial lead time not only to test new proposed changes, but also to procure the equipment and materials needed to conduct the count. However, the timing of this panel has allowed it to meaningfully contribute to the 2005 and 2006 census tests through discussions at our public meetings, and our report lays out a broader agenda of testing and experimentation, some of which can and should be conducted in 2010 and some of which is intended to provide a knowledge base to inform residence rule considerations for the 2020 census and beyond.

1–D OVERVIEW OF THE REPORT

This report is structured in three parts. Following this introduction, Chapter 2 rounds out Part I by examining the meaning and development of residence rules and residence-related concepts in the decennial census context. We discuss the form of the residence rules used in the 2000 census and the basic difficulties faced by both census respondents and the Census Bureau itself in defining residence.

Part II describes numerous living situations for which specifying a "usual residence" is not straightforward. Chapter 3 begins by describing major segments of what the Census Bureau has traditionally termed the "group quarters" population. These groups—students in dormitories, health care patients, and prisoners, along with parts of the military population—are major potential sources of census error. Chapter 4 focuses on general living situations and social and demographic trends that make specification of a single residence extremely difficult. The chapter also includes discussion of some population groups that may be missed, or not well handled, by current census procedures. Chapter 5 draws some basic conclusions and directions for specific research from the overview in the preceding two chapters.

The panel's core findings and recommendations are detailed in a sequence of chapters contained in Part III. Chapter 6 argues for a reconceptualization of residence rules as they have been used in the past, suggesting instead the development of a core set of residence principles. We discuss the nature of the census questionnaire itself in light of this approach and recommend that the Bureau move toward a question-based (rather than instruction-based) method of collecting residence information. Chapter 7 recommends improvements in group quarters and nonhousehold enumeration and Chapter 8 suggests guidance on other residence-related census operations. In addition, Chapter 8 discusses the Census Bureau's research and testing program, directed in part at specific concepts that should be tested as part of the 2010 census, but also focused on a broader research agenda to improve the collection of basic census data in the future.

The Census Bureau's residence rules for the 2000 census are reproduced in Appendix A. The residence concepts and questionnaire structures used in selected foreign censuses are outlined in Appendix B. Finally, though our charge precludes consideration of whether American citizens living overseas should or should not be included in the census, that population—including military and federal government personnel stationed overseas—has been at the core of several residence rules revisions; Appendix C explores that history.

– 2 –

Residence Rules: Development and Interpretation

EVERY 10 YEARS, the decennial census data provide the means to assess the size and dynamics of communities and social groups. The data are put to myriad uses every day. However, as complicated as the census is and as varied as its uses are, it is not an exaggeration to say that the census relies fundamentally on the core concept of residence. In the end, each census stands or falls on its ability to gather accurate information on how many people live in particular structures at specific geographic locations; all other questions in the census and the information they elicit are secondary to getting residence information right.

Residence rules form a crucial connective link in the census process, taking the data and attributes of each American resident and producing results that can be tabulated by whatever geographic boundaries may be needed. Residence rules are critical to assigning each person to a "correct" address; a second key linkage—between the address and a specific geographic location—is provided by the Census Bureau's Master Address File (MAF) and geographic database systems. Properly understood and executed, residence rules provide structure to the highly complex task of census data collection.

The specific residence rules used in the census have changed with time; so, too, has the role of residence rules changed, as the census has come to rely on mail-based, self-administered forms in recent censuses. In this chapter, we discuss the broad context of residence in the census, discussing the nature and scope of residence rules (Sections 2–A and 2–B). We also discuss the general problem of why the seemingly simple topic of residence can be so dauntingly complex to both census designers and respondents (Sections 2–C and 2–D). We outline some of the consequences of difficulties in defining residence in Section 2–E, and close in Section 2–F with a description of the Census Bureau's preliminary residence rules for 2010.

2–A WHY ARE RESIDENCE RULES NEEDED?

The explicit constitutional mandate for the decennial census is the generation of counts used to apportion the U.S. House of Representatives. Through this action—assignment of the number of seats in the House and, accordingly, the number of votes in the electoral college for president—representative power is distributed among the states. Hence, the accurate placement of residents by their geographic location (at least to the state level) has always been key to the accuracy of the census.

A related use of census data—and arguably as primary a use as apportionment—is the division of states into legislative and voting districts. The use of the census for redistricting greatly heightens the need for accurate links between people and geographic locations: the redrawing of districts demands data at the fine-grained resolution of blocks and tracts. According to McMillen (2000b), in the early days of the country, the division of states into legislative areas (redistricting) was left primarily to the states. In 1842, Congress enacted a law requiring states with more than one legislative district to divide the state into districts with one representative per district. However, nothing was said about district size until the latter half of the 1800s, when geographical compactness and equal population emerged as requirements. In 1929, Congress passed a permanent apportionment act that contained no directives on the geographic size or population of districts, beginning a period of great variability in legislative districts. This period lasted until the U.S. Supreme Court's ruling in 1962 (*Baker v. Carr*, 369 U.S. 186) that required legislative districts within states to be drawn to include equal numbers of people (McMillen, 2000b). In the wake of that ruling, and other cases that reinforced the "one person, one vote" standard, legislative redistricting is now based entirely on the most recent decennial census counts, and districts are held to exacting standards of numerical equivalence in population. Since 1965, enforcement of the Voting Rights Act and protection of minority rights have also depended on census counts.

Another important use of census counts is in the distribution of funds by the federal government to the states and substate units. In 1998, $185 billion in federal aid was distributed to states and substate areas based in whole or part on census counts (U.S. General Accounting Office, 1999).

In all of these uses of census data, the need to count each person once, only once, and in the right place, is crucial. Yet the ideal goal—unequivocally linking each person's census record to a single specific geographic location—is difficult to achieve in practice. Accordingly, the census relies on a set of residence rules to define what "in the right place" means for various census respondents, so that their census returns can be accurately tabulated. Because they are meant to establish the "correct" location of a person in the census context, residence rules can help alleviate problems of duplication; however, as we will describe, residence rules alone can not solve all problems of census error.

2–B WHAT ARE THE RESIDENCE RULES?

As they have developed over time, the residence rules for the decennial census are a formal list of clarifications and interpretations, indicating where people in various residence situations should be counted in the census. In recent censuses, the actual list of residence rules has been an internal Census Bureau document, although a somewhat edited version of the rules was posted on the Census Bureau Web site during the 2000 census.[1] The rules were also incorporated in some form into the training materials for census enumerators. The formal residence rule list is used to answer questions, both inside and outside the Census Bureau, on residence questions (U.S. Census Bureau, Population Division, 2004).

To understanding residence rules, it is important to remember what the residence rules are *not:*

- Most fundamentally, the residence rules are not the specific instructions on the census questionnaire—the guidance to census respondents on who should be included or excluded on the census form, such as those on the 2000 census questionnaire (see Figure 2-1). Confusion on this point may arise because the instructions and questions on the census form are the general public's primary point of interaction with the census residence rules. However, the instructions are only an extract from the full set of residence rules.
- Residence rules are not the link between a housing unit and a specific geographic location; rather, they provide the link between an individual person (and data about that person, on the questionnaire) and a specific housing unit. The specific geographic referencing between housing units and geographic locations is done through the Census Bureau's

[1]The presence of these rules online was indicated in a press release detailing the mass mailing of census questionnaires: "a complete set of residency rules telling where students, nursing home residents, military personnel, 'snowbirds' and others are counted can be found on the Census Bureau's Internet site at http://www.census.gov/population/www/censusdata/resid_rules.html" (the link was still functional as of 6/1/06); see U.S. Department of Commerce (2000).

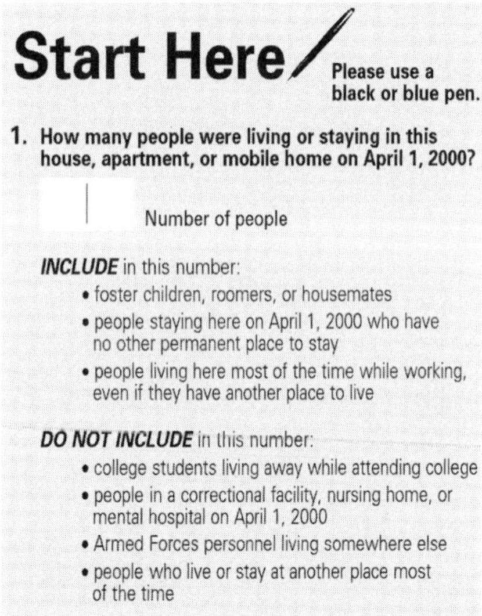
Start Here Please use a black or blue pen.

1. How many people were living or staying in this house, apartment, or mobile home on April 1, 2000?

Number of people

INCLUDE in this number:
- foster children, roomers, or housemates
- people staying here on April 1, 2000 who have no other permanent place to stay
- people living here most of the time while working, even if they have another place to live

DO NOT INCLUDE in this number:
- college students living away while attending college
- people in a correctional facility, nursing home, or mental hospital on April 1, 2000
- Armed Forces personnel living somewhere else
- people who live or stay at another place most of the time

Figure 2-1 Basic residence question (Question 1), 2000 census questionnaire

MAF and geographic reference database, which we discuss in greater detail in Section 8–A.

- Despite their name, the residence rules are not rules in any legal or regulatory sense. Residence rules are not written into census law; indeed, as we discuss below, not even the term "usual residence" (much less its definition) is written into active census law. Rather, the residence rules are guidelines, internal to the Census Bureau, on how certain living situations should be handled in terms of defining "usual residence."

2–B.1 Historical Development

While the U.S. Constitution specifies that a census be conducted every 10 years for the purpose of reapportioning the U.S. House of Representatives, it offers no further guidance on exactly how the count is to be performed.[2] The first U.S. Congress faced the practical problem of performing a count through the Act of March 1, 1790. That act authorized marshals to carry out

[2] Clemence (1987:16–17) summarizes the evolution of the Constitution's census clause from its first draft (apportioning by "the quotas of contribution or the number of free inhabitants, [as] may seem best in different cases") through several revisions, including one, and only one, instance where "citizens" were distinguished from inhabitants or residents.

a count such that "every person whose usual place of abode shall be in any family on [August 1, 1790,] shall be returned as of such family." In addition, "every person, who shall be an inhabitant of any district, but without a settled place of residence, [shall be counted] in that division where he or she shall be on" August 1, and "every person occasionally absent at the time of the enumeration [shall be counted] as belonging to that place in which he usually resides in the United States" (1 Stat. 101, §5).[3] The legislative text of 1790 outlined the rules for defining residence for census purposes, with the goal of counting each resident of the United States once and only once and in the correct location; the first census residence rules, like the underlying census goal, have guided every subsequent census.

For most people—in 1790, as in 2006—the meaning of "usual residence" is clear.[4] These people are affiliated with only one address or household, and have no difficulty identifying it. Indeed, the inaugural census of 1790 presumed that the concept was sufficiently self-evident that the marshals charged with obtaining the counts through personal contact with residents were provided with no written rules or instructions.

Yet even in the earliest days of the census, conceptual problems with "usual residence" were evident; as Clemence (1987:18) comments, "the [first] census law contained about 1,600 words and not a single definition." Moreover, ambiguous residential situations were as plentiful in those days as in modern times. Notably, Clemence (1987:12–14) cites the cases of George Washington and Thomas Jefferson, who served as president and secretary of state during the official period of conduct for the 1790 census (August 2, 1790–April 1791). In those 36 weeks, Clemence observes that Washington "was on the road for 16 weeks, visiting every State in the Union from Rhode Island to Georgia; 15 weeks at the seat of Government, and only 10 weeks at his home in Mount Vernon, Virginia," yet he was almost certainly counted as head of family at Mount Vernon. Likewise, Jefferson spent most of that period in Philadelphia: "he, like many others, was following the seat of Government around, which had no settled place of residence for itself." His name appears twice in 1790 census records—once at his Monticello home in Virginia (where he seems to have been tallied) and once in Philadelphia, where both he and attorney general Edmund Randolph signed a census schedule posted on a tavern wall posted (as the law directed) so that people who believed they were not listed at home could still be registered. More generally, Clemence (1987:19) concludes:

[3]Records of floor debate from the first Congress suggest that the census legislation was enacted rather swiftly. Though James Madison's suggestion that residents be listed by occupation was challenged, "no one on the floor spoke a word about place of abode" (Clemence, 1987:18).

[4]The schedule used by enumerators in the 1940 census was the first to include the terminology "usual place of residence"; the 1930 and several preceding census continued to use the phrase "place of abode" (Gauthier, 2002; Mills, 1993).

> In 1790, there were at least a few Americans abroad, including Thomas Jefferson until his summer voyage home from France. There were people in institutions, people with two homes, riverboat captains with no settled place of residence, militia on post duty in the territories outside of any State, and college students attending, for example, Yale, Harvard, Princeton, and William and Mary.

The first few decennial censuses did not include definitions or residence rules; indeed, even the exact layout of the schedule used by marshals to conduct the count varied by state. The first set of detailed instructions that touched on residence concepts accompanied the 1850 census.[5] The 1850 census instructions were the first to specify residence rules (albeit not formally listed or labeled as such) in order to adapt to changing social conditions. Among the emergent residence rules developed for the 1850 census was a determination on where to count college students (Clemence, 1987:21):

> Students in colleges, academies, or schools, when absent from the families to which they belong, are to be enumerated only as members of the families[6] in which they usually boarded and lodged on [Census Day].

"Because no uniform rule was adopted for the whole United States" regarding the counting of crews of marine vessels, the 1850 instructions continued, "errors necessarily occurred in the last census in enumerating those employed in navigation"; accordingly, the instructions laid out new counting rules that "assistant marshals are required to be particular in following" (Gauthier, 2002:10). Later censuses continued to add rules and instructions for the counting of different groups, typically in response to questions and ambiguities experienced in the field.

Though specific residence rules have shifted over the years, as has the date of the census (see Box 2-1), the "usual residence" benchmark inherited from 1790 has generally prevailed as the underlying residence concept for the census. Even though the concept is enshrined in census practice, there exists no definition of "usual residence" in current census law (Title 13 of the U.S. Code). Moreover, Title 13 does not directly specify what residence standard—a *de jure* enumeration based on usual residence or a *de facto* count based on current residence or where a person is found on Census Day—should apply to the decennial census.

---

[5]Charged with conduct of the 1820 census during his service as secretary of state, future president John Quincy Adams did provide detailed instructions—and, for the first time, a printed list of questions to be asked by the census—but did not expand on the definition of residence. However, his instructions did ask for information about each person's settled place of residence and family members temporarily absent; "all of the questions refer to the day when the enumeration is to commence," and enumerators were cautioned to include family members who had died and exclude babies born after Census Day (Clemence, 1987:21).

[6]As discussed in Section 2–C.1, censuses of this period interpreted "family" as any collective of people in a "dwelling house," without regard to kinship. Hence, the wording of this rule is tantamount to counting students at their college locations.

Box 2-1 Why Is April 1 "Census Day"?

As Anderson (1988:44) notes, "census takers always knew that the count could be affected by the month of the year it was taken." Early American censuses had to balance the difficulty of making personal contact with residents (slowly, with transportation by foot or horseback) with the prospects of duplicate counting or omissions that would follow from allowing the enumeration to run too long. The first four decennial censuses all used an early August date as the reference, since the summer and fall months were judged to be the best time to find people in what was still an agricultural society. For 1830, "Congress also moved the date of the census ahead two months, to June 1 instead of August [7] as in 1820, on President Adams's suggestion that this would permit a longer stretch of good weather for the house-to-house enumeration" (Cohen, 2000:121). June 1 remained the census date for the 1840 through 1900 censuses.

As the census became more routinized and professional, officials worked to shorten the time of the count and shift the census date earlier in the year. The increasing urbanization of the United States prompted the shift from June 1 to April 15 in the 1910 census; June was "unsatisfactory . . . because some city dwellers were already out of town for summer vacations, and farmers did not remember enough about the previous year's crop for the agricultural census" (Anderson, 1988:44). The Census Bureau tried moving Census Day back further still in the 1920 census, to January 1, "at the request of the Department of Agriculture, and also because it was contended that more people would be found at their usual place of abode in January than in April" (Steuart, 1921:571). However, this change "ran into trouble because the winter weather impeded the enumeration and rural leaders complained that many people were working in the city community and hence were not properly counted" (Anderson, 1988:45).

After the 1920 experience, census officials were "convinced . . . that a more nearly perfect and a more rapid count of the people can be made in April than in January." However—presaging the continuing problem of counting seasonal residents—they acknowledged that this represented a tradeoff (Steuart, 1921:572):

> It is true that during April and June, when the enumeration has heretofore been in progress, large numbers have been at summer resorts. But at [the January 1920] enumeration it was found that surprisingly high numbers were at winter resorts. Thousands who have their usual places of residence in the northern states spend the winter months in California, Florida, and other southern states. Some of them live in the south several months of each year, and it was difficult to determine their usual places of abode. In this respect the change complicated the work; certainly it did not simplify it.

Hence, the 1930 census set April 1 as Census Day, and that date has since been written into Title 13 of the U.S. Code for subsequent censuses. One exception in the 2000 and other recent censuses is the enumeration of remote villages in Alaska, which are rendered unreachable by weather conditions in March and April. In 1930, the count there only began in October; recent censuses have tallied those areas in January or February.

2–B.2 The Changing Role of Residence Rules: From Enumerator Interviews to Self-Response

The earliest decennial censuses were conducted by marshals on horseback; though the federal agency charged with conducting the census varied, the method of collecting census information by face-to-face interviewing remained the norm well into the 20th century. The fact that the census was administered in person meant that the field enumerators, ultimately, were responsible for explaining and deciding who should be counted on a "usual residence" standard. This role of enumerator as residence adjudicator was emphasized in the 1880 instructions to enumerators:[7]

> The census law furnishes no definition of the phrase, 'usual place of abode,' and it is difficult, under the American system of a protracted enumeration, to afford administrative directions which will wholly obviate the danger that some persons will be reported in two places and others not reported at all. Much must be left to the judgment of the enumerator, who can, if he will take the pains, in the great majority of instances satisfy himself as to the propriety of including or not including doubtful cases in his enumeration of any given family.

Arguably, the most significant change in residence rules and their role in the decennial census was brought about by a major paradigm shift in census operations: the switch from an enumerator-conducted census to a mailed-questionnaire, self-administered response model of census data collection. The 1960 census was the first to move significantly toward this model;[8] in that year, households were mailed an "Advance Census Report," which they were asked to fill out but *not* return by mail. Instead, enumerators visited the household to collect the forms and transcribe the information onto forms more conducive to the optical film reader then used to process census data. If a household did not complete the advance form, the residents were interviewed directly by the enumerator. Subsequently, legislation passed in 1964 (P.L. 88-530) eliminated the requirement that decennial census enumerators personally visit every dwelling place, enabling broader change in census methodology.

---

[7]The enumerator instructions and forms for censuses dating back to 1850 are very helpfully archived as part of the documentation of the Integrated Public Use Microdata Series in the "Counting the Past" section of http://www.ipums.umn.edu/usa/doc.html [8/1/06]. Gauthier (2002) also comprehensively lists enumerator instructions and census schedules from 1790 to 2000.

[8]However, the 1960 census was not the first to use the mail in census data collection. The Census Bureau's procedural history of the 1970 census (U.S. Bureau of the Census, 1970) indicates that mail was used in specialized operations as early as 1890, when questionnaires concerning residential finance were mailed to households with a request for mail return (the same was repeated in 1920, and a similar mail-based program on income and finance was used in 1950). Supplemental information on the blind and deaf was requested by mail in 1910, 1920, and 1930, and an "Absent Family Schedule" was used for some follow-up in 1910, 1930, and 1940. Presaging the 1960 approach, an "Advance Schedule of Population" was delivered to households in 1910; farm households also received an advance copy of an agriculture questionnaire administered as part of that census. Prior to implementation of large-scale mailout/mailback in 1970, experiments and tests of the method were conducted in 1948, 1950 (as a census experiment), 1957, 1958, 1959, 1960 (some of the Advance Census Report responses were requested by mail), 1964, and 1965.

The 1970 census took the mailout-census model a step forward: census questionnaires were mailed to all households in major urbanized areas (which were thought to include about 60 percent of all housing units), and respondents were asked to mail them back to their local census office on Census Day. This mailout/mailback methodology has been expanded and used in the subsequent censuses; by 2000, 82 percent of the population was covered by mail. The balance of the country was enumerated through a mix of approaches including enumerator visits; Box 2-2 lists the nine "type of enumeration areas" used in 2000.

This major change in census data collection techniques dramatically changed the role of residence rules in the census. Prior to the mail-based model, residence rules had been incorporated in the instructions to census enumerators. The Census Bureau's task in administering residence rules was relatively small: only enumerators had to be trained in the rules, and enumerators could answer questions from respondents, to the best of their ability, during the census interview. But mail-based methods changed the nature of the exercise: the Census Bureau could develop its own interpretation of "usual residence," but it now had to try to get every census respondent to grasp the Bureau's definition based only on information included in the census questionnaire.[9] Because only so much information can be included on a single piece of paper, the application of residence rules became trickier—trying to find the right combination of words and cues to induce respondents to make their concept of "usual residence" square with the Bureau's. (We discuss the evolution of the mail-based census instruments in Chapter 6, illustrating the approaches used in the 1960–1990 censuses in Figures 6-1–6-4.)

2–B.3 Assessment of the 2000 Census Residence Rules

By 2000, the Census Bureau's internal list of residence rules grew to 31 specific rules, plus a related statement on the meaning of time cycles (e.g., daily, weekly, or monthly) in determining usual residence. This set of rules is reprinted in Appendix A.

The first impression that comes from reviewing the 2000 census residence rules is that they are not organized in a way that a general reader can follow; in large part, this is attributable to the internal (not for public consumption) nature of the full residence rules document. Designed for a more general audience, the residence rules might be structured by major group type (e.g., "students" or "military personnel") or by the approximate population of the group (so that a living situation in which a person is likely to find oneself comes earlier in the list than rare groups). Instead, the organizational structure of the

[9]Of course, the temporary enumerator hired to do field follow-up would also need to develop a solid understanding of the "usual residence" concept; how well training materials achieved that is an open question.

Box 2-2 Types of Enumeration Areas (TEAs), 2000 Census

(1) **Mailout/mailback** In areas with predominantly city-style addresses, U.S. Postal Service carriers delivered an address-labeled advance letter to every housing unit on the MAF the week of March 6. In mid-March the carriers delivered address-labeled questionnaires, followed 2 weeks later by a reminder postcard. Households were instructed to fill out the questionnaire and mail it back.

(2) **Update/leave** In areas with predominantly rural route and post office box addresses that could not be tied to a specific location, census enumerators dropped off address-labeled questionnaires to housing units in their assignment areas; respondents were instructed to return the completed form by mail. While delivering questionnaires, enumerators updated address entries to include new units not on the list, noting for each its location on a map (map spot), so that follow-up enumerators could find units that did not mail back a questionnaire.

(3) **List/enumerate** In remote, sparsely populated, and hard-to-visit areas, census enumerators combined address listing and enumeration. There was no MAF for these areas created in advance. The enumerators searched for housing units, listed each unit in an address register (also its map spot), and enumerated the household at the same time.

(4) **Remote Alaska** The enumeration procedure in remote areas of Alaska was similar to list/enumerate. It was conducted in February, before ice break-up and snow melt.

(5) **Rural update/enumerate** The Census Bureau determined that some blocks originally planned to be enumerated by update/leave would be better handled by a procedure in which address list updating and enumeration were conducted concurrently. Areas covered by the operation included resort areas (believed to have high concentrations of seasonally vacant housing units), some American Indian reservations, and *colonias* on the U.S.-Mexico border. "Rural" refers to the source of the address list, which were operations focused on areas with mainly non-city-style addresses.

(6) **Military** Mailout/mailback procedures were used for all residential blocks on military bases (excluding group quarters). Such blocks in type 2 enumeration areas (but not those in type 1 enumeration areas) were assigned an enumeration area code of 6 because there was no need to update the address list or provide map spots.

(7) **"Urban" update/leave** It was determined that some blocks originally planned to have questionnaire delivery by the Postal Service would be better handled by having census enumerators follow an update/leave procedure. Such blocks contained older apartment buildings that lacked clear apartment unit designators, or they had many residents, despite having city-style addresses, who received their mail at post office boxes.

(8) **"Urban" update/enumerate** Some American Indian reservations contained blocks in more than one TEA. In these instances, all blocks in the reservation were enumerated using update/enumerate methods (type 5). However, those blocks for which the mailing list was developed using "urban" procedures and for which no map spotting was required were made type 8.

(9) **Mailout/mailback conversion to update/leave** Late reexamination of planned TEA 1 areas, conducted in 1999, suggested that some blocks contained a significant number of non-city-style addresses. These were converted to update/leave but treated as a separate type.

SOURCE: Adapted from National Research Council (2004c:Box C.2).

2000 census residence rules is both weak and operational in nature. Its major headings are obscure—"household population," "group quarters population, [usual home elsewhere (UHE)] allowed," "group quarters population, UHE not allowed," "overseas population," and "do not list population"—and use jargon (e.g., "group quarters" and "do not list") and acronyms (UHE) that are opaque to a lay audience. Further, major population groups are not addressed coherently. The rules for counting college students are dispersed into rules 5, 6, and 25 (although that group of rules does, helpfully, provide explicit directions to look at the other related rules); newborn babies are mentioned almost as an afterthought in rule 3 and lost in technical detail of counting hospital patients; military personnel are divided across rules 4, 13, 26, and 27.

As we discuss further in Chapter 6, census residence rules serve several different purposes, and would thus be better handled by crafting different products to fill those various purposes. The 2000 census residence rules document takes an omnibus approach and tries to satisfy all the needs. By straining too hard to be an operational blueprint (e.g., categorizing by "household" versus "group quarters" operations, complete with specific group quarter code numbers in rules 13–19), the document becomes less effective in its primary purpose of clarifying the meaning of "usual residence."

Finding 2.1: As developed and used in the 2000 census, the residence rules for the decennial census were too complicated and difficult to communicate. The set of 31 formal residence rules was not organized for ease in comprehension, and instead seemed to be a loose amalgamation of previously encountered problematic residence situations. The sheer number and redundancy of the rules detract from their effectiveness in training temporary census enumerators.

2–C WHY IS MEASURING RESIDENCE DIFFICULT FOR THE CENSUS BUREAU?

In this and the next section we discuss some of the difficulties associated with measuring residence, and "usual residence" in particular, from two basic viewpoints. In this section we focus on the challenges faced by the Census Bureau in specifying what it means by residence; in Section 2–D, we consider the difficulties faced by respondents in answering residence questions.

2–C.1 Definitional Challenges

An inherent problem with the Census Bureau's "usual residence" approach—particularly when the primary mode of data collection is self-response by individual persons—is that it requires respondents to interpret and apply "usual residence" as the Bureau does. Yet "usual residence" is not everyday parlance for most people; respondents instinctively need to reconcile complex and technical terms into concepts more familiar to them.

"Residence," "abode," "home," "domicile," "household," "lodging," "dwelling"—even the basic semantics of describing living situations to census respondents present difficulties because these words have different connotations. For instance, the overtones of "domicile" strongly suggest emphasis on a legal address; "home" suggests the presence of family and could also connote a place of origin or birth; "lodging" suggests temporary or current location. At the most mechanical level, "address" is arguably the best word to describe the entity with which the Census Bureau would like respondents to associate themselves—a geographic reference that can readily be located in a particular area. But that word, too, is problematic, given its near-automatic association with "mailing address" and mail delivery; one usually does not think of a post office box as a place of residence, but it would be the natural "address" to report in rural areas without household mail delivery.

The term the Census Bureau uses as its basic unit of measurement is the "household"; operationally, the Bureau draws distinctions between the household population and the "group quarters" (e.g., prisons, dormitories, and hospitals; see Box 2-3) and "service-based" populations (e.g., shelters and soup kitchens). Like the other possible terms, "household" can be a difficult concept to grasp: the word can be "associated with a physical structure, a co-resident social group, a consumption unit, and a kinship group, usually thought to be the family" (Hainer, 1994:337).

The definition and implementation of the basic unit of measurement in the census has varied over time. The 1850 census—seminal in many respects regarding the topic of residence because it was the first to provide detailed enumerator instructions—attempted a sharp delineation. The grouping of people the 1850 census was concerned with was the "family"—and actual kinship had nothing to do with the definition. "A widow living alone and separately providing for herself, or 200 individuals living together and provided for by a common head," or the "resident inmates of a hotel, jail, . . . or other similar institution"—these cases "should each be numbered as one family" (Gauthier, 2002:9).[10] The 1860 enumerator instructions used similar examples to define "family," concluding (Gauthier, 2002:14):

> Under whatever circumstances, and in whatever numbers, people live together under one roof, and are provided for at a common table, there is a family in the meaning of the law.[11]

---

[10]Likewise, the term "dwelling house" was used to describe any "separate inhabited tenement, containing one or more families under one roof." Even if used partly for business, or an institution like a hotel or prison, the "dwelling house" label still applied.

[11]See Smith (1992) for a review of the evolving definitions of "family" and "head of household" in the census.

Box 2-3 Group Quarters Categories for the 2000 Census

1. **Correctional institutions** Federal detention centers; federal prisons; state prisons; local jails and other confinement facilities; halfway houses (correctional); military disciplinary barracks and jails; other
2. **Juvenile institutions** Homes for abused, dependent, and neglected children (subdivided by ownership—public, private, or unknown); residential treatment centers (for emotionally disturbed children); training schools for juvenile delinquents (subdivided by ownership—public, private, or unknown); detention centers (diagnostic centers and short-term care facilities); other
3. **Nursing homes** Seven subdivisions by ownership (federal, state/local, public, private nonprofit, private for profit, private undetermined, and other)
4. **Hospitals/wards, hospices, and schools for the handicapped** Hospitals and wards for drug/alcohol abuse; hospitals or wards for chronically ill (military or civilian hospitals); hospices; mental (psychiatric) hospitals; schools, hospitals, or wards for the mentally retarded; institutions for the deaf; institutions for the blind; orthopedic wards and institutions for the physically handicapped; wards in general hospitals for patients who have no usual home elsewhere; wards in military hospitals for patients who have no usual home elsewhere
5. **College dormitories (includes college quarters off campus)** College dormitories, fraternity and sorority houses
6. **Military quarters** Barracks and unaccompanied personnel housing; transient quarters for military personnel; military ships
7. **Service-based facilities** Emergency and transitional shelters (with sleeping facilities); shelters for children who are runaways, neglected, or without conventional housing; shelters for abused women (shelters against domestic violence); soup kitchens; regularly scheduled mobile food vans; targeted nonsheltered outdoor locations
8. **Group homes/halfway houses** Homes or halfway houses for drug/alcohol abuse; homes for the mentally ill; homes for the mentally retarded; homes for the physically handicapped; other group homes (communes, foster homes, homes for unwed mothers)
9. **Dormitories and other group quarters** Crews of maritime vessels; agriculture workers' dormitories; other workers' dormitories; Job Corps and vocational training facilities; dormitories for staff at military institutional group quarters; dormitories for staff at civilian institutional group quarters; religious group quarters; other nonhousehold living situations (includes hostels, YMCAs, YWCAs); natural disaster emergency shelters; residential facilities providing "protective oversight"

In the parlance used in the 2000 census, one or more group quarters make up a *special place*: special places are administrative units, while group quarters are the actual living and sleeping facilities. For instance, a university would be considered a special place; its individual dormitories are each group quarters.

SOURCE: Jonas (2003).

Over time, the concept of a "dwelling-house" (1900) evolved into the contemporary terms "household" (1940) and "housing unit" (1960)[12] in census instructions. These housing units would be defined by the presence of some factors, including separate front doors (1900), two or more rooms (1950), and kitchen facilities (1950–1970). Similarly, "households" came to be distinguished from what would be called "group quarters" by set characteristics or counts—for instance, the threshold of 5 (1950–1970) or 10 (1980–1990) unrelated persons living together being a "group quarters" rather than a "household." The 2000 census moved away from a size criterion to differentiate housing units from group quarters; as we discuss further in Chapter 7, the sharp operational distinction between the two concepts raised problems because the Bureau's housing unit and group quarters listings were maintained separately.

Defining the unit of measurement is difficult in itself; defining "usual residence" at that unit is harder still. In the absence of a universally clear and applicable word to describe what is meant by "usual residence," another approach is to try to find the actions and activities that respondents are most likely to associate with what the Census Bureau deems to be the "usual residence." This is the approach taken in most recent censuses, trying to find the bundle of activities—"living and sleeping"? "living and staying"? "eating"?—that most people would identify as corresponding to their usual residence. The 1960 and 1970 censuses, with their explicit definitions of housing units as containing kitchen facilities, clearly put stock in "eating" as being a usual residence activity. Earlier, the enumerator instructions for the 1890 census bluntly sided with "sleeping" as the defining activity (Gauthier, 2002:26):

> A person's home is where he sleeps. There are many people who lodge in one place and board in another. All such persons should be returned as members of that family with which they lodge.[13]

Instructions in 1950 were equally blunt: "as a rule [the place of usual residence] will be the place where the person usually sleeps." Respondent sensitivity to these particular key words has been the focus of cognitive testing by the Census Bureau, including those conducted as part of the 1993 Living Situation Survey (we discuss the Living Situation Survey in Chapter 5 and cognitive tests in Chapter 8).

An additional complication in specifying a standard for "usual residence" is that the word "usual" demands reference to a period in time, in ways that are not always easy to determine:

---

[12]These were still called "dwelling units" in 1950, but assumed the final name "housing units" thereafter.

[13]The first detailed set of enumerator instructions in 1850 suggested the importance of sleep as a criterion for defining residence; marshals were directed to "make inquiry at all stores, shops, eating houses," and other establishments to record "every person who usually slept there, provided such person is not otherwise enumerated."

- *"Usual," relative to how long a time period?* Is there some minimum amount of time necessary to establish a place as a "usual" residence? If so, specifying too fine or too coarse a time reference period when asking about residence could yield different answers.
- *Retrospective or prospective?* There is also a basic question of whether the time frame should be retrospective (e.g., where a person lived during the 12 months prior to Census Day), prospective (e.g., where a person intends to live for the next 3 months), or centered (an interval around Census Day). A person who has just completed a lengthy business trip or vacation could potentially be tripped up if asked to consider the place where they lived and slept most often during the past week, while asking a person who has moved to a new residence in the past few months where they lived and slept most often during the previous year is problematic.
- *How much weight to put on "intent"?* A key factor in distinguishing between a "usual" and a "current" residence location is intent—intent to eventually return to a place even though one is temporarily somewhere else, or intent to remain at a place even though one may have just moved in. Left ambiguous are questions of how highly intent should be considered in determining a usual residence: Can "temporary absences" be so long that they break the tie to a usual residence (such as a 2- or 30-year prison term, a displacement of indeterminate length due to a natural disaster, or near-continuous long-haul truck driving)?

Though the formal residence rules of the 2000 census included an attachment that tried to outline weekly, monthly, and annual time "cycles," the 2000 census form and some of its predecessors have taken the approach of leaving the time frame decision to the respondent. Respondents are asked to consider where they live "most of the time," allowing flexibility in interpretation.

2–C.2 Discrepant Standards

Etymology and syntax aside, determining "usual residence" can be difficult because the general concept of residence can be approached through any number of standards, some of which conflict directly with each other and some of which clash with residence standards encountered by people in everyday life. The residence requirements for voter registration in state and local elections may differ from those that federal, state, and local tax authorities use. Likewise, residence for purposes of qualifying for in-state tuition at state colleges and universities can vary substantially from those required to obtain a driver's license or identification card.

Box 2-4 describes the range of residence definitions and standards that are used in these kinds of everyday applications in the state of California; other states could be used to illustrate the same kind of diversity. Although none of the applications defines residence in exactly the same way, the definitions do connect in limited ways: for instance, the motor vehicle code cites payment of resident tuition at a California college or voting in a state election as a means of qualifying as a resident for obtaining a driver's license.

Residence for purposes of registering to vote in elections is a particularly important application in the U.S. system. At the federal level, the 1970 amendments to the Voting Rights Act of 1965 prohibited states from imposing duration requirements of longer than 30 days for eligibility to vote in presidential elections. Individual states vary in their residence guidelines for general state and local elections; as illustrated in the California example, states typically invoke a type of *de jure* standard ("fixed," "permanent," or "usual" residence), though an explicit time period is often stated. The U.S. Supreme Court's 1972 ruling in *Dunn v. Blumstein* (405 U.S. 330) struck down long-term residence requirements for general state and local elections as discriminatory; the standard in question in the case was Tennessee's requirement of 1 year residence in the state and 3 months in the county. The court suggested that residence in the state 30 days before an election was a more appropriate benchmark.[14] The 1993 enactment of the National Voter Registration Act (better known as the "Motor Voter" Act) requires states to offer opportunities for voter registration in conjunction with applying for or renewing driver's licenses or other identification cards. The act links the processes of voter registration and driver certification, even though the residence qualifications for both purposes may differ in state law or regulation.

Enduring Ties

As discussed above, residence can also be seen as an act of belonging. The adage "home is where the heart is" rings true for many; people may consider their family home to be their real home or "usual residence," even if that location is not what a strict majority-of-nights-stayed or other measure might suggest. Family and kinship ties may be important in determining where people say that they usually live; so too may be the presence of friends or affiliation in community organizations.

The notion of "usual residence" as being defined by feelings of connectedness to a place, for any number of reasons, may usefully be called an "enduring ties" standard. The standard takes its name from the U.S. Supreme Court's ruling in *Franklin v. Massachusetts* (505 U.S. 788, 1992); in that case, the majority opinion commented that usual residence "has been used broadly enough to include some element of allegiance or enduring tie to a place." *Franklin v.*

[14]The 30 days specification was not meant as an absolute requirement; the Court later ruled that Arizona's decision to adopt a 50-day durational requirement was justified (*Marston v. Lewis*, 410 U.S. 679, 1973).

Box 2-4 State Definitions of Residence: California

Residence for In-State College Tuition: With slight variations, the campuses of the University of California, California State University, and California Community College systems define residency consistent with "Uniform Student Residency Requirements" (California Education Code, Part 41 and §68062). These requirements hold that:

(a) There can only be one residence.

(b) A residence is the place where one remains when not called elsewhere for labor or other special or temporary purpose, and to which he or she returns in seasons of repose.

(c) A residence cannot be lost until another is gained.

(d) The residence can be changed only by the union of act and intent.

(e) A man or woman may establish his or her residence. A woman's residence shall not be derivative from that of her husband.

(f) The residence of the parent with whom an unmarried minor child maintains his or her place of abode is the residence of the unmarried minor child. When the minor lives with neither parent his or her residence is that of the parent with whom he or she maintained his or her last place of abode, provided the minor may establish his or her residence when both parents are deceased and a legal guardian has not been appointed.

(g) The residence of an unmarried minor who has a parent living cannot be changed by his or her own act, by the appointment of a legal guardian, or by relinquishment of a parent's right of control.

(In particular, minors under the age of 18 who may be attending college are considered by the state definitions to be resident at their parental home.)

Consistent with these guidelines, the state colleges and universities generally require that residence be demonstrated by all three of the following criteria:

1. A **physical presence** must be established for at least 366 days ("more than one year") prior to the residence determination date for a term (California Education Code, §68017). "Continuous physical presence is not mandatory," suggests the University of California, Santa Barbara (UCSB), "but a student who leaves California after establishing residence has the burden of demonstrating that he/she intended to remain a California resident, and that his/her principal place of residence has been in California."
2. Physical presence must be accompanied by **intent** to remain in California. UCSB guidelines caution that "intent is . . . demonstrated by establishing residential ties in California, and relinquishing ties to the former place of residence." University of California, Berkeley (UCB), instructions suggest that intent can be shown through documentation of such activities as voter registration, "designating California as your permanent address on all school and employment records," obtaining a driver's license or identification card, and paying taxes as a California resident. UCB warns that "your intent will be questioned if you return to your prior state of residence when the University is not in session."
3. **Financial Independence:** Students who do not have a California resident parent or have not previously been enrolled at the campus are required to be "financially independent." The UCSB documentation notes that "this requirement makes it extremely difficult for most undergraduates who do not have a parent living in California, including transfer students from community colleges and other post-secondary institutions within California, to qualify for classification as a resident at a University of California campus."

Links to the individual UC campuses' interpretation of residence can be found at http://www.universityofcalifornia.edu/admissions/undergrad_adm/ca_residency.html [6/1/06].

Box 2-4 (continued)

Residence for Voting Purposes: California election law defines a person's residence for voting purposes as their "domicile:" "that place in which his or her habitation is fixed, wherein the person has the intention of remaining, and to which, whenever he or she is absent, the person has the intention of returning" (California Elections Code, §349). Generally, the section defines a person's residence as "that place in which the person's habitation is fixed for some period of time, but wherein he or she does not have the intention of remaining." Hence, at a given time, a person may have more than one "residence" but only one "domicile."

Domicile status in California is voided with a move to another state, with the intention of either "making it his or her domicile" (§2022) or "remaining there for an indefinite time. . .notwithstanding that he or she intends to return at some future time" (§2023). However, "a person does not gain or lose a domicile solely by reason of his or her presence or absence from a place" while in military service, "nor in navigation, nor while a student of any institution of learning, nor while kept in an almshouse, asylum or prison" (§2025). "If a person has a family fixed in one place, and he or she does business in another, the former is his or her place of domicile," unless the person intends to remain at the business location (§2028). Further, "the domicile of one spouse shall not be presumed to be that of the other, but shall be determined independently" (§2029).

Residence for Taxation Purposes: The Franchise Tax Board's "Guidelines for Determining Resident Status—2004" (FTB Publication 1031) indicates that California tax law follows a pragmatic definition of resident: someone who is "in California for other than a temporary or transitory purpose" or who is "domiciled in California, but outside California for a temporary or transitory purpose." However, in asking taxpayers to classify their own residency, the tax standards invoke enduring ties. "The underlying theory of residency is that you are a resident of the place where you have the closest connections," says the board. The publication lists several factors that can determine residency:

- Amount of time you spend in California versus [time spent] outside California;
- Location of your spouse and children;
- Location of your principal residence;
- Where your driver's license was issued;
- Where your vehicles are registered;
- Where you maintain your professional licenses;
- Where you are registered to vote;
- Location of the banks where you maintain accounts;
- Location of your doctors, dentists, accountants, and attorneys;
- Location of the church, temple or mosque, professional associations, or social and country clubs of which you are a member;
- Location of your real property and investments;
- Permanence of your work assignments in California; and
- Location of your social ties.

"In using these factors, it is the strength of your ties, not just the number of ties, that determines your residency." [As a more concrete suggestion, the document notes that "you will be presumed to be a California resident for any tax year in which you spend more than nine months in this state."]

Residence for Obtaining a Driver's License: The California Department of Motor Vehicles (http://www.dmv.ca.gov/dl/dl_info.htm) advises that "residency is established by voting in a California election, paying resident tuition, filing for a homeowner's property tax exemption, or any other privilege or benefit not ordinarily extended to nonresidents."

Massachusetts is described in more detail in Box 2-5; we will discuss the case further in Sections 3–F and 3–D.1.

Box 2-4 provides a thorough example of an "enduring ties" standard in the suggested definition of residence used in California for purposes of taxation.

Residence in Administrative Records

Thus far, we have focused on the Census Bureau's basic problems in defining a residence concept so that it can be used to try to elicit accurate information from census respondents. However, a potential mismatch in residence standards arises in situations when the Bureau has to blend data gathered directly from respondents (or through enumerator follow-up interviews with respondents) with data from administrative records. As we discuss below (in Chapter 7 and Table 7-1), roughly half of the data collected on group quarters residents in the 2000 census was obtained through use of administrative records maintained by facilities where personal interviewing or questionnaire distribution was not feasible or not permitted. Thus, the quality of residence information and the definition under which it is collected could vary greatly: prisons might have records of inmates' sentencing jurisdiction but not detailed information on preincarceration or family addresses, and records available for military personnel might list a deployment location or a home base or port but not detailed address information.

In terms of final counts, discrepancies in administrative records' residence standards were likely not a major problem in the 2000 census. Only a few group quarters types were eligible for "usual home elsewhere" reporting—that is, the respondent could indicate that they did not usually live in the group quarters facility and could instead identify the address which they considered their usual residence. The highest-frequency group quarters types—college housing, prisons, and health care facilities—were not eligible for this reporting; regardless of what residence information might be coded in administrative records, people at those facilities were counted at the facility location.

2–C.3 Changing Norms and Living Situations

From the Census Bureau's perspective as data collector, residence information can be difficult to obtain accurately simply because of the demographic and social diversity of the American population. Over the course of the past few decades, living situations have taken different and more fluid forms. For instance, greater rates of cohabitation of couples—living together, but unmarried—and increased cultural acceptance of cohabitation arrangements challenge traditional definitions of family and, with it, "usual" residence. The prevalence of divorce and joint physical custody arrangements for children of divorced couples creates conundrums in identifying a single place as a child's

Box 2-5 *Franklin v. Massachusetts* (1992)

One of several legal challenges to arise from the 1990 census, the case eventually decided as *Franklin v. Massachusetts* (505 U.S. 788), targeted the Census Bureau's procedure of allocating overseas employees of the U.S. Department of Defense (DoD) to individual states for purposes of apportionment. The commonwealth of Massachusetts, and two of its registered voters, brought the suit, arguing that this allocation may have deprived Massachusetts of a congressional seat that was ultimately awarded to the state of Washington. The district court sided with Massachusetts, directing the Secretary of Commerce to remove the overseas employees from apportionment counts.

In July 1989, then-commerce secretary Robert Mosbacher decided to allocate overseas federal employees to their home states, citing growing sentiment in Congress (as evidenced by a number of introduced, but not passed, bills) in favor of their inclusion. Moreover, the Mosbacher decision was buoyed by DoD's announced plans to poll its employees to determine "which State they considered their permanent home" (505 U.S. 788, §I). Ultimately, though, DoD scrapped the proposed survey, and still later DoD was unable to provide data on employees' last 6 months of residence within the United States. Instead, the Census Bureau allocated DoD employees by the "home of record" indicated in their personnel files.

Sections I and II of Justice Sandra Day O'Connor's opinion for a 5–4 court majority focused on the legal underpinnings of Massachusetts' claim that the decision of the president and the commerce secretary to include overseas federal employees was inconsistent with the Administrative Protection Act. Section III of the O'Connor opinion spoke to Massachusetts' standing to bring the case on constitutional grounds, and was only joined by three other justices; opinions filed by Justices John Paul Stevens and Antonin Scalia explained disagreements, in part or in whole, with the O'Connor opinion's conclusions on the constitutional standing arguments.

Section IV of the O'Connor opinion, on the merits of a constitutional challenge, is the most relevant to discussion of census residence issues, and was joined by eight justices (with only Justice Scalia, having concluded that Massachusetts lacked standing, declining to join an argument on the merits). Referring to the Act of March 1, 1790, O'Connor wrote that "'usual residence' was the gloss given the constitutional phrase 'in each State' by the first enumeration Act and has been used by the Census Bureau ever since to allocate persons to their home states. . . . The term can mean more than mere physical presence, and has been used broadly enough to include some element of allegiance or enduring tie to a place." The opinion further noted cases in which "usual residence" had been broadly defined, such as the pre-1950 placement of college students in the state of their parents' residence.

The opinion concluded that Mosbacher's decision was "consonant with, though not dictated by, the text and history of the Constitution, that many federal employees temporarily stationed overseas had retained their ties to the States and could and should be counted toward their States' representation in Congress." Indeed, the allocation to the employees' home states "actually promotes equality [of representation]," assuming that the employees have legitimately retained ties to their home states. Thus concluding that the Massachusetts case failed on its merits, the district court judgment was reversed and the Bureau's inclusion of overseas federal employees in apportionment totals was upheld.

"usual" residence; joint custody arrangements also heighten the potential for duplication of census records for children, in cases where neither parent (often without any discussion in the matter) cedes a child's "usual" residence to their ex-partner. Shifting economic tides and general public attitudes affect immigration patterns and the tendency for foreign workers to migrate from place to place to work, further complicating the meaning of usual residence.

The past several decades have also seen changes of technology that can reduce a person's physical tie to one "usual residence." Of course, ongoing development of transportation systems make it easier for people to be more mobile; recreational vehicles make it possible for people to literally live on the road for weeks or months at a stretch, and air travel makes it possible for employers to deploy staff across the nation (or around the world) for extended periods. More significantly, cellular phones and e-mail truly untether people from a single physical address.

Other broader social and demographic trends that complicate the definition of residence are more subtle and technical, but affect major groups of interest. In particular, across many fields, the very notion of what the Census Bureau has come to identify as "group quarters" has been upended. Medical advances that promote longevity, coupled with the changing economics of health care, have led to a diversity of health care services that do not square with traditional notions of "hospitals" and "nursing homes," among them the increased number of assisted living options and the presence of semiresidential long-term care options in some hospital wards. To attract student interest, some colleges and universities have modified their campus housing stock, providing more complete apartment-style living communities that are virtually indistinguishable from regular (nonuniversity) households.

2–C.4 Inherent Tie to Geography

On a practical level, residence can also be a difficult issue for the Census Bureau because residence is inextricably linked to geography. A perfect set of residence rules, which could flawlessly guide respondents through the process of identifying themselves at the place the Census Bureau considers the person's "usual residence," is ultimately futile if the Bureau's geographic resources are not in order.

Inclusion on the Bureau's MAF is central to inclusion in the census; the MAF is the source of mailing addresses for the mass mailout of questionnaires to most of the country and is the basis for follow-up with nonresponding households. Just as social and demographic trends affect respondents' notion of usual residence, so, too, can they affect the ability to put together and maintain a comprehensive address list. We have already mentioned the increasing blurriness between what has typically been dubbed a "group quarters" location and the general household population. Other examples include the dense concentration of new immigrants in unconventional housing stock in urban centers (making it difficult to maintain a complete housing inventory and to define what constitutes a housing unit), the changing nature of migrant workers camps in the agriculture industry, the use of subletting and renting of finished basements and attics of houses to nonfamily members (who might not be thought of as belonging to the "household"), and the rapid pace of new construction in large suburban and exurban subdivisions and areas. Finally, even if an address is known on the MAF, it still needs to be accurately geocoded (that is, linked to a fixed geographic location, such as latitude and longitude coordinates) for accurate tabulation. Gaps and inaccuracies in the Census Bureau's geographic reference database, the Topologically Integrated Geographic Encoding and Referencing (TIGER) system, can lead—and did lead, in 2000—to geocoding failures and, possibly, exclusion from the census or gross geographic misallocation.

In advance of the 2010 census, the Census Bureau has embarked on its MAF/TIGER Enhancements Program in order to modernize its geographic resources; we discuss this effort, and the role of the MAF and TIGER, in Section 8–A.

2–D WHY IS DEFINING RESIDENCE DIFFICULT FOR RESPONDENTS?

There are several reasons that respondents themselves can find it difficult to follow residence rules and, in some cases, to resist them entirely. First, respondents have their own notions of what constitutes residence; these are often consistent with the Census Bureau's rules, but not always. Gerber (2004) summarizes some basic factors that can shape a respondent's concept of residence:

- *Social ties:* People without close ties may not naturally be included in a person's accounting of who "lives" at their home; a houseguest who stays for an extended period of time who legitimately has no other place to stay may not be dubbed a resident by a census respondent, even though they would be considered residents under Census Bureau rules.
- *Kinship and economic contribution:* Family membership can be critical to some respondents' concept of residence: the person renting a finished room in a basement or attic is not family, and thus may not be tallied as a resident. Conversely, the strength of economic ties may be essential to some people: if an extended family member is staying in the house for some time, between jobs or residences, but is not being charged rent, the lack of economic contribution may trigger some respondents to exclude them from the count.

- *Discounting "just for work" living situations:* People may discount residences that are arranged just to facilitate work. Examples include an apartment close to work that a person uses during the week rather than the family home he or she returns to on the weekend, the long-term lodging used when a person is stationed for work at another site for several weeks or months, or rooms for live-in employees. The same reasoning may apply to college students and their "just for school" residences.
- *Varying legal standards:* As described in the previous section, residence is defined in myriad ways in state and local law; people encounter these definitions in their day-to-day lives and may be confused as to what exactly constitutes residence in a particular context.

Second, respondents' concepts of "household" may vary, which in turn can alter their notion of who lives in their "household." Hainer (1994:337) argues that the traditional Census Bureau interpretation of "household" as a "discrete monothetic unit useful for description and comparison, and one that is presumed to be universally appropriate for the accurate counting of all populations," is flawed. Rather, "household" living arrangements can be very fluid—unrelated persons living together in groups, extended family members moving in and out, and so forth. In short, "household is a polythetic category" and its definition can vary sharply "between various social groups and within them."

Third, the instinctive desire to preserve the family unit runs counter to the Census Bureau's "usual residence" principle in several major cases, and can lead to noncompliance with residence rules. Prominent among these cases is the situation of children at college; regardless of how clear instructions may be, or how logical a "usual residence" ruling that college students be counted at school may be, some parents will undoubtedly consider it anathema to count their children at any location other than the family home. Likewise, a spouse in a nursing home or long-term hospital stay may instinctively be considered part of the household, even if his or her "usual residence" is at the facility, and families with members serving in the military may include those absent service members in their household count.

In some part, the notion of preserving family structure in the census may arise from the emerging role of the census as a social research tool rather than being viewed strictly as a head count. Specifically, the popularity of genealogical searches using census records—released 72 years after the census, as was done most recently for the 1930 census—to trace family ties may in some way lead people to expect that future generations may use modern census data in the same way.

Fourth, as we discuss further in Chapter 6, some respondents will not follow residence instructions because they are likely to ignore instructions, generally. Some people will reason that they already know how to fill out a form and decide that they do not need to look at the instructions. Particularly if instructions appear long and complex, they may determine that they can figure out the questions on their own—and on their own terms.

Even for respondents who do read the instructions, there is limited space on the form to provide residence information and instructions, and not all concepts can be adequately explained in that space. If there is an easy mechanism for respondents to obtain help on the question through other means (e.g., looking at a Web site or calling a help center), some may take advantage of those options. But most will not: if the provided instructions do not address their own situation, they will answer as best they can—which may not be strictly correct, by the Census Bureau standards.

2–E CONSEQUENCES OF RESIDENCE COMPLEXITIES

The basic consequence of difficulties with census residence rules—either in their definition or in their interpretation by census respondents—is spotty census coverage. That is, some people will be omitted from the census entirely, while others will be counted multiple times. Others may be counted only once, but in the wrong place.

2–E.1 Omission and Duplication

Since the 1940s the Census Bureau has published evaluations of the census, showing that the census has undercounted several groups. Though census coverage was always of academic interest, undercount or overcount in the census was not perceived as a major political issue until the U.S. Supreme Court's 1962 ruling in *Baker v. Carr* reinforced the "one person, one vote" principle. In the altered political landscape that followed, with its increased attention to strict mathematical equality in legislative districts and judicial invalidation of districting plans with even tiny amounts of variability, the "exactness" of the census count became ever more important and contentious.

Planning for the 1980 and 1990 censuses featured lengthy debates over the prospective undercount of certain groups, particularly in urban areas and among minority groups.[15] Anticipating undercount and other coverage problems in 2000, the Census Bureau put in place an Accuracy and Coverage Evaluation (A.C.E.) Program. However, the A.C.E. analysis ultimately spotlighted an unexpected problem: compared with a separate analysis using demographic analysis, the A.C.E. suggested an overall census *overcount* driven by an esti-

[15]Statistical adjustment for undercount—using dual-systems estimation based on a follow-up survey—was considered in 1990 but ultimately ruled out by the U.S. Department of Commerce, the Census Bureau's parent agency.

mated 5.2 million duplicate persons in the census.[16] The estimates of net undercount and overcount in the 2000 census are described further in Box 2-6.

Duplication in the census has always had a certain "advantage" over census omission in that it is easier to grasp and conceptualize; at least in theory, data can be reanalyzed and cross-checked to try to identify duplicate records. Until 2000, though, census data capture lacked critical information to make such a records check possible: the 2000 census was the first to use optical character recognition to capture and store the names of persons on census records. This advance permitted A.C.E. researchers to perform matches by name and date of birth. This was first done in the Further Study of Person Duplication conducted as part of the Census Bureau's Executive Steering Committee on A.C.E. Policy research program on the possible statistical adjustment of 2000 census totals; in its first stages, the A.C.E. follow-up survey was matched by name and date of birth against the census to try to find duplicate records (Mule, 2002). Later, as the method was refined, the complete set of census records was matched against itself, using a model based on name, geographic distance, age group, and status as a housing unit or group quarters to estimate the probability of duplication (Fay, 2004).[17] The success of the methodology has raised the strong prospect of real-time unduplication procedures in the 2010 census—an ongoing set of internal matches as census returns are being received and processed that could spotlight potential duplicates and help census administrators send enumerators to those households to try to obtain the correct information.

The Fay (2004) analysis of the full census-to-census person match suggests empirical results that are consistent with prior impressions of the nature of duplication. Age distributions of estimated duplication rates suggest relatively high levels of duplication among children around 10 years of age (possibly suggestive of children in joint custody situations). Young women in their early to mid-20s, and to a lesser extent men in the same age range, also show elevated duplication rates, "possibly associated with forming new households while being reported by a previous one" or with double counting of college students (Fay, 2004:2). Focusing on those cases where duplicates

---

[16]As we discuss in more detail in Chapter 8, duplicate housing units can occur as well as duplicate persons. Indeed, the first indication that duplication would be a major story of the 2000 census came as the 2000 census was in progress; a comparison with the count of housing unit addresses on the MAF with another estimate of the size of the housing stock suggested a potentially large duplication problem. The Census Bureau mounted a special ad hoc unduplication program in summer 2000; based on that operation, 2.4 million housing units (comprised of 6 million people) were flagged as potential duplicates. After review, 1 million housing units (2.4 million people) were included in the census but the others were permanently deleted (National Research Council, 2004c:137–138).

[17]Fay (2004:1) notes that the models take into account "the frequency of the name in the geographic area in question. In practice, the models indicate that exact matches on name and date of birth are almost always true duplicates if they occur in the same county, but the models have an important effect in estimating the number of between-county and between-state duplicates."

Box 2-6 Undercount and Overcount in the 2000 Census

Prior to the April 2001 deadline for delivery of redistricting data, the Census Bureau's Executive Steering Committee for A.C.E. Policy (ESCAP) weighed a set of highly discrepant results. Relative to the census count of 281.4 million, estimates derived using the A.C.E. suggested a 1.18 percent net undercount at the national level (an improvement from the estimated 1.61 percent net undercount in the 1990 census). Moreover, the A.C.E. estimates suggested a reduction in net undercount among the black and Hispanic populations, from 4.57 percent and 4.99 percent in 1990 to 2.17 and 2.85 percent in 2000, respectively. However, results from demographic analysis (DA)—essentially, adding estimates of births and immigrants to the previous census count and subtracting estimated deaths and emigrants—told a different story. The hard-to-estimate size of the illegal immigrant population is important to a DA count; depending on which estimate was used for this subgroup, DA suggested either a much smaller national net undercount (0.32 percent) or an unexpected net *overcount* (−0.65 percent). In light of these discrepancies, ESCAP concluded that there might be a significant flaw in the A.C.E. methodology; it directed a program of further research and recommended that the census data not be adjusted, a recommendation later approved by the commerce secretary.

Further refinement of the A.C.E. methodology was conducted during summer 2001, in advance of an October decision on whether adjusted census data should be used as the basis for census totals used for such purposes as fund allocation. The new studies suggested that use of the A.C.E. to adjust census results would overstate the population because the census itself had more errors of erroneous enumeration (overcount) than originally indicated. A careful evalution follow-up study, combined with methods for detecting duplicates by matching by name and date of birth (discussed further in Box 8-1), found an estimated 2.9 million erroneous enumerations that were not discovered in the original A.C.E. Bureau analysts and a National Research Council (2004c:180) panel conjectured that problems in defining census residence might explain many of these duplicates: cases seemed to include college students double-counted at their parental homes and at school, children in joint custody counted at the homes of both parents, and seasonal residents with more than one house. The Bureau once again recommended against statistical adjustment of the 2000 census data for fund allocation and other purposes.

For a third and final time, the Bureau recommended against adjustment in March 2003; this time, the decision involved whether adjusted census data would be used as the basis for postcensal population estimates. Between October 2001 and March 2003, Census Bureau staff engaged in a further major set of evaluation studies, producing a final set of results dubbed A.C.E. Revision II; these results indicated a 0.5 percent net overcount of the population. Final DA results suggested a very small, 0.1 percent net undercount.

SOURCE: National Research Council (2004c:Tables 5.1, 5.2).

appear to exist between persons counted in both a group quarters and a regular housing unit, sharp duplication rate peaks occur in the college-age population and for people starting around age 70, suggesting the importance of college students and hospital and nursing home patients as potential sources of duplication. Analyses of the 1990 and 2000 censuses, as well as their predecessors, suggest that census omissions—the undercount—are heavily concentrated among young black males, children under 10, people living in homes where they are unrelated to others, movers, and those who rent rather than own their homes (National Research Council, 2004c).

Census residence rules that are clearly conveyed are a key part of a strategy to combat person duplication; given the prominence of groups like college students and nursing home patients as potential census duplicates, a more effective way of ensuring that these groups are counted at one place is certainly needed. But residence rules alone can not solve the entire problem of census omission and unduplication. As was also learned in 2000, even census programs designed to counter census errors can sometimes serve to exacerbate them. For example, the coverage edit follow-up (CEFU) operation both added and subtracted people from the census. This telephone follow-up operation was intended to collect data on people in large households (those that indicated more than 6 people living there, even though the census form only allowed for collection of data on 6 residents) and to resolve count discrepancies between the reported household population count and the actual number of data-defined[18] people recorded on the census form. The operation was applied to all mail returns, including both census short and long forms, as well as certain Be Counted[19] forms and Internet data collection responses processed before June 8, 2000. About half of all CEFU cases were not completed for various reasons, including the lack of a telephone number, but some decision had to be made on the household size for all of them. Since 6.5 percent of the cases that were completed were found to be duplicates, the failure to complete CEFU for all cases (sticking with the original determined count) undoubtedly added some people to the count who were duplicates (Sheppard, 2003).

As our predecessor census panels have noted (National Research Council, 2004c:Finding 1.10), the Census Bureau's coverage evaluation research based on the 2000 A.C.E. is insightful and of high quality, and the subsequent work that has been done on the complete matching of census records against themselves is similarly commendable. Although the work has yielded some better glimpse at the nature of duplication in the census, it does not speak as strongly to the nature of census omission. As the next census draws near and becomes the Census Bureau's overriding operational goal, it can be difficult to devote resources to continuing to mine the data from the previous census. However, we believe that a continuous process of developing hypotheses from those data and using those lessons for future census planning is absolutely essential.

[18]A household is data-defined if at least one member has reported values for at least two complete-count items, including name. A data-defined person has at least two complete-count items, including name.

[19]Be Counted forms were unaddressed census short-form questionnaires that could be picked up in local post offices or other locations, and were intended to be returned by people who felt that they had otherwise been missed in the census. The 2000 "Be Counted" program was the successor to the similar "Were You Counted?" program in 1990.

In that spirit, we suggest that evaluation of the 2000 census not be considered to be a completed process.

> **Finding 2.2:** Although the research done to date does provide some information on the nature of omissions and duplicates in the 2000 census, the analyses are not sufficient to fully sort out important effects, and the data that have been collected need further analysis.

We note that the person-matching routines face a significant data limitation, which is the lack of coded information on "any residence elsewhere" reported by census respondents. Combining name and date of birth is certainly better than matching on name alone, but very common names (e.g., "Bob Smith") will still create matching difficulties. Information on "any residence elsewhere" could augment search capabilities by refining the geographic scope. We discuss this point further in Section 8–B.

2–E.2 Group Quarters Enumeration

The group quarters population is a particularly challenging one for census residence rules. Difficult as the concept may be to work with, though, it is important to keep the general nature of group quarters data collection in mind when thinking of the consequence of definitional and operational aspects of census residence rules. It is important for census residence concepts to deal with group quarters as accurately as possible because the decennial census has historically been the only comprehensive data source on characteristics of the group quarters population. Regardless of its flaws, the decennial census serves as the best—and sometimes the only—window on this small (2.7 percent) but significant part of the population.

The Census Bureau's fiscal 2006 budget request included funds that would add group quarters into data collection for the American Community Survey (ACS). The quality of group quarters data collection is uncertain—whether improved group quarters definitions and more highly skilled ACS interviewers will be able to offset the disturbingly high missing data rate (and, correspondingly, the rate with which those missing data had to be imputed) for many census long-form data items in 2000 (National Research Council, 2004c). Other federal and private surveys probe parts of the group quarters problem, but the decennial census remains unique in its comprehensive nature. As in previous years, the 2010 census will be examined as a source of benchmark data on the size, growth, and nature of the population living in prisons, college dormitories, nursing homes, and other group quarters.[20]

[20]That said, the ability to measure change from the previous census may be affected by changes to the definitions of group quarters; see Chapter 7.

2–F PLANS FOR 2010

2–F.1 One Rule: Proposed Residence Rules Revision

In preparation for the 2010 census, the Census Bureau established an internal working group to refine residence rules and concepts, with the objective of testing revisions in the 2006 census test and the 2008 dress rehearsal, and, ultimately, using them in the 2010 census. This panel's work is intended in part to review this initial work and provide guidance.

One of the panel's regular meetings in December 2004 was devoted almost exclusively to a comprehensive walk-through of the residence rules of the 2000 census. Census Bureau staff briefly reviewed historical highlights of each of the 31 formal rules; "straw man" suggestions for changes to the rules were advanced, and the ensuing discussion provided both constructive criticism of the existing rules as well as possible directions for improvement. In particular, the public discussion at that meeting yielded in raw form the basic argument that we underscore and elaborate on in this report—namely, that there is a need for a clearly articulated set of residence *principles*, and that it is from these principles that other products (including analogues of the current census residence rules) should be developed.

Based on that feedback, the Census Bureau residence rules staff continued work in advance of a March 2005 meeting with the panel. At that meeting, the Census Bureau presented a draft recommendation that replaced the 31 formal rules of 2000 with a single residence rule for 2010; a supporting document, similar to the 2000 census residence rules list, described how this single residence rule should be applied in a variety of living situations. This "one rule" approach and the major proposed differences are summarized in Box 2-7.

To its credit, the Census Bureau has also conducted parallel work on redefining group quarters, with the objectives of creating an integrated address file (rather than a separate MAF and group quarters roster) and of testing revised definitions in 2006 and 2008. We discuss the group quarters redefinition efforts in Section 7–B and Box 7-1.

2–F.2 Assessment

The Census Bureau's progress to date in revising the census residence rules has been highly commendable. With its draft revision, the Bureau has shown a willingness to make broad changes; we encourage this reconsideration of core census concepts and urge that old conventions be substantially revised, if not fully for 2010 then for future censuses.

In this chapter, we have reviewed the nature of census residence rules and the general issues of residence in the context of the decennial census. The chapters that follow expand our comments and present our recommendations

Box 2-7 Census Bureau's Proposed 2010 Census Residence Rule

The 2010 Census Residence Rule

The 2010 Census residence rule is used for determining where people should be counted (which means tabulated) in the census.

Residence Rule: Count people at their usual residence, which is the place where they live and sleep most of the time. People in certain types of group quarters (GQs) on Census Day should be counted at the GQ. These GQ types are listed in the box below. People who do not have a usual residence or cannot determine a usual residence, and who are not in one of the GQs types listed below, should be counted where they are on Census Day.

(As of the Census Bureau's March 15, 2005, draft, the specific "Group Quarters Where People Are Counted at the Group Quarters" had not yet been determined.)

Proposed Changes to Residence Situation Applications

The Census Bureau's proposed "one rule" is accompanied by a listing of how the rule applies to a variety of residence situations. Most of these are adapted from the 2000 census residence rules, with one change and several formal additions:

- *Boarding school students:* The Census Bureau proposes changing the interpretation of "usual residence" for boarding school students (below the college level) to the "residence at the boarding school where they live and sleep most of the time"; previously, it had been the parental home.
- *Births and deaths on Census Day:* The Census Bureau proposes adding three interpretive statements. "Babies born on or before 11:59:59 p.m. on Census Day" are to be counted at the "residence where they will live and sleep most of the time." Both "babies born after 11:59:59 p.m. on Census Day" and "people who die before 12:00:00 a.m. on Census Day" are not counted in the census.
- *Movers on Census Day:* The Census Bureau proposes that
  - "People who move into a residence on Census Day who have not been listed on a questionnaire for any residence" should be counted at the residence they move into on Census Day;
  - "People who move out of a residence on Census Day and have not moved into a new residence on Census Day who have not been listed on a questionnaire for any residence" should be counted at the residence they move out of on Census Day; and
  - "People who move out of a residence or move into a residence on Census Day who have been listed on a questionnaire for any residence" are not to be counted again.

on specific issues, but we first end this chapter with our assessment at the macro level.

First is a basic but surprising fact arising from examining the legal mandates of the census.

> **Finding 2.3:** Though the concept inherits from a long tradition of practice dating to the 1790 census act, active census law and regulation do not define the residence standard for the decennial census (*de jure* or *de facto*), nor do they define what constitutes "usual residence."

To be clear, the panel believes that it is very much for the best that Title 13 is not highly prescriptive of the exact mechanics of the decennial census; the open-endedness gives the Census Bureau much-needed latitude to develop and continue to refine its craft. Nor do we relish or advise a full congressional review and reopening of Title 13. What we do find is that, in contrast with the enabling laws for international censuses (see Appendix B), it is somewhat unusual for core census residence concepts in the United States to be as "hidden" as they have been in the past, deriving from the spirit of a centuries-old statute and kept in full form as a purely internal Census Bureau document. Accordingly, we suggest that census residence rules and concepts be made more transparent. Approaches to promote the transparency of census residence concepts to the public and decision makers could include posting notice of, and inviting comment on, residence standards through promulgation in the *Federal Register*, as the Bureau does with other basic operational plans. Information on how the Bureau defines residence should also continue to be posted to the agency's Web site.

The second point of the assessment follows from the first: the base residence standard of the decennial census—whether it is a *de jure* type or a *de facto* type—is not explicitly written into law. As a result, an evaluation of census residence rules necessarily prompts some consideration of this fundamental question. Since its inception, the U.S. census has followed a *de jure*-type model with its "usual residence" standard; we believe that it is important to consider whether a change in the residential standard—to more of a *de facto* orientation—is warranted. Moreover, it is very appropriate that the choice of residence standard continue to be periodically revisited in ensuing decades.

The choice of a *de jure* or *de facto* approach is particularly salient for the U.S. census because of the adoption of the ACS as a replacement for the census long form. While the decennial census follows a "usual residence" standard that can be considered a *de jure* style, the ACS relies on "current residence"—defined by a "two-month rule"—as its benchmark, establishing it as a *de facto* type. Having just begun full-scale data collection in 2005, the ACS is still 2 years away from producing a steady stream of estimates for many geographic areas. When that production begins, however, the Census Bureau will be in the position of having two related flagship products—the decennial census and the ACS—follow two very discrepant residence models. How this discrepancy will affect estimates of population characteristics, and the extent of the gap that may arise between ACS and census estimates in decennial census years, is unknown. However, the coexistence of the census and the ACS makes the choice of residence standard as an open and important issue; we discuss the ACS and the census residence standards in detail in Chapter 8.

Both the *de jure* and *de facto* ideals have strengths and weaknesses. One is not clearly superior to the other; rather, the choice between which standard to follow depends critically on the uses of the resulting data. The case can be made that a pure *de facto* count is the least ambiguous concept for respondents to understand, and it may be the simplest to describe: "Who is here right now?" As a result, "current residence" models based on a *de facto* approach are generally considered easier to explain and implement; they have become the standard for many surveys and polls, as their point-in-time snapshot orientation makes them well suited for measuring the characteristics of a population (rather than an absolute count). However, for purposes of reallocating political representation based on population counts, the emphasis in a pure *de facto* model on counting people exactly where they are found would emphasize people's stays in situations that are temporary or transitory: away on trips, in hotels or motels or staying as houseguests, or in transit at the time of the count. As a result, a *de facto* approach would arguably be less appropriate than a *de jure*-type rule that counts people where they belong, in some legal or other sense.

Some nations strive for the *de jure* ideal by maintaining population registers or using counts derived from administrative records in lieu of a census. Appendix B summarizes residence concepts used in several foreign nations that—like the United States—conduct regular population censuses rather than compile records-based counts. Of these countries, most adhere to a *de jure*-type concept: Austria, Canada, Finland, Israel, Italy, Portugal, Switzerland, and the United Kingdom (the United Kingdom switched to a *de jure* count in 2001 after using a *de facto* standard since 1801). However, that choice is not universal. Australia, Estonia, Ireland, New Zealand, and South Africa use a strict *de facto* standard; Japan uses a 3-month hybrid rule akin to the ACS rule. Several nations, among them Japan and New Zealand, ask both usual residence and current residence questions; perhaps more interesting are cases like Austria and Finland, which ask their respondents to identify their "main" or "permanent" residence but also ask them for a "secondary" or "temporary" residence if one exists. Likewise, *de facto*-type rules are common in general surveys and polls, but some major federal household surveys vary and follow a usual residence rule: prominent among these are the Census Bureau-conducted Current Population Survey and the Survey of Income and Program

Table 2-1 Residence Rules for the Current Population Survey

Living Situation	Include in Household?
A. PERSONS STAYING IN SAMPLE UNIT AT TIME OF INTERVIEW	
Person is member of family, lodger, servant, visitor, etc.	
1. Ordinarily stays here all the time (sleeps here)	Yes
2. Here temporarily—no living quarters held for person elsewhere	Yes
3. Here temporarily—living quarters held for person elsewhere	No
Person is in Armed Forces	
1. Stationed in this locality, usually sleeps here	Yes
2. Temporarily here on leave—stationed elsewhere	No
Person is a student—Here temporarily attending school—living quarters held for person elsewhere	
1. Not married or not living with immediate family	No
2. Married and living with immediate family	Yes
3. Student nurse living at school	Yes
B. ABSENT PERSON WHO USUALLY LIVES HERE IN SAMPLE UNIT	
Person is inmate of institutional special place—Absent because inmate in a specified institution regardless of whether or not living quarters held for person here	No
Person is temporarily absent on vacation, in general hospital, etc. [including veterans' facilities]—Living quarters held here for person	Yes
Person is absent in connection with job	
1. Living quarters held here for person—temporarily absent while "on the road" in connection with job (e.g., traveling salesperson, railroad conductor, bus driver)	Yes
2. Living quarters held here and elsewhere for person but comes here infrequently (e.g., construction engineer)	No
3. Living quarters held here at home for unmarried college student working away from home during summer school vacation	Yes
Person is in Armed Forces—was member of this household at time of induction but currently stationed elsewhere	No
Person is a student in school—away temporarily attending school—living quarters held for person here	
1. Not married or not living with immediate family	Yes
2. Married and living with immediate family	No
3. Attending school overseas	No
4. Student nurse living at school	No
C. EXCEPTIONS AND DOUBTFUL CASES	
Person with two concurrent residences—determine length of time person has maintained two concurrent residences	
1. Has slept greater part of that time in another locality	No
2. Has slept greater part of that time in sample unit	Yes
Citizen of foreign country temporarily in the U.S.	
1. Living on premises of an embassy, ministry, legation, chancellery, or consulate	No
2. Not living on premises of an embassy, ministry, etc.	
a. Living here and no usual place of residence elsewhere in the U.S.	Yes
b. Visiting or traveling in the U.S.	No

SOURCE: U.S. Census Bureau (2002:Figure 7-5).

Participation (see Table 2-1 for a summary of the Current Population Survey's residence rules).

As discussed above, data from the U.S. decennial census have several principal uses, and it is relative to those uses that the adequacy of either a *de jure* or a *de facto* approach must be assessed. If the principal use of the decennial census data was the allocation of federal and state funds, a *de facto*-type approach emphasizing current physical presence (and hence demand on resources) in an area would likely be preferable as a base for postcensal population estimates. However, the co-primary uses of census data—the reallocation of political power through reapportionment and redistricting—argue for a *de jure* approach that puts less weight on short-term or transitory residence situations.

In short, we know of no pressing reason for a major change in the residence standard for the decennial census from the *de jure*-based "usual residence" concept to a *de facto* model. The balance of this report offers recommendations predicated on the assumption that a *de jure* approach will continue to be the standard. However, we reiterate that this basic question is one that should be revisited from time to time, particularly following further experience with the ACS's "current residence" standard and the completion of research on respondent difficulty in interpreting the two types of residence standards; we discuss both of these items in more detail in Chapter 8.

Third, we offer a recommendation that underlies the balance of the report.

> *Recommendation 2.1:* **The residence rules for the 2010 and future censuses should be substantially rewritten (relative to those used in 2000), and the Census Bureau should make a concerted effort in 2010 to improve the communication of residence rules. Core concepts should be expressed as a small number of concise residence principles. These residence principles should then be used to develop other products, such as any instructions or cues to respondents on the census questionnaire, training materials for enumerators, census processing and editing routines, and a "frequently asked questions" list for enumerator and respondent reference and posting on the Internet.**

In offering this recommendation, we reiterate that the Census Bureau has made some good first strides in this direction with its proposed revisions for the 2006 census test and 2008 dress rehearsal. We applaud and endorse the Bureau's continued effort, and urge that further work be given high priority commensurate with the importance of residence concepts to the accuracy of the census. We also wish to acknowledge and emphasize, in this context, a tension that will recur throughout our detailed discussion in later chapters. Some principles, and strict application of principles, will necessarily run counter to the most optimal or cost-effective census operations, and some may be extremely difficult to implement in practice. Striking a balance between pure principle and effective operations may not be easy, but the accuracy of the census will be improved by the development of basic residence principles and their use in developing implementation plans.

Part II

Residence Rules Meet Real Life: Challenges in Defining Residence

– 3 –

The Nonhousehold Population

AN INEVITABLE TRUTH in every decennial census is that there are groups of people who are extremely difficult to count. In some cases, their living situations make it difficult to accurately gather data from them by standard enumeration techniques or even to locate them at all. In other cases, they may simply be unwilling or unable to provide accurate information even if a questionnaire reaches them. This is the first of two chapters in which we focus on groups of people who may have multiple residences (making it difficult to specify one "usual residence") or whose ties to any fixed residence are ambiguous. In addition, we identify groups that are not explicitly covered by current census residence rules or that have historically proven difficult to count by the standard census methods and questionnaires. This listing of complex living situations is by no means exhaustive, but is intended to provide concrete examples of the breadth of difficulties in defining residence.

In each case we attempt to give some indication of what is known about the size of the group; this is important because not all the groups are the same in terms of the magnitude of the problems they present to the census count. Ultimately, as the Census Bureau and other agencies work on approaches to reach these problematic groups, some sense of prioritization is needed in order to make effective use of time and resources. However, there are cases in which no real quantitative assessment of a group's size is possible; instead, we rely on qualitative impressions. For each group, we also indicate how the group was handled under the 2000 census residence rules as well as past censuses.[1]

[1]Our focus throughout this report, particularly in these two descriptive chapters, is on the

We begin in this chapter with some major segments of what the Census Bureau has traditionally termed the "group quarters" population (Section 3–A). Among these groups are students at colleges and boarding schools (3–B), patients of health care facilities (3–C), and persons serving terms in correctional facilities (3–D). As we argue in more detail in Chapter 7, we believe that term "group quarters"—indeed, the very concept—deserves reconsideration and revision. We prefer the nomenclature "nonhousehold population," but still use "group quarters" in this and later chapters for consistency with past work. We also include in this chapter two groups that blend elements of group quarters and standard household enumeration. Children in foster care (3–E) are predominantly placed in individual family homes but may also reside in group home or semi-institutional settings, and they may transition between these settings with some frequency. Likewise, the domestic military population (3–F) includes both barracks housing on military bases as well as housing in surrounding communities. (Of course, the general issue of counting of the military population raises issues related to the enumeration of American citizens stationed or living overseas; we discuss the overseas aspects in Appendix C.)

3–A THE CONCEPT OF "GROUP QUARTERS"

Before discussing large classes of group quarters, it is useful to first describe how the general concept has evolved in past censuses.

Some tabulations of the 1900 census drew a distinction between "private families" and "families not private," and the latter category was subdivided into the categories "hotel," "boarding," "school," "institution," and "other."[2] The 1930 census was the first to make a firm differentiation between places like institutions, prisons, and boarding houses as distinct from more conventional households. Dubbed "quasi-households" in that census and in 1940, and "nondwelling-unit quarters" in 1950, this segment of the population was subsequently renamed "group quarters": the label has endured since then, even though the exact definition and distinction between group quarters and the household population have evolved from census to census.

Most censuses since 1930 have used some numerical standard to define group quarters: the 1930 and 1940 censuses defined 12 unrelated people living

census as it is conducted in the 50 U.S. states and the District of Columbia. The parallel set of census processes conducted in the Commonwealth of Puerto Rico raise unique challenges of their own, from demands on Spanish-language resources to differences in street and mail address formatting.

[2]"Other," in turn, "includes groups of laborers at work on farms and plantations, railroads, roads, etc.; groups of miners and lumbermen in camps, etc.; crews of boats and vessels; soldiers and sailors at military posts and stations and on naval vessels; and miscellaneous groups of persons lodging together but having no family relationship in common" (see U.S. Census Office, 1902:clviii–clix; Tables 95 and 96).

in the same unit as a group quarter, the 1950–1970 censuses used 5 or more unrelated people, and the 1980 and 1990 censuses used 10 or more unrelated people (Ruggles and Brower, 2003:75). However, the 1980 census also began the practice of declaring some types of living facilities—notably, college dormitories, as well as some hospitals and missions or flophouses—as "noninstitutional group quarters" regardless of the number or relationship of the people within those units. This approach was used in the 2000 census: types of places were designated as group quarters in advance, without any numerical criterion in the definition of "group quarters."

Table 3-1 lists the 2000 census totals for the group quarters population, divided by group quarters type. The major groups in that listing—college students, patients in health care facilities, and persons in correctional facilities—are also significant cases where residential ambiguity and census error are potential problems. Before discussing issues specific to each group, we note five issues that are common across the groups.

The first general issue is competing claims as to where facility population should be counted. The size of the population of group quarters facilities makes the geographic placement of the group in census tabulations a sensitive issue, politically and economically. Colleges and universities, prisons, and military bases (with on-base housing) can account for large shares of the overall population and job base of cities, towns, and counties, and—in the quest for allocated state and federal funds—areas that house facilities have a strong incentive to have the facilities counted in their population tallies. However, the various facility types are not the same in how they fit the specifications of funding programs: for example, by virtue of their ability to move around and seek employment outside the facility, the populations of colleges and military installations are arguably more service needy and factor heavily into local areas' planning of almost all activities. By contrast, the inherently less mobile and more insular populations of prisons are less likely to play a role in, say, local transit funding and education decisions; however, the prison population may need to be covered by local fire and emergency response personnel, and so be accounted for in those allocations.

The second broad issue involves gradations in the length of stay. Large group quarters segments differ greatly in the expected length of stay of their residents. College and boarding school students live in dormitories during the 8–10 months of the academic year, and the dormitories are largely vacant during the summer months. Health care facilities like hospitals and nursing homes can hold patients for stays ranging from hours (emergency room visits and admissions) to weeks (short-term physical rehabilitation stays at nursing facilities) to years (long-term nursing home assignment), and individual patients may cycle back and forth between home and hospital stays. Likewise, local jails may hold inmates for a matter of hours but may have to hold some for weeks or months (e.g., incarceration during a trial), and prison sentences

Table 3-1 Group Quarters Population by Group Quarters Type, 2000 Census

Group Quarters Type	Population
Institutionalized Population	4,059,039
Correctional institutions	1,976,019
Federal prisons and detention centers	154,391
Halfway houses	30,334
Local jails and other confinement facilities (including police lockups)	623,815
Military disciplinary barracks	20,511
State prisons	1,077,672
Other types of correctional institutions	69,296
Nursing homes	1,720,500
Hospitals/wards, hospices, and schools for the handicapped	234,241
Hospitals/wards and hospices for chronically ill	40,022
Hospices or homes for chronically ill	7,682
Military hospitals or wards for chronically ill	1,944
Other hospitals or wards for chronically ill	30,396
Hospitals or wards for drug/alcohol abuse	19,029
Mental (psychiatric) hospitals or wards	79,106
Schools, hospitals, or wards for the mentally retarded	41,857
Schools, hospitals, or wards for the physically handicapped	20,765
Institutions for the deaf	2,325
Institutions for the blind	1,885
Orthopedic wards and institutions for the physically handicapped	16,555
Wards in general hospitals for patients who have no usual home elsewhere	23,538
Wards in military hospitals for patients who have no usual home elsewhere	9,924
Juvenile institutions	128,279
Long-term care	71,496
Homes for abused, dependent, and neglected children	19,290
Residential treatment centers for emotionally disturbed children	12,725
Training schools for juvenile delinquents	39,481

Short-term care, detention or diagnostic centers for delinquent children	29,437
Type of juvenile institution unknown	27,346
Noninstitutionalized Population	3,719,594
College dormitories (includes college quarters off campus)	2,064,128
Military quarters	355,155
On base	310,791
Barracks, unaccompanied personnel housing (enlisted/officer), and similar group living quarters for military personnel	281,416
Transient quarters for temporary residents	29,375
Military ships	44,364
Group homes	454,055
Homes or halfway houses for drug/alcohol abuse	94,243
Homes for the mentally ill	63,912
Homes for the mentally retarded	148,043
Homes for the physically handicapped	16,119
Other group homes	131,738
Religious group quarters	78,102
Dormitories	97,697
Agriculture workers' dormitories on farms	47,664
Job Corps and vocational training facilities	32,143
Other workers' dormitories	17,890
Crews of maritime vessels	4,523
Other nonhousehold living situations	95,565
Other noninstitutional group quarters	570,369
Total	7,778,633

SOURCE: 2000 Census Summary File 1, tabulated through http://factfinder.census.gov.

can range from months to life. This gradation in the possible length of stay at facilities can create residence conundrums for census respondents and their families—if a household member is expected to return home from a hospital stay in a few days/weeks/months, should they be counted at their home or at the nursing facility? And, as we discuss in more detail later, should prisoners be counted at a "home of record" rather than a prison location and, if so, should remaining time left in a prisoner's sentence (e.g., 3 months or natural lifetime) be a factor in making that determination?

The third general issue concerns the continuum of attachment to a facility. Consistent with the variation in lengths of stay at facilities, the group quarters population spans a continuum of levels of attachment to the facility. The population blends groups of people who are unequivocally in the facility as a usual residence (they have no other residence outside the facility) with those who have strong ties to a usual residence elsewhere.

In a different vein, the fourth general issue is the presence of a "gatekeeper" or other barriers to direct enumeration. Residents of large group quarters also vary greatly in the degree to which direct enumeration—distribution of questionnaires to be filled out directly by residents—is either feasible or permissible. Some college campuses, eager to get as complete a count as possible, may encourage such direct contact. Others—citing resource constraints or the magnitude of the job—may insist that enumeration not be done directly and that university records be used. Hospital and group home residents may be physically incapable of completing forms on their own, requiring special enumeration methods or reliance on administrative records. And, finally, prison administrators may cite safety concerns in prohibiting any direct contact with their inmate populations.

Lastly, there are complicated address listing issues that affect all groups. As examined in greater detail in Chapter 7, it can be difficult to maintain an accurate and up-to-date roster of group quarters units, and—if the lists of group quarters and "regular" housing units are kept separate—duplication or omission can occur if the two lists are not reconciled. An analysis of the 2000 census public-use microdata sample (Ruggles, 2003:481) concluded that the Census Bureau's decision to scrap a rule that would define a group quarters on the basis of a set number of unrelated individuals living together was largely inconsequential because only 42 households in the sample would have been declared group quarters using such a rule (using the 1990 cutoff). However, many of these stray households appear to be clear cases in which buildings were misclassified and errors were made in listing group quarters facilities: these "large households" consisted entirely of college students, farm laborers, or construction laborers.[3]

[3]Among these is "one unit with 53 female housemates aged 18 to 24, all of whom were attending a public university" (Ruggles, 2003:481).

Large institutional group quarters may also have housing units that should rightly be contacted by usual census methods like mailout/mailback: college dormitories may include rooms for resident assistants who maintain a permanent residence there; boarding schools may likewise have permanent on-site resident faculty and staff quarters; and contemporary nursing homes may blend apartment-style assisted living units with more institutional, hospital-like nursing beds in the same facility.

It should also be noted that the questionnaire used to enumerate group quarters residents differs somewhat from the general household questionnaire. Group quarters resident information was to be collected on Individual Census Reports (ICRs), or customized versions for the military and shipboard populations; see Box 3-1. In addition to the ICRs—one per person—enumerators of group quarters facilities were expected to complete a general roster of residents. The general roster and every ICR were supposed to be marked with the same group quarters identification number (the space for which is shown on page 2 of the ICR in Box 3-1).

3–B STUDENTS

3–B.1 Colleges and Universities

As Lowry (1987:15) observed, "[college] students are individually transient—each year many leave never to return—but as a class they form a permanent component of the local population that must be served in various ways by local government." As a large group of people with strong ties to two locations, college students have understandably been prominent in census residence issues from the outset. College students were the focus of one of the first formal census residence rules (see Section 2–B.1), and were the focus of one of the most significant changes to the rules in recent decades.

Enumerator instructions for 1850 directed that college students be "enumerated only as the members of the family in which they usually boarded and lodged on the 1st day of June." Instructions for the 1870 census clarified that "children and youth absent for purposes of education on the 1st of June, and having their home in a family where the school or college is situated, will be enumerated at the latter place." Whether the student was to be counted at home or school would depend on the course of study and whether students returned home at the end of the academic calendar. The 1880–1900 census instructions signaled a clearer inclination for counting students at their parental homes, although that specific language was not explicitly used. A college student was identified by the 1880 instructions as a person for whom the enumerator "can, by one or two well-directed inquiries, ascertain whether [there] is any other place of abode within another district at which he is likely to be reported." (Similar text is included in the subsequent census instructions.) The

Box 3-1 Individual, Military, and Shipboard Census Reports

The census questionnaire used to gather data from people in group quarters facilities is similar to, but not exactly the same as, the questionnaire mailed to the general household population. In the 2000 census, four separate forms were used for group quarters. The *Individual Census Report (ICR), short form,* was a two-page questionnaire, shown here. A six-page *ICR, long form,* was administered to those people included in the long-form sample; it presented the ICR short-form questions (though in less space), along with the social and demographic long-form questions. The *Military Census Report (MCR)* and *Shipboard Census Report (SCR)* were based on the ICR long form, for enumeration of military personnel and maritime vessel crew, respectively. The MCR and SCR forms begin by asking for context-specific information—installation/base name and unit name for the military, name and operator of ship for the shipboard population—after requesting the respondent's name in question 1.

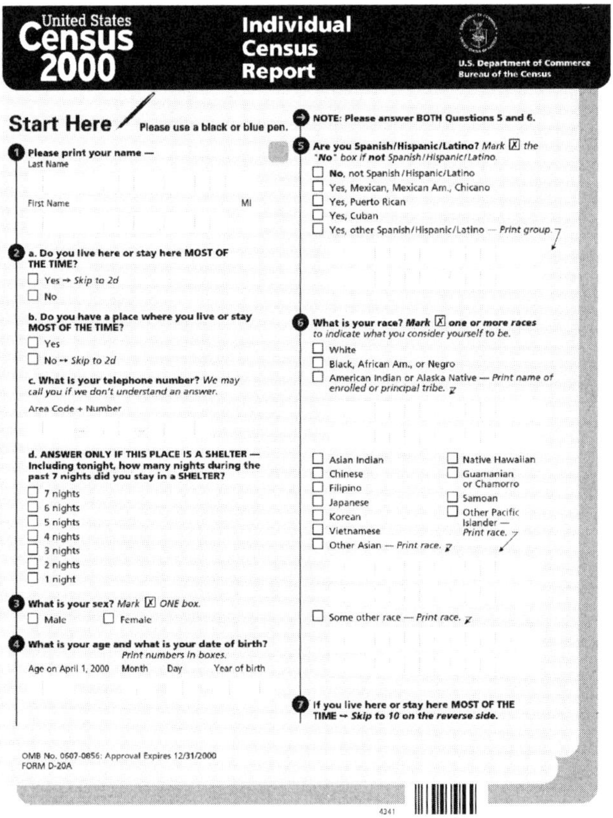
United States Census 2000

Individual Census Report

U.S. Department of Commerce
Bureau of the Census

Start Here Please use a black or blue pen.

1 Please print your name —
Last Name
First Name MI

2 a. Do you live here or stay here MOST OF THE TIME?
☐ Yes → *Skip to 2d*
☐ No

b. Do you have a place where you live or stay MOST OF THE TIME?
☐ Yes
☐ No → *Skip to 2d*

c. What is your telephone number? *We may call you if we don't understand an answer.*
Area Code + Number

d. ANSWER ONLY IF THIS PLACE IS A SHELTER — Including tonight, how many nights during the past 7 nights did you stay in a SHELTER?
☐ 7 nights
☐ 6 nights
☐ 5 nights
☐ 4 nights
☐ 3 nights
☐ 2 nights
☐ 1 night

3 What is your sex? *Mark* ☒ *ONE box.*
☐ Male ☐ Female

4 What is your age and what is your date of birth? *Print numbers in boxes.*
Age on April 1, 2000 Month Day Year of birth

→ NOTE: Please answer BOTH Questions 5 and 6.

5 Are you Spanish/Hispanic/Latino? *Mark* ☒ *the "No" box if not Spanish/Hispanic/Latino.*
☐ **No**, not Spanish/Hispanic/Latino
☐ Yes, Mexican, Mexican Am., Chicano
☐ Yes, Puerto Rican
☐ Yes, Cuban
☐ Yes, other Spanish/Hispanic/Latino — *Print group.*

6 What is your race? ***Mark*** ☒ ***one or more races*** *to indicate what you consider yourself to be.*
☐ White
☐ Black, African Am., or Negro
☐ American Indian or Alaska Native — *Print name of enrolled or principal tribe.*
☐ Asian Indian
☐ Chinese
☐ Filipino
☐ Japanese
☐ Korean
☐ Vietnamese
☐ Other Asian — *Print race.*
☐ Native Hawaiian
☐ Guamanian or Chamorro
☐ Samoan
☐ Other Pacific Islander — *Print race.*
☐ Some other race — *Print race.*

7 If you live here or stay here MOST OF THE TIME → *Skip to 10 on the reverse side.*

OMB No. 0607-0856: Approval Expires 12/31/2000
FORM D-20A

4341

Box 3-1 (continued)

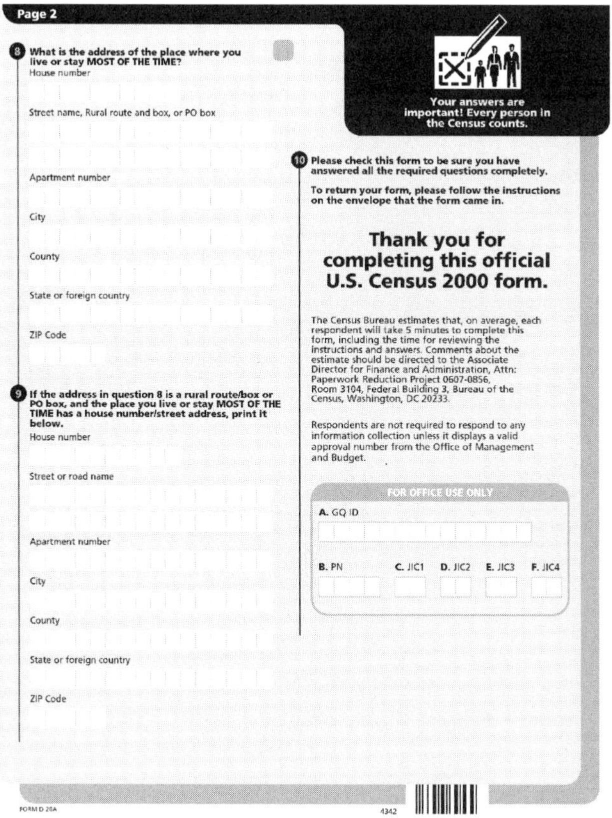

Page 2

8 What is the address of the place where you live or stay MOST OF THE TIME?
House number
Street name, Rural route and box, or PO box
Apartment number
City
County
State or foreign country
ZIP Code

9 If the address in question 8 is a rural route/box or PO box, and the place you live or stay MOST OF THE TIME has a house number/street address, print it below.
House number
Street or road name
Apartment number
City
County
State or foreign country
ZIP Code

Your answers are important! Every person in the Census counts.

10 Please check this form to be sure you have answered all the required questions completely.

To return your form, please follow the instructions on the envelope that the form came in.

Thank you for completing this official U.S. Census 2000 form.

The Census Bureau estimates that, on average, each respondent will take 5 minutes to complete this form, including the time for reviewing the instructions and answers. Comments about the estimate should be directed to the Associate Director for Finance and Administration, Attn: Paperwork Reduction Project 0607-0856, Room 3104, Federal Building 3, Bureau of the Census, Washington, DC 20233.

Respondents are not required to respond to any information collection unless it displays a valid approval number from the Office of Management and Budget.

FOR OFFICE USE ONLY
A. GQ ID
B. PN C. JIC1 D. JIC2 E. JIC3 F. JIC4

FORM D-20A 4342

A particular feature of note in the ICR is that it asks for information on a "usual home elsewhere" ("What is the address of the place where you live or stay MOST OF THE TIME?"), even if the respondent lived in a group quarters type for which the Census Bureau did not allow "usual home elsewhere" information to be used. The MCR adds a filtering question and is more specific in defining a time period: "Is the place where you stay at least 4 nights a week a barracks, BOQ [bachelor officers' quarters], disciplinary barracks, hospital, etc., or a house, apartment, or mobile home?" If the answer is "house, apartment, or mobile home," the person is asked for the address; if not, he or she is routed to a question asking for a telephone number. The SCR asks, "Do you have a residence (house, apartment, or mobile home) where you usually stay when off duty?" If yes, the address is requested.

exact resolution if such a home elsewhere was found (the parental home) was not made clear, but the implication is that students were not to be counted if found at school if they were already counted at home. No specific mention of college students (distinct from general boarders) are made in enumerator instructions between 1910 and 1940, preserving the notion of counting college students at their parental homes.[4]

In 1950 the Census Bureau reversed the rule and concluded that the students should be counted at the place they were living while attending college. Bureau documentation ascribes the change to the desire for consistency with "usual residence"—"most students live in college communities for as much as nine months of the year, so the college is their usual residence"—but also because it was believed that the count would be more accurate since students "were often overlooked in the enumeration of their parental homes" in past censuses (U.S. Bureau of the Census, 1955:11). In part, this shift owed to the evolution of the "usual residence" concept, but it also reflected a significant demographic fact of the times—following World War II, large numbers of returning veterans became college students with assistance from the GI Bill. Accordingly, the basic demography of college campuses shifted from a children-away-from-home model to an adult education model, and considering students' school residences as their "usual" residences was consonant with that shift. The rule change in 1950 created a substantial shock in demographic data series that future analysts would have to consider: "63.7 percent of students aged 18 to 22 resided without family in 1950, compared to just 7.0 percent in 1940" (Ruggles and Brower, 2003:82).[5]

The size of the college population, as a whole but particularly as a share of many cities and towns, makes the question of where college students should be counted a contentious one. In this respect, the college population is similar to the military and incarcerated populations: they are all instances where slight variation in the census residence rules can dramatically alter the demographic portrait of local areas (and the state and federal fund allocations to those areas). The counting of the college population has thus been challenged over time, but the U.S. Supreme Court's 1971 decision in *Borough of Bethel Park v. Stans* (see Box 3-2) upheld the Census Bureau's right to make a rational determination as to where college students and other institutional populations should be counted.[6] For purposes of constructing its legislative districts and allocating state funds, the state of Kansas disagrees with the Census Bureau's

[4]Gauthier (2002) does note an exception invoked in 1930: cadets at the U.S. Military Academy and the U.S. Naval Academy were to be counted at the academies.

[5]Ruggles and Brower (2003:83) note that "the number of students was still small in 1950," so "the consequences for the population as a whole were small." Considering the whole population aged 15 and older, for instance, inference on the proportions residing with or without family would not be greatly altered by using either the 1940 or the 1950 rule, but the change is significant for analyses focused more squarely on the college-age population.

[6]Legal and political pressures on the counting of college students continue; we discuss the

interpretation; the state fields its own decennial survey to "adjust" where college students (and military personnel) in the state are counted. We discuss Kansas' approach in Box 7-2 in Chapter 7.

In the 2000 census, college students were the explicit focus of 3 of the 31 formal residence rules. Students who live at home while attending college are to be counted at home (Rule 6); those who live away from the parental home while attending college and are "only here during break or vacation" are to be counted at their residence at college (Rule 5); and students living in a group quarters (such as a dormitory or a fraternity or sorority house) are to be counted at the group quarters (Rule 25). Two other rules address college students indirectly; foreign students studying in the United States on Census Day are to be counted at their household location in the United States (Rule 11), while U.S. students studying abroad are excluded from the census (Rule 29).

The standard of counting college students at the college location had been revisited by Bureau staff working groups prior to the 2000 census, and the decision to retain it was not a unanimous recommendation. As Rolark (1995:5) comments, "some members of [the Bureau team] felt very strongly that we should allow all college students to claim a [usual home elsewhere] in the 2000 census"—presumably, students could claim and be counted at their family home. However, the Bureau's steering committee for the census rejected the change. Instead, "there was general agreement [on] the need to educate respondents on the correct procedure for reporting students away at college," and the committee indicated that it would be important "to collect 'hard' data on the reporting habits of college students and their parents" (Rolark, 1995:Comments on Rule 5).

College students are a challenging population to count accurately because they present difficulties on many different levels, both in terms of definitions and operations and because of potential misunderstandings by both students and parents. Three major difficulties are students' independence versus their familial ties, the variation in kinds of college residences, and the academic calendar.

Student Independence Versus Parents' Enduring Ties

For most college students, aged 18–24, the college years are an important time of transition from dependence on parents to independent adulthood. In

---

analogous arguments on the counting of prisoners elsewhere in this report. With specific respect to college students, the adequacy of the Census Bureau's procedures for generating intercensal population estimates for including college students has drawn some fire. Most recently, the state of Massachusetts has contemplated legal action, contending that failure to include on the order of 30,000 college students—along with recent immigrants and others who may well not be fully included in federal tax records, which are used in part to derive population estimates—deprive the state of "millions of dollars in federal aid" (Lewis, 2005).

Box 3-2 *Borough of Bethel Park v. Stans* (1971)

Plans to include overseas U.S. military personnel and their dependents in the state-level apportionment counts in the 1970 census, based on "home of record," occasioned a major challenge to the Census Bureau's practices in counting other special populations. Several Pennsylvania municipalities and other parties brought suit against the U.S. Secretary of Commerce and the census director, arguing that the Bureau's residence standards for college students, military personnel, and the institutionalized population violated the Constitution and the Census Act. Specifically, the Pennsylvania municipalities argued that their populations were undercounted because college students, domestic military personnel, and institutional inmates were counted at the location of the facilities rather than the place that those individuals might consider their "legal residence for all purposes other than the census." The appellants approved of the Bureau's new policy of siting federal and military personnel and dependents stationed overseas at their "home of record," but argued that they should be assigned to specific addresses (and not just included in the state-level apportionment totals).

After a district court ruling sided with the Census Bureau, the city of Philadelphia and Rep. James Fulton (R-Pennsylvania) appealed the case; the U.S. Court of Appeals for the Third Circuit rendered its judgment in *Borough of Bethel Park v. Stans* (499 F.2d. 575) on September 30, 1971. Although the Court of Appeals noted that the current U.S. Code contains no attempt of definition of "usual residence" akin to the Census Act of 1790, the court said that "it has been stipulated that the following criterion was used by the Bureau of the Census to determine usual place of residence for the 1970 Census: 'Persons are enumerated at the place in which they generally eat, sleep and work, with persons who are temporarily absent for days or weeks from such usual place of abode being counted as residents of their usual place of abode.'" The Court held that this criterion is a "historically reasonable means of interpreting" the Constitutional and legislative mandate of the census.

The 1970 census was only the third in which college students were counted at their schools rather than parental homes. "Once a person has left his parental home to pursue a course of study at a college in another state which normally will last for a period of years, it is reasonable to conclude that his usual place of abode ceases to be that of his parents." The Court of Appeals was not swayed by the appellant's argument that, "if an individual college student indicates that he feels a particular connection or attachment to the state of his parental home, registers to vote in that state, and accordingly regards it as his home, the Bureau should consider these facts in determining his residence for the purpose of the census enumeration." Instead, the Court ruled that "the Bureau is entitled to limit its inquiry to the objective facts as to where the student chooses to generally eat, sleep, and work—the state of his college rather than the state where his contacts are substantially reduced." This gives the Bureau a "definite, accurate and verifiable standard." [In a footnote, the Court also accepted the Bureau's practice of reversing course and counting boarding school students at the parental home.]

Using similar logic, the Court also found that "appellants have failed to impugn the Bureau's exercise of discretion" regarding the military and institutional populations. Accepting the Bureau's assertion that allocating overseas federal and military personnel to specific addresses or to levels of geography below the state "would be an impossible task," the Court held that the Bureau's decisions "possessed a rational basis." Likewise, "persons confined to institutions where individuals usually stay for long periods of time" (as contrasted with short hospital stays) are counted "as residents of the state where they are confined" under census rules. "We think that the decision of the Bureau as to the place of counting institution inmates has a rational basis," concluded the Court.

many cases, college represents the first extended period of being away from their family home; the peak college age group straddles important legal and cultural thresholds of adult responsibilities at ages 18 and 21 (e.g., voting, Selective Service enrollment, and legal ability to consume alcohol). We also noted in Chapter 2 the intricacies of qualifying for resident status for purposes of college tuition and how that may conflict with other residence standards. College students are also of the age where they can begin to take on financial burdens in their own right (e.g., taxes on employment income or student loans). These and other factors can lead to contradictory impulses in college students' minds when they are asked where they reside—the new independence (and burden) of their college life may lead them to identify with their college community, but the tie to family homes may be just as strong.

At the same time, parents may have strong impulses to list their college student sons and daughters when asked who lives at their household—whether or not that coincides with Census Bureau notions of residence and regardless of clear instructions on how students are to be counted. In many cases, it is the parents who bear the substantial costs of college tuition; strength of kinship ties aside, the financial tie of paying for student tuition alone may induce some parents to count college students as part of their household. The "enduring ties" philosophy makes it counterintuitive to many parents to exclude their children from a list of household members. Therefore, there is a conflict between the subjective feelings of both parents and students and the census residence rules, which can lead to both omission (college students assuming that they are counted at "home" and not completing a form at their school location) and—particularly—duplication (being counted at both the college and home locations).

Variation in College Housing Types and Options

College housing options can be divided into on campus (dormitories, graduate and married student housing, residential colleges, and fraternity or sorority houses) and off campus, for which the basic distinction is whether the school serves as the effective landlord. Both these broad categories present their own challenges.

First, students living in college-owned on-campus facilities are generally tied to the academic calendar; they live at the on-campus location during the academic year but do not (and most often cannot) live there during the summer months. Colleges and universities vary in the extent to which they require students to live on campus; some require college freshmen to live on campus and others may require students to live on campus during part or all of their college career. However, they need not live in the same specific room or hall during the entire time; from year to year, the population in residence calls can be highly dynamic. More salient for purposes of a census, on-campus living arrangements vary significantly in the style of housing, from traditional dormitories with shared bathroom facilities, common dining halls, and with postal boxes in the building or at a student union to rooms and suites that are indistinguishable from regular apartments.[7] (Indeed, in attracting students, colleges and universities have diversified their housing stock—adding more apartment-style living, with ready access to fitness facilities and other amenities.) It is possible that more apartment-style and "independent" living may encourage students to more naturally think of the college residence as their usual residence (even if that does not square with their parents' notions).

Second, off-campus housing options for college students include apartments, either alone or in conjunction with other students, or cooperation in leasing a home with peers. In principle, these students should be enumerated in the same manner as the general population, through receiving a form in the mail and returning it. While the act of leasing an apartment or part of a house creates a strong link to the college location, an enduring tie can remain to the family home. In many cases, students' parents may be cosignatories on leases and may contribute part or all of rent payments, further leading parents to count their at-college students as part of their household.

Of course, some students may be able—and choose—to attend college without leaving their parents' home, depending on the location of the school.

As shown in Table 3-1, the 2000 census counted 2,064,128 students in dormitories or college-owned quarters on or off campus. With data from the National Postsecondary Student Aid Study, a survey conducted every 4 years by the National Center for Education Statistics, Table 3-2 shows students' living arrangements while enrolled in college. Among full-time enrollees at 4-year universities (public and private), the data suggest that about 40 percent of undergraduates live in on-campus housing; 42 percent live in off-campus housing; and about 18 percent live with their parents.

By comparison, for 2-year (community or junior) colleges and private for-profit schools, the combination of off-campus housing and living with parents accounts for about 97 percent of full-time enrollees. On-campus housing in residence halls is not the exclusive province of 4-year institutions; some community colleges (particularly in rural areas, with large service districts) offer on-campus housing, and other community colleges may view adding housing options (either residence halls or partnerships with off-site apartments) as means for attracting students.

---

[7]At some urban universities, it is not uncommon for the school to acquire apartment buildings or hotels for conversion to residence halls (see, e.g., http://www.bizjournals.com/washington/stories/1999/05/31/story4.html [8/1/06] on George Washington University's conversion of the 193-room Howard Johnson Premier Hotel—best known as the lookout point during the 1972 Watergate break-in—to a student residence hall). These types of conversions further complicate the task of making sure unit listings for those buildings are listed correctly and handled by the proper enumeration method.

Table 3-2 Undergraduate College Housing, 2003–2004

Student and Institution Characteristics	Housing			
	On Campus	Off Campus	With Parents/ Family Home	Total
Class Level				
Freshman/1st year	959.6	3,403.0	1,967.4	6,330.0
Sophomore/2nd year	666.0	2,548.5	1,270.6	4,485.2
Junior/3rd year	484.1	1,518.5	503.2	2,505.8
Senior/4th year	435.6	1,625.8	424.6	2,486.0
5th Year+ or unclassified	86.1	1,428.0	302.9	1,817.0
Student Enrollment Status and Institution Type				
Full-time/full-year, public 4-year	1,166.3	1,629.0	660.0	3,455.4
Full-time/full-year, private 4-year	876.3	518.6	247.1	1,642.0
Full-time/full-year, public 2-year	63.0	814.2	922.9	1,800.1
Full-time/full-year, private for-profit	9.7	321.3	121.2	452.2
Part-time/part-year, single institution	516.8	7,240.3	2,517.2	10,274.3
Lived with Parents While Not Enrolled				
Yes	1,914.5	1,180.0	3,534.1	6,628.6
No or independent	716.9	9,343.8	934.6	10,995.3
Attend Institution in State of Legal Residence				
Yes	1,900.4	9,488.0	4,254.4	15,642.8
No	677.8	858.7	153.4	1,689.9
Foreign or international student	53.6	176.7	60.9	291.2
Institution Distance from Home (miles)				
0–30	605.3	7,492.1	3,888.1	11,985.5
30–100	745.5	1,567.3	442.2	2,754.9
100-500	946.5	989.1	82.7	2,018.3
More than 500	311.3	407.2	42.4	760.9
Total	2,631.5	10,524.3	4,468.1	17,623.9

NOTES: Cell counts calculated from row-wise weighted population estimates (in thousands). Tabulations exclude students who attended more than one institution during the 2003–2004 academic year.

SOURCE: Data from National Center for Education Statistics, National Postsecondary Student Aid Study, 2004 Undergraduates. Tabulation from DAS-T Online, accessed through http://nces.ed.gov/das/.

Census Schedule Versus School Schedule

Depending on the academic calendar at particular schools and the calendar structure (e.g., semesters or quarters), the peak of decennial census enumeration activity may coincide with particularly bad times to obtain an on-campus count: spring break or a change in term. Moreover, nonresponse follow-up activities can begin in earnest at the time that the school year winds down and students either graduate or return home. Later still, postenumeration surveys to assess coverage may occur when students are absent from the college location. (The short-term population shifts associated with college students' spring break is also discussed briefly in Section 4–A.1.)

3–B.2 Boarding Schools

Boarding and preparatory schools pose the same basic issues and challenges to enumeration as colleges and universities. They also conform to an academic calendar and can be largely vacant during the summer months. Likewise, not all boarding school students actually live on school grounds, as some boarding schools take "day students" who live in the area and only come to school for classes. However, boarding school students differ from college students significantly in their age: because boarding school students are minors, they are legally and financially dependent on their parents and are almost certain to return to their family homes when school is not in session. Moreover, given the students' youth, parents' instinctive urge to count them at home is perhaps stronger than for older college students.

Data from the National Center for Education Statistics suggest that the total enrollment at boarding schools in the United States is approximately 100,000 students at 1,500 schools, including both day and boarding students.[8] As of 2003, the Bureau of Indian Affairs operated 66 boarding schools with approximately 9,500 students; some of these students are known to live at the school location year-round, staying with a state- or tribe-appointed guardian during vacations (U.S. Census Bureau, Population Division, 2004).

In the 2000 census, Rule 7 of the formal residence rules said that "a student attending school away from home, below the college level, such as a boarding school or a Bureau of Indian Affairs boarding school" should be counted at the parental home. Accordingly, it stands as a key exception to a strictly applied "usual residence" concept. The Census Bureau's internal committee that formulated the rules for 2000 elected to count boarding school students at home due to the "inherent dependency that boarding school students have on

[8]Particularly challenging to count (for census purposes) are those students who are enrolled at U.S. boarding schools but whose parents live overseas. If the students are to be counted at the school location, they would be included in the domestic population count; if they are to be counted at the parental home, they would be excluded from census totals altogether.

their parents;" the analysts worried about accurately operationalizing a count of boarding students at the school because "there would most certainly be a greater possibility of misreporting if we were to leave the enumeration of this population to the students themselves or to school administrators" (Rolark, 1995:6). That said, boarding schools were not specifically mentioned in the include/exclude instructions at the beginning of the 2000 census questionnaire. If respondents (parents) strictly followed the questionnaire guidance, the advisory to exclude "people who live or stay at another place most of the time" would lead them to exclude boarding school students, even though the formal rule is to count them at home.

Other commentators who have looked at recent census residence rules (Lowry, 1987; Hill, 1987; Mann, 1987; Sweet, 1987) have advocated that the rule be changed to count students at the boarding school location, for consistency with true "usual residence."[9] Reflecting the desire for consistency, the Census Bureau's draft discussion points on rules for the 2006 census test (shared with the panel) suggested that boarding school students be counted at the school location, not at the parental location, for consistency with "usual residence" (U.S. Census Bureau, Population Division, 2004).

3–C HEALTH CARE FACILITIES

Another major class of group quarters are those that provide health care and related services. Of particular interest in the census context are long-term care facilities, such as nursing homes and other residential communities for senior citizens, although health care settings where potential ambiguity in defining residence may occur include hospitals and rehabilitation and treatment centers. Health care-related group quarters are sufficiently varied, and living situations so potentially complex, that 4 of the 31 formal residence rules for the 2000 census dealt explicitly with them. Rule 3 says that a person who "lives in this household, but is in a general or Veterans Affairs hospital on Census Day," should be counted at "this household, unless in a psychiatric or chronic disease hospital ward, or a hospital or ward for the mentally retarded, the physically handicapped, or drug/alcohol abuse patients. If so, the person should be counted in the hospital." Rules 21 and 23 indicate that persons who are, on Census Day, "under formally authorized, supervised care or custody," in a "nursing, convalescent, or rest home for the aged and dependent" or a "home, school, hospital, or ward for the physically handicapped, mentally re-

[9]These papers, from the 1986 Council of Professional Associations on Federal Statistics (COPAFS) residence rules workshop, argued for counting boarding school students at the school. However, in synthesizing the results of the workshop, CEC Associates (1987:24) make the treatment of boarding students one of their formal recommendations but, in the text, give greater weight to those reasons that "argue for counting boarding school students at the parental home, as has been done in the past."

tarded, or mentally ill," are to be counted at the facility. Likewise, Rule 24 indicates that persons at emergency or other shelters on Census Day should be counted at the shelter.

As shown in Table 3-1, the 2000 census tallied 1,720,500 people in nursing homes. It included in its "institutionalized" count 234,241 people in hospitals and other health care facilities (including schools for the handicapped), the largest share of which (79,106) were in mental (psychiatric) hospitals or wards. Group homes, including those connected to drug abuse recovery programs and homes for the mentally ill or disabled, accounted for 454,055 of the noninstitutionalized population.

The distinction drawn in the census tabulations between institutionalized and noninstitutionalized health service facilities is a telling one, because it speaks to what has been pointed out as a problem in the Bureau's current handling of these types of group quarters. An important trend that has emerged over the past decade is a blurring of distinction between medical care facilities and what might ordinarily be considered an institutional group quarter. In the past, nursing homes were typically inhabited by long-term residents needing care and supervision, with some people staying at the facility for a 30-day-or-less rehabilitation stay after a hospital procedure. More contemporary practice in health care suggests at least a three-level hierarchy in the type of care needed by sick and elderly patients:

- *residential care* for persons who are physically capable of maintaining a large measure of independence in day-to-day living but who do require some level of assistance;
- *intermediate care* for persons with conditions that render them unable to live independently, but not to the point that they need constant intensive care (instead, for example, they may need rehabilitation therapy to regain daily living functions); and
- *skilled care* for persons who need constant attention from nursing personnel.

To satisfy these care needs, providers have developed an array of living settings and situations. McCormick and Chulis (2003:143) observe that the "proliferation of facility-like residential alternatives to nursing homes"—going by "various names including assisted living facilities, continuing care facilities, retirement communities, staged living communities, age-limited communities"—over the past 10 to 15 years is a major reason why 1980s-era projections of looming shortages of traditional nursing home beds "did not materialize." These "life care communities," which McCormick and Chulis (2003:143) dub elderly group residential arrangements, have added to the complexity in defining residence.

Some "life care communities" have three levels of care and housing. One level consists of people in independent living facilities, usually an apartment that they own or rent. Staff may come in to assist with some housekeeping tasks and the residents can cook for themselves or eat in a central restaurant. In the second level, assisted living, people still live independently, but they require more assistance with daily activities. The third level is for people unable to care for themselves and that need continuous daily care and supervision. Yet all three groups may live on the same campus under the same auspices. Are all of these people to be counted as institutionalized populations?

At the same time that health care services have diversified the type of housing and care options available to a longer lived (and longer active) elder population, the levels and types of care provided at hospitals—usually thought of as acute, short-term care facilities—has also changed in important respects. As health care providers have become more vertically and horizontally integrated, the physical structure that used to be only a "hospital" may now include such components as nursing home care, substance abuse treatment, psychiatric care, physical rehabilitation, and assisted living in separate wings (Drabek, 2005). Many hospitals offer extended care units (ECUs), which can offer home-like settings for lengthy periods of time while still permitting 24-hour observation and care. These ECUs are often connected to psychiatric services provided for both adult and adolescent patients, but can be more general; extended-term rehabilitation and nursing units may be embedded in a hospital system (and indeed within hospital buildings). Still other hospitals have cited the cost and liability of maintaining ECUs and have dissolved them, transferring patients to dedicated facilities elsewhere.

A series of surveys conducted by the Census Bureau on behalf of the National Center for Health Statistics (NCHS) offers empirical evidence of some of the dynamics of the long-term care population. Since 1973, NCHS has periodically contracted with the Census Bureau to conduct the National Nursing Home Surveys (NNHS).[10] In the most recent round for which data are available (1999), a sample of 1,500 facilities was drawn, each was mailed an advance letter, and an interviewer set up an appointment with the administrator. The interviewer administered a questionnaire on facility characteristics to the administrator. They then asked for a list of both current residents and residents who had been discharged during a specified month (one chosen between October 1998 and September 1999); up to six current residents and six discharged patients were randomly selected from these rosters. No resident—current or discharged—was interviewed directly; instead, the interviewer collected information from facility staff members suggested by the

[10]NNHS rounds were fielded in 1973–1974, 1977, 1985, 1995, 1997, and 1999; a new version was to be fielded in late 2004, but data from that round have not been released. See http://www.cdc.gov/nchs/about/major/nnhsd/nnhsdesc.htm [8/1/06].

administrator. The current patient questionnaire[11] directs the interviewers to ask the staff member if they have the medical file and records for the selected patient; "if no record is available for a resident, try to obtain as much information as possible from whatever administrative records are available and/or from the respondent's memory."

Analyzing data from the 1977, 1985, and 1999 NNHS rounds, Decker (2005) estimates that the total number of residents receiving care at nursing facilities in 1999 was 1.63 million, a 27 percent increase from 1977; these were distributed across 18,000 facilities. However, this growth in total residents served was accompanied by a higher rate of discharge and shorter lengths of stay. The number of discharges per 100 nursing home beds grew from 86 and 77 in 1977 and 1985, respectively, to 134 in 1999. This escalation in the discharge rate was concentrated almost entirely among cases in which the patient's length of stay was less than 3 months; see Table 3-3. The growth in short-term stays resulted in an overall drop in the mean length of stay from 398 days in 1985 to 272 days in 1999. Decker (2005:3) comments that "the increase in the number of nursing home residents requiring short stays coincides with the implementation in 1983 of the hospital prospective payment system. This system shortened hospital stays and increased Medicare-funded postacute care in nursing homes."

Still, "the nursing facility was in 1999, as it was in 1977, a place where many of the residents had been in the facility for substantial durations since their admission"—27 percent of current residents in 1999 had been in the facility for 3 years or more since admission, and 30 percent had stays of 1–3 years (see Table 3-3) (Decker, 2005:3–4).

A second survey covering part of the long-term care population—specifically, transitions between short-term hospitalization and long-term care—is the National Hospital Discharge Survey (NHDS). Every year since 1965, the NHDS has examined the characteristics of inpatients discharged from non-Federal short-stay hospitals, based on sample of approximately 270,000 patient records across about 500 hospitals. Only hospitals with an average length of stay of less than 30 days are eligible for inclusion in the sample. The NHDS collects data from a sample acquired in two ways—manual transcription from hospital records to survey forms, by hospital staff or by Census Bureau staff (on NCHS' behalf) or through purchase of electronic medical records data. Analysis of data from the NHDS (Kozak, 2002) shows 2.8 million transfers of patients from hospitals to long-term care institutions in 1999, up from 1.6 million in 1990. These transferred patients tended to have had shorter hospital stays in 1999 than in 1990: nearly half stayed in the hospital 5 days or less in 1999. Kozak (2002) concludes that these trends

[11]The questionnaire is viewable at http://www.cdc.gov/nchs/data/nnhsd/nnhs99_3.pdf [8/1/06].

Table 3-3 Patient Discharges and Distribution of Current Nursing Home Residents, by Length of Stay (in percent)

Length of Stay	1977	1985	1999
Patient Discharges per 100 Beds			
Less than 3 months	46.1	40.3	91.7
3 months to less than 1 year	19.3	17.2	20.0
1 year to less than 3 years	12.0	11.6	12.5
3 years or more	8.7	8.5	10.0
Percent Distribution of Current Residents			
Less than 3 months	14.4	12.8	17.8
3 months to less than 1 year	22.1	23.9	25.0
1 year to less than 3 years	32.8	31.4	30.1
3 years or more	30.7	31.9	27.1

SOURCE: Decker (2005:Figures 3 and 4), from National Nursing Homes Surveys, 1977, 1985, and 1999.

suggest that skilled, subacute care in long-term care facilities is increasingly a substitute for hospital care.

Thus, overall, trends in the industry make it difficult to accurately develop address lists for health care facilities, and the internal diversity in living situations in facilities labeled "hospitals" or "nursing homes" exacerbate the problem. Address listing efforts that rely on the name of an institution, or the classification derived from state licensing regulations, may be out of step with the specific unit-by-unit duration of stay and level of care or oversight provided in the facility. For example, the unit occupied by "a person with Alzheimer's [disease] who lives in a residential care setting that provides 24 hour, 7 day a week oversight could be counted as residing in a non-family housing unit, an institutional setting, or a [group quarters] non-institutional setting depending upon the name of the place (assisted living, nursing home or group home, respectively)" in which the unit is nested (Drabek, 2005:1).

Noting this limitation of facility-based analysis, McCormick and Chulis (2003:143) pursued another route by analyzing the Medicare Current Beneficiary Survey, whose sample of respondents is drawn from the Medicare Enrollment Files (person-based), and thus can be contacted wherever they are found. The authors estimate that 30 percent of Medicare recipients living in long-term care facilities lived in elder group residential arrangements in 2001, a near doubling from the 16 percent estimated in 1996; the percentage living in Medicare- or Medicaid-certified nursing homes (institutional) dropped from 70 percent in 1996 to 59 percent in 2001 (McCormick and Chulis, 2003:144).

In addition to the problems associated with simply constructing a list of the units in health care group quarters are the no-less-tricky concerns of collecting information from their patients, once and only once:

- Within the facilities, people's willingness and basic capability to respond to a census questionnaire can vary greatly; for the sake of patients under care and to ease disruption, then, facility administrators bar direct access to patients in many cases. In the 2000 census, just 5 percent of people counted in nursing homes responded to the census questionnaire on their own, and only 8.8 percent of those in hospitals responded on their own; administrative records were used to fill in information on 72.8 and 65.8 percent of those groups, respectively (see Chapter 7 and Table 7-1).
- Much as is the case with college students or on-duty military personnel, the family members remaining at a household when a loved one enters a hospital or nursing facility may not think of the facility as their loved one's "usual residence." This is particularly the case if the move is seen as limited in term (e.g., physical rehabilitation) rather than long-term care.

3–D CORRECTIONAL FACILITIES

As shown in Table 3-1, the 2000 census counted nearly 2 million people at correctional facilities of some sort; about 55 percent of that total was recorded in state prisons. The vast majority of the incarcerated population is housed in *prisons* or *jails*, which differ principally in the expected length of stay. Sickmund (2004:20) distinguishes between the two types of facilities:

- Prisons "are generally state or federal facilities used to incarcerate offenders convicted in criminal court. The convicted population usually consists of felons sentenced to more than a year."
- Jails "are generally local correctional facilities used to incarcerate both persons detained pending adjudication and adjudicated/convicted offenders. The convicted population usually consists of misdemeanants sentenced to a year or less. Under certain circumstances, they may hold juveniles awaiting juvenile court hearings."

The number of persons actually incarcerated in prisons or jails is a relatively small part of the overall population in the criminal justice system. "Of all offenders under correctional care on any given day, nearly three-fourths are under some form of community supervision" (Clear and Terry, 2000:517), which is to say probation or parole arrangements.

Bureau of Justice Statistics (BJS) tabulations of the correctional population at the end of calendar year 2004 put the total incarcerated population at

2,267,787 (Harrison and Beck, 2005). Of these, 1,421,911 were housed in federal and state prisons and 713,990 in local jails (15,757 were housed in prisons in U.S. territories). In addition, 9,788 were housed in facilities maintained by the Bureau of Immigration and Customs Enforcement and 2,177 in military disciplinary facilities. Estimated population counts were available for slightly different time vintages for two other groups, those in Indian tribal jails (1,826 as of midyear 2003) and in juvenile correctional facilities (102,338 as of October 2002).

Federal prisoners constitute a small share of the overall prison count; as of December 8, 2005, weekly population data from the Federal Bureau of Prisons[12] indicated 188,288 federal inmates total. Of these, 85 percent were housed in prisons maintained by the Bureau of Prisons, 10 percent in privately managed secure facilities, and 5 percent in community correction settings (most of these in designated halfway houses or community correction centers).

The incarcerated population shares many of the same challenging features as other institutional group quarters, but in most respects the conceptual problems they raise are even more complicated. Prisoners are arguably tied to more than one residence location and yet defining them as resident of any location is troublesome. They do not—and cannot—live day-to-day in the communities from which they were sent to prison, and yet their possible eventual return creates demands for such local services as parole monitoring, substance abuse rehabilitation, and job counseling social services. They also do not live day-to-day in the communities in which the prisons are located, in that they do not drive on the roads or use other services. Yet they may be counted in those locations for purposes of legislative representation—even though they may be prohibited from voting for said representatives. Like the other institutional group quarters, prisoners may also have family members "back home" who may be reluctant to exclude them from a household count; alternatively, the stigma of incarceration may lead family members to sever ties with the convicted.

Given their size relative to other categories, we focus in this discussion on prisons and jails. However, we note that non- or semi-incarcerative correctional programs may have residential components and so can present challenges to accurate counting. "Halfway houses" or community correctional centers provide room and board (and possibly other services, such as counseling) for those reentering society after serving their sentences; depending on sentence and parole terms, prisoners may spend up to 1 year (and sometimes more) during the transition from prison to society. Work-study release centers typically "allow a person to be free during working hours in order obtain or keep a job but require a return to the facility overnight" (either a local

[12]See http://www.bop.gov/locations/weekly_report.jsp [12/8/05].

jail or other location); "boot camp" facilities put inmates through a short-term (90–180-day) quasi-military program to instill discipline, while "shock probation" sentencing arrangements can pair a short-term incarceration with a probationary period (with the intent of deterring future criminal activity) (Clear and Terry, 2000:518–519).

3–D.1 Prisons

Rules and procedures for counting prisoners have been a long-standing concern in the census. The enumerator instructions for the 1850 census were the first to discuss the issue directly, stipulating that "all landlords, jailors, [and superintendents of institutions] are to be considered as heads of their respective families, and the inmates under their care to be registered as members thereof," with "convict" status noted in another column (Gauthier, 2002:10–11).[13] The census schedule (enumerator form) further directed that "the crime for which each inmate is confined, and of which each person was convicted," be recorded, but urges that information on recent prison releases might best be gathered by reference to administrative records:

> When persons who had been convicted of crime within the year reside in families on [Census Day], the fact should be stated, as in the other cases of criminals; but as the interrogatory might give offence, the assistants had better refer to the county record for information on this head, and not make the inquiry of any family. With the county record and his own knowledge he can seldom err.

Similar procedures held through 1880; the 1890 census still considered inhabitants of a prison as a family but directed that prisoners be listed on a separate schedule from the warden or other resident staff.

The 1900 census included a special schedule (enumerator form) for questions on crime. The instructions for that schedule made the first mention in census materials of the ambiguity of prisoners' residence—"many prisoners are incarcerated in a state or county of which they are not permanent residents. In every case, therefore, enter the name of the county and state in which the prisoner is known, or claims, to reside." In the main census schedule, enumerator instructions read that "all inmates of . . . institutions are to be enumerated; but if they have some other permanent place of residence,

[13]The concept of treating incarcerated groups as families was consistent with practice dating to colonial times; "unlike contemporary institutions, [jails] had no special prison architectural design but resembled regular households, with the keeper and his family typically living on the premises." Local town jails were only conceived as being short-term facilities "to detain people only until they could be tried and then, if convicted, sanctioned or punished shortly thereafter." It was only over the course of the 19th century, with evolving debate on the appropriateness of capital punishment and the potential for criminals to reform, that the concept of a long-term "penitentiary" came into being (Cullen and Sundt, 2000).

write it in the margin of the [population] schedule on the left-hand side of the page." No such provision was made in 1910 or subsequent censuses, and the practice of counting prisoners at the institutional location was adopted. Accordingly, rule 20 of the 2000 census residence rules indicated that a person "under formally authorized, supervised care or custody, in a correctional institution, such as a federal or state prison" should be counted at the prison, with no provision made to claim a usual home elsewhere.

The Census Bureau has long argued that considering the prison the "usual residence" of an incarcerated person is a reasonable choice, and that other interpretations would raise further problems. Testifying before Congress in 1999, Census Bureau director Kenneth Prewitt summarized the Bureau's major arguments, suggesting that creating an exception to usual residence for prisoners "may open a Pandora's box of pressures for other exceptions to our residence rules" (U.S. House of Representatives, Committee on Government Reform, 2000).[14] Subsequently, the Bureau has continued to evoke the "Pandora's box" argument in justifying its position on prisoner counting; for instance, commenting in a Texas newspaper article in late 2005, a census official noted that, in changing the rule, "you open up a Pandora's box, a free-for-all census, where there's no principle for where people are counted" (Price, 2005).[15]

Five years later, former director Prewitt took the opposite stance on prisoner counting, writing in a foreword to a Brennan Center for Justice report that "changes in the criminal justice system over the last three decades call into question the fairness of counting persons where they are imprisoned" and that "current census residency rules ignore the reality of prison life." "Counting people in prison as residents of their home communities offers a more accurate picture of the size, demographics, and needs of our nation's communities," he continued, "and will lead to more informed policies and a more just distribution of public funds" (Allard and Levingston, 2004).

What makes the issue of prisoner counting in the census a significant one, empirically, is a major sociological trend—namely, the major growth in the incarcerated population since the mid-1970s through 1990s, following decades of remarkable stability in the incarceration rate from 1925 to 1975 (for reviews, see, e.g., Blumstein, 1995; Cullen and Sundt, 2000; King, 1998; Zimring and Hawkins, 1991). Analysts from the Brennan Center for Justice at New

[14]In addition to substantive issues, Prewitt also challenged specific features of H.R. 1639, the bill in question. Specifically, the bill would have counted prisoners at another location depending on which state bore most of the costs for housing the prisoner; Prewitt questioned the adequacy of records needed for this determination. The bill was also vague as to whether it applied only to state-run prisons, or to private and federal prisons as well. He noted a final, major factor in opposing the proposed change—the then-short time (just less than a year) between the hearing and Census Day, 2000.

[15]The same source was quoted as saying that "if you've been ordered by a judge to be in prison, that's where you're living at the time of the census."

York University School of Law summarize this expansion by noting that the incarcerated population grew fourfold between 1980 and 2002, from about 500,000 to more than 2 million people (Allard et al., 2004).

Prisoner counting in the census is especially controversial (and will likely remain so) because it lies at the intersection of two of the most potent and enduring social struggles in the American experience: race and the urban-rural divide. The incarcerated population is markedly different from the general population in that it is more than 90 percent male and includes a disproportionate share of racial minorities. Using 1998-vintage estimates, Cullen and Sundt (2000:483) found that blacks represent 12.7 percent of the general population, compared with 41.3 percent of the state and federal prison population. Lotke and Wagner (2004:593–594) find that "nearly 9% of all African American men in their twenties and thirties live in prison."

The second core political tension at the heart of the prisoner counting issue is the tension between urban and rural areas for equitable shares of representation and resources. Prisons are frequently sited in rural areas rather than large urban centers. "[Prisoners] are not generally from the county where the census has them placed," note Lotke and Wagner (2004:592); "they were imported from other counties for purposes of confinement" and "if the [prison] doors were opened, few would stay" in the vicinity of the prison. For example, Heyer and Wagner (2004) estimate that 99 percent of Illinois' prison cells are not located in Cook County (Chicago), although 60 percent of the state's prisoners come from the county, and Los Angeles County, California, is the source of 34 percent of the state's prisoners, but only 3 percent of the state's prison population is housed there. Similar arguments can be made for other urban centers; notably, Philadelphia (which functions jointly as a city and a county) "is the legal residence for 40 [percent] of Pennsylvania's prisoners, but the County contains no state prisons." In comparison with the general population, Allard and Levingston (2004) suggest that 40 percent of incarcerated persons nationwide are in rural facilities, while rural residents make up 20 percent of the U.S. population.

As a share of the total U.S. population, prisoners (like the rest of the group quarters population) are a relatively small group, but at the local level prisons can be very significant. Lotke and Wagner (2004:603) cite the extreme example of Florence, Arizona, which "has a free population of roughly 5,000 plus another 12,000 living under lock and key," as it is the location of four prisons.[16] Six state prisons are located in Gatesville, Texas, and about 58 percent of Gatesville's 2000 census population of 15,591 were listed as coming from institutional group quarters. Moreover, all but one of the Gatesville facili-

[16]Florence (about 60 miles southeast of Phoenix) houses two state prisons (Arizona State Prison Complex–Florence, Arizona State Prison Complex–Eyman), the privately operated (but state-contracted) Arizona State Prison–Florence West, and the U.S. Immigration and Customs Enforcement Florence Service Processing Center.

ties are female only, heavily skewing the city's overall demographic portrait.[17] More generally, Lotke and Wagner (2004) observe that 197 counties out of 3,141 (6.3 percent) have more than 5 percent of their population in prison. Eighteen counties (13 of which have majority rural populations) have more than 20 percent of their population in prison. Heyer and Wagner (2004) observed that 56 counties that reported growth during the 1990s (comparing the 2000 census to 1990) would have actually posted declining populations, save for the addition of new prison beds to their jurisdictions.

A major use of decennial census data is for allocation of state and federal funds: such allocations—and whether the monies should be directed at the communities housing prisons or those which prisoners leave and reenter—are a source of much contention. "Where prisoners are counted—in penitentiaries usually in remote areas far from home—effectively shifts political power, taking federal and state dollars, and social services, from urban areas to rural ones" (Price, 2005): hence, it could be argued that this shift "skews a state's public policy agenda." Analyzing 1992–1994 data, Beale (1996:25) found that 60 percent of new prison construction is in rural, nonmetropolitan counties, even though those counties account for only 20 percent of the national population; he concluded that, "whether through unsought placement of facilities or aggressive local bidding for them, prison construction and employment have become economically important for many rural areas." The potential effects on available revenues associated with prison location can be major. For Florence, Arizona, Lotke and Wagner (2004:603) estimate the influx of annual funds tied to the various prison facilities at \$4 million; other state and local funds account for \$1.8 million, and local measures produce \$2.3 million.[18]

Arguments about the fairness of fund allocations, and the influence of prisoner counts on them, can be made by both the source and destination communities of prisoners. Proponents of change to the Census Bureau's current policy of counting prisoners at the facilities argue that the policy disadvantages prisoners' home communities by denying them resources they need to facilitate effective prisoner reentry to the community (e.g., occupational

[17]The male-only exception is the 2,900-inmate-capacity Hughes prison, established in 1990. The city's Gatesville, Hilltop, Mountain View, Murray, and Woodman prisons are women-only; the oldest of the facilities came on line in 1975, while three of the women's prisons have been established since 1990.

[18]Knowledge of these funds was not lost on other Arizona communities; "such lucre tempted the Arizona town of Buckeye to annex nearby Lewis State Prison, population 4,600, though first it had to defeat a matching attempt by neighboring Gila Bend. The mayor of Buckeye, with a population 5,038 before the annexation, promised to use the expected \$1.3 million to upgrade parks and family services, and assured everybody that it would more than pay for the additional burden on fire and police" (Lotke and Wagner, 2004:603). While some rural communities may engage in bidding wars to host prison facilities, correctional facilities remain a quintessential "not in my back yard" proposition; some local officials see increased funding as "the one positive" to hosting a prison, as a counterbalance to the "stigma" and "ill effects" of prison placement (*East Oregonian* and Associated Press, 2006).

counseling, substance abuse treatment) and prevent recidivism. It is these "home communities," the argument follows, that educational and transportation systems will have to meet the needs of returning prisoner populations; by nature of their confinement, prisoners do not use those resources in the prison host community. Acknowledging that the prison site "may lose some formula funding for water or sewerage," proponents of a change in counting rules for prisoners suggest that these costs could be recouped in other ways, including revised contracts with state governments and departments of correction (Lotke and Wagner, 2004:606).

However, prison sites argue just as vigorously that they need the resources due to the large immediate burden placed on them by prisoner. For instance, the city of Pendleton, Oregon, "receives around $35,000 a year from the state's liquor and cigarette taxes and 9-1-1 fees" and "about $75,000 annually from the state's gas tax," all based on residency counts. These figures are swayed in part by the location of the 1,600-inmate Eastern Oregon Correctional Institution in the city. Local officials concede that the inmates are not active participants in the community, but they argue that additional funds are justified by the demands on city water, sewer, and gas services by the prison, as well as city fire and police emergency response (*East Oregonian* and Associated Press, 2006).[19]

Two federal court precedents are central to the current Census Bureau handling of prisoners in the census. We discussed the Third Circuit Court of Appeals' 1971 ruling in *Borough of Bethel Park v. Stans* in Box 3-2. The *Bethel Park* plaintiffs claimed that institutionalized persons (including prisoners) were one of three major groups that were misallocated under the Bureau's residence rules. The *Bethel Park* court concluded that counting "persons confined to institutions where individuals usually stay for long periods of time as residents of the state where they are confined . . . has a rational basis." Thirty years later, the counting of prisoners in the census was directly challenged by the District of Columbia, which then housed all of its felony prisoners at the Lorton Correctional Complex in Virginia. The U.S. District Court for the District of Columbia held that the Census Bureau is not compelled to consider factors like the impact of residence rule determinations on fund allocation; they are required to make a "rational determination" in defining residence, and the court judged that the Bureau's policy of counting the prisoners in Virginia had a rational basis. Box 3-3 summarizes the case in more detail.

[19]The surrounding Umatilla County also contains the Two Rivers Correctional Institution in the city of Umatilla. That city's finance director estimates that the city would lose about $117,000 annually if Two Rivers inmates were not counted in Umatilla (*East Oregonian* and Associated Press, 2006).

In the rest of this section, we consider the major threads of argument that have been advanced concerning whether prisoners should be counted at prison or at some other location.

Implications for Districting—The Potential for Intrastate Distortion

A fundamental use of census data is for the construction of congressional and state legislative districts. Regardless of where they are counted, prisoners present an immediate complication for this use of census data. Only two states permit prisoners to vote during their periods of incarceration, but all prisoners count toward the composition of districts—for purposes of ensuring equal district population size.[20] This source of distortion is compounded by the size and general location of correctional facilities; particularly for state legislative districts, the siting of prisons in rural areas can draw representation away from urban centers.

Both the Prison Policy Initiative (see, e.g., Wagner, 2002; Heyer and Wagner, 2004; Lotke and Wagner, 2004)[21] and the Brennan Center for Justice at New York University School of Law (Allard and Muller, 2004) offer clear and compelling empirical evidence of the distortion that can be induced by prisoner counts. Much of the work has focused on the state of New York, where prisons are generally located upstate while the majority of incarcerated persons come from the New York City area. Pendall (2003) found that the upstate New York region achieved only very mild population growth over the 1990s, but a large share of that was attributable to prisons; 30 percent of new "residents" to the upstate counties in the decade were prisoners. More anecdotally, Lotke and Wagner (2004) cite the admission of a New York state senator (whose upstate district includes six prisons) that "it's a good thing his captive constituents can't vote, because if they could, 'They would never vote for me.' "

As a result, Lotke and Wagner (2004:593–594) argue that the incarcerated population—including the "nearly 9% of all African American men in their twenties and thirties" who are imprisoned—"[are] apportioned to legislative districts that do not reflect their communities of interest or their personal political concerns. Whether they can or do vote is irrelevant; their bodies still count in the prison district. A more refined analysis shows that the impact is modest in U.S. Congressional Districts but more significant in state legislative districts."

[20]It should be noted that distortion in voting influence may also arise because the population counts used to determine district size include other large population segments that do not have the right to vote, most notably, children under voting age and noncitizens. In addition, population counts also include those people who choose not to vote or register to vote.

[21]See also the state-level analyses posted at http://www.prisonersofthecensus.org/impact.shtml [8/1/06].

Box 3-3 *District of Columbia v. U.S. Department of Commerce* (1992)

In November 2001—pursuant to congressional legislation enacted in 1997—the Lorton Correctional Complex in Northern Virginia's Fairfax County ceased operations. The facility's closure ended nearly 90 years of a unique arrangement, in that the Lorton facility was purchased at the direction of Congress to house felony prisoners from the District of Columbia; though located in Virginia, the complex was operated and managed by the District of Columbia Department of Corrections.

The unique situation at Lorton occasioned a challenge to the 1990 census, with the potential of setting a precedent for the counting of prisoners housed out of their home state. Specifically, the District of Columbia filed suit against the U.S. Department of Commerce and the Census Bureau, arguing that the Bureau's "inclusion of Lorton Correctional Facility inmates in the 1990 Census as residents of Virginia rather than of the District of Columbia violates the Constitution and the Census Act," a decision which the District alleged would mean a loss of $60 million in federal fund allocations over the 1990s. In particular, the District cited the unique ceding of Virginia land by Congress for purposes of housing District of Columbia prisoners as support for counting Lorton prisoners in the District. The Census Bureau countered that "the Bureau's application of the usual residence rule to Lorton inmates is a rational decision that is not arbitrary and capricious," that the counting of Lorton prisoners in Virginia was consistent with the definition of usual residence, and that Lorton inmates had been counted as Virginia residents since the opening of the prison in 1916. The U.S. District Court for the District of Columbia ruled on *District of Columbia v. U.S. Department of Commerce* (789 F.Supp. 1179) in April 1992.

The court was unconvinced that Lorton's management by the District of Columbia should have any impact on where its residents are counted. "Retention of control and management, despite United States ownership, does not automatically qualify a property as 'unique' such that it deserves different rules for the Census enumeration;" all federal prisons which hold inmates from outside the state where the prison is located are subject to the same arguments, as are military installations. That said, the court observed that the District of Columbia did make a solid point when arguing that the District "pays all of the costs of maintaining Lorton, including water and electricity," and that inmates retained eligibility for "health, social and educational benefits" paid for the by District. In short, the District bore all costs for Lorton, while "Virginia's only connection to the facility is that it is within the geographic boundaries of the state and Virginia can not collect taxes from businesses or residences that otherwise might have been there."

"In one light, this would appear to be a convincing argument," the court held; indeed, "it appears that all that separates Lorton residents from being counted as District of Columbia residents is a mere vagary of the District of Columbia's strange position as a city without a state." However, the court concluded that the paramount goal of the census is the production of a count for the purposes of apportionment and that, "however rational it may seem to examine the source and nature of fiscal support" won or lost by residence definitions like the Lorton case, "the Census Bureau is not required to do so." Rather, the court held that the Census Bureau's interpretation of usual residence, "based on geography" and "developed and consistently applied," constitutes a "rational determination." In short, the court ruled that, "although including Lorton within the District of Columbia population may be more equitable, we cannot say that the Bureau acted without a rational basis.. . .The solution to the District's problems lies not in adjusting the Census count, but in changing the way funds are distributed."

Box 3-3 (continued)

(In a footnote, the court raised another problem with diverging from the usual residence rule: namely, the conceptual problems that would arise. "Where would inmates actually be counted, where they lived prior to incarceration? And then what of the current residents of that address? How would they be counted for representation purposes?")

After the closure of the Lorton Correctional Complex, prisoners housed there were to be transferred to facilities of the Federal Bureau of Prisons or to private correctional facilities. In most cases, these new transfer locations were substantially further away from the District of Columbia than Lorton.

Custody Versus Jurisdiction

In a correctional setting, the distinction between the agency or location of jurisdiction (the one with legal responsibility for a prisoner) and the agency or location of custody (where the prisoner is actually housed) is an important one, because they are not always identical. The nature of incarceration is such that divergence in the locations of custody and jurisdiction creates the possibility for distortion in population counts, both across and within states.

The massive growth in the prison population also drove an increase in the construction of prison space, but the supply of prison beds does not always keep pace with demand. To accommodate prisoner flows, several states with tight prison space began contracting with other states (with surplus prison beds) to house prisoners. Physical custody of prisoners was transferred to the other state, but the origin state retained jurisdiction and paid most if not all of the costs of housing the prisoners out of state. Under census "usual residence" standards, prisoners would be counted at the state of custody, not the state of jurisdiction.

In 1999, Rep. Mark Green (R-Wisconsin) introduced H.R. 1639 in the 106th Congress. At the time, Wisconsin was said to be the "largest exporter of prisoners in the United States," sending more than 3,500 prisoners to Tennessee, Oklahoma, Texas, and Minnesota.[22] The Green bill would have compelled the Bureau to count a prisoner in the state of jurisdiction if that state pays "greater than half the costs associated with such individual's incarceration." The bill was the topic of a hearing of the House Subcommittee on the Census, but never advanced beyond subcommittee consideration.

The Wisconsin example is a telling one for another reason—namely, the remedy the state ultimately pursued. Beginning in 2003, Wisconsin began entering into partnerships with county sheriffs to house state prisoners in local

[22]See "Green Introduces Census Bill Today" [press release], Office of U.S. Rep. Mark Green, April 29, 1999; http://www.house.gov/markgreen/PRESS/1999/April99News/NRCensusPrisoners.htm [8/1/06]. Green further claimed that "the number of Wisconsin prisoners sent to other states is expected to grow to nearly 10,000 over the next two years."

jail cells; as of May 2005, the state no longer housed prisoners in other states.[23] These kinds of arrangements, in which overflow state prisoners are housed in local facilities, are a possible technique for dealing with prison space shortage without shifting prisoners out of state. However, in the census count context, the practice presents at least two major difficulties. First, long-term housing of prisoners in local jails makes it difficult to assign enumeration strategies based solely on facility type (e.g., it can not always be assumed that local jails are short-stay-only, no-usual-residence facilities). Second, while working with in-state jurisdictions ensures that prisoners are allocated to the "right" state, intrastate distortion can occur; prisoners housed in another county are still not located in the place of jurisdiction.

In recent years, the distinction between jurisdiction and custody has also been complicated by increased privatization. Faced with the expense of creating prison space for the growing correctional population, governments also faced the question of privatizing prisons—that is, permitting local companies to construct and operate correctional institutions. The first modern private prison opened in 1984, and 91 private prisons were operational by 1995 (Cullen and Sundt, 2000:485). More recent statistics by BJS indicate 98,901 prisoners being housed in privately owned prisons, a slight increase from 95,707 the year before.

Private prisons present the quintessential case of the divide between custody and jurisdiction: persons housed in private prisons are under the jurisdiction of the government correctional agency but the physical custody of the private prison operator. With respect to inclusion in the census, private prisons represent a different set of actors who have to be involved in order to obtain a count; private vendors may provide different levels of access to either prisoners or to administrative records than government agencies, and private prisons may also be difficult to tally accurately in address listings.

Sentence Length

The duration of time actually spent in prison is an important question in considering whether the facility is the appropriate place to count prisoners. Table 3-4 details the mean and median sentences imposed on new state prisoners in 1993 and 2002. The sentences imposed are fairly consistent between the two years, and fairly long (median 48 months in both years for all offenses, and higher—72 and 60 months, respectively, in 1993 and 2002—for violent offenses). But advocates of counting prisoners outside the prisons correctly note that actual time served is shorter, and certainly shorter than the "lock them up and throw away the key" model that is sometimes assumed by

[23]See "Governor Doyle Announces Return of Last Remaining Wisconsin Inmates Housed in Out-of-State Facilities" [press release], Wisconsin Department of Corrections, May 10, 2005; http://www.wi-doc.com/index_news.htm [8/1/06].

the public. Rather, the median time served among those prisoners released in 2002 was just 17 months, a level that has been fairly constant since the mid-1990s (see Table 3-5). For those serving time for violent offenses, the mean sentence served was tugged higher, to 30 months, and both median and mean time served are markedly higher. Table 3-6 details the average time served by offense for 1993 and 2002; the relatively low overall mean and median time served amounts are driven by the large bulk of lighter-sentence property and drug offenses. The tables listed here cover state felony convictions only; local jail sentences and sentences for misdemeanors are, of course, much shorter.

The census is meant to be a resident count at a particular point in time; in itself, it is not meant to be a population projection for some future time. Still, average sentence length prompts the question of the appropriate time reference for a rule governing residence assignment in a census. On the one hand, it makes intuitive sense that a 17-month stay at a particular location is eminent justification for declaring that place to be where a person is "resident." But census counts can affect the drawing of legislative districts and allocation of funds over 10 years or more: in light of the possible effects over a window of 120 months, a sentence of 17 months could be viewed as a more temporary arrangement.

Where Is "Home" for a Prisoner?

While prisoners are obviously in physical custody at the prison location during their term of incarceration, the "enduring ties" argument of *Franklin v. Massachusetts* can be invoked to suggest that prisoners may remain linked to family members and a residence beyond the prison walls. Allard et al. (2004:1) argue that counting prisoners at the prison "disserves people returning home from prison because they are subtracted from the populations in which their interests, if not their bodies, permanently remain." College students may build ties with their surrounding communities, but—by the nature of incarceration—prisoners "do not become active members of the communities in which their prisons are located. Their only options are to maintain enduring ties with their home communities, or maintain no ties at all" (Allard et al., 2004:2).

The "enduring ties" argument for prisoners' "homes" centers around the family members left behind, and in particular on children. In support of their argument, Allard et al. (2004) and Allard and Levingston (2004) cite the analysis by Mumola (2000), who estimated that 55 percent of state prisoners and 63 percent of federal prisoners in 1999—721,500, in total—were parents of minor children. Accordingly, Mumola (2000:1) estimates that 336,300 households and 1,498,800 minor children were "affected by the imprisonment of a resident parent." The estimates were generated based on the 1997 wave of the

Table 3-4 Sentence Length for Most Serious Individual Offense, New Court Commitments to State Prisons, by Offense, 1993 and 2002

	Sentence Imposed (months)					
	1993			2002		
Most Serious Offense	% of Releases	Median	Mean	% of Releases	Median	Mean
Violent Offenses	29.3	72	107	28.6	60	95
Homicide	4.2	240	188	2.9	240	205
Kidnapping	0.6	108	133	0.7	60	108
Rape	2.1	120	141	1.6	120	145
Other sexual assault	4.0	72	108	4.6	72	105
Robbery	9.9	72	102	7.9	60	95
Assault	7.7	48	76	9.4	48	63
Other violent	0.8	48	58	1.5	36	56
Property Offenses	30.5	48	58	27.5	36	51
Burglary	12.9	48	69	10.4	48	61
Larceny/theft	8.0	36	46	6.7	36	43
Motor vehicle theft	2.4	36	44	2.1	32	39
Arson	0.6	60	89	0.5	55	75
Fraud	3.8	36	52	4.8	36	46
Stolen property	1.9	48	52	2.0	36	45
Other property	0.9	36	48	1.1	36	43
Drug Offenses	29.9	48	62	30.5	48	58
Possession	6.5	48	60	9.8	36	47
Trafficking	18.5	48	65	14.5	48	63
Other/unspecified drug	4.8	48	54	6.2	48	65
Public-Order Offenses	9.3	30	42	12.8	36	46
Weapons	2.5	36	49	3.3	36	48
Driving while intoxicated	2.7	24	32	4.0	36	45
Other public-order	4.1	31	45	5.6	36	47
Other Offenses	1.1	24	45	0.5	60	81
All Offenses	100.0	48	71	100.0	48	65

NOTES: Per original source notes, data are based on new court commitments with a total sentence of more than 12 months for which the most serious offense is reported.

SOURCE: National Corrections Reporting Program, 1993 and 2002; tables posted at http://www.ojp.usdoj.gov/bjs/prisons.htm [1/26/06].

Table 3-5 Time Served by Newly Released State Prisoners, 1993–2002

		Time Served to First Release from State Prison (months)			
		All Offenses		Violent Offenses	
Year	Releases	Median	Mean	Median	Mean
1993	229,098	12	21	23	36
1994	227,217	13	22	24	36
1995	234,360	15	23	26	37
1996	255,220	16	25	27	39
1997	248,789	17	27	29	41
1998	259,284	18	28	30	43
1999	245,444	18	29	31	45
2000	236,733	18	29	32	47
2001	271,147	18	31	33	49
2002	288,883	17	30	32	49

NOTES: Per original source notes, "data are based on release type with a total sentence of more than a year for whom the most serious offense and time served were reported. All data exclude persons released from prison by escape, death, transfer, appeal, or detainer."

SOURCE: National Corrections Reporting Program, 1993–2002; tables posted at http://www.ojp.usdoj.gov/bjs/prisons.htm [1/26/06].

Survey of Inmates in State and Federal Correctional Facilities, administered by the Census Bureau under contract by BJS.

Ties between prisoners and the family members and locations located outside the prison may be enduring throughout a prison term; so, too, can they be diminished or severed. Mumola (2000) also finds that less than half—46.8 percent—of imprisoned parents lived with their children prior to incarceration; this total includes 43.8 percent of imprisoned fathers and 64.3 percent of imprisoned mothers. Although 62.4 percent of male prisoners and 78.4 percent of female prisoners reported having at least monthly contact with their children while in prison, about half (57 percent of fathers and 54 percent of mothers) had received no personal visit from their children since being incarcerated. About 20 percent of parents in state prisons and 8 percent of parents in federal prisons reported no contact of any kind with their minor children since their imprisonment.

Fundamentally, "enduring ties" between a prisoner and his or her "home" may also be irrevocably strained or severed by the nature of the offense or the offender's traits. As Travis et al. (2001:39) put it bluntly: "There are situations

Table 3-6 Time Served by Newly Released State Prisoners, by Offense Type, 1993 and 2002

	Time Served in State Prison to First Release (months)					
	1993			2002		
Most Serious Offense	% of Releases	Median	Mean	% of Releases	Median	Mean
Violent Offenses	24.6	23	36	27.3	32	49
Homicide	2.4	42	60	2.3	82	93
Kidnapping	0.5	28	41	0.6	34	55
Rape	1.6	44	57	1.6	68	81
Other sexual assault	3.4	25	33	3.8	33	45
Robbery	8.8	25	38	8.1	41	55
Assault	7.2	15	23	9.5	20	32
Other violent	0.7	14	20	1.4	17	24
Property Offenses	34.5	11	17	28.1	14	24
Burglary	14.6	13	21	11.2	19	32
Larceny/theft	9.2	9	13	6.8	12	19
Motor vehicle theft	2.8	11	14	2.1	12	21
Arson	0.6	16	25	0.6	25	36
Fraud	4.3	9	14	4.4	11	17
Stolen property	2.0	9	14	2.0	12	19
Other property	0.9	8	14	1.0	10	16
Drug Offenses	30.1	11	16	31.9	15	23
Possession	6.5	9	14	9.9	11	18
Trafficking	18.6	13	17	15.8	17	26
Other/unspecified drug	5.0	10	14	6.2	15	22
Public-Order Offenses	9.7	8	13	12.3	13	19
Weapons	2.2	10	14	3.0	16	23
Driving while intoxicated	3.0	7	10	4.2	13	16
Other public-order	4.6	9	15	5.0	12	20
Other Offenses	1.1	10	15	0.4	16	23
All Offenses	100.0	12	21	100.0	17	30

NOTES: Per original source notes, "data are based on release type with a total sentence of more than a year for whom the most serious offense and time served were reported. All data exclude persons released from prison by escape, death, transfer, appeal, or detainer."

SOURCE: National Corrections Reporting Program, 1993–2002; tables posted at http://www.ojp.usdoj.gov/bjs/prisons.htm [1/26/06].

where families are better off without a neglectful or abusive parent or partner in their lives. Some individuals may have been convicted of a crime of violence or abuse in the home, while others were convicted of different crimes but may exhibit a pattern of abuse." In such cases, it would be difficult to justify an "enduring ties" case; a spouse or ex-spouse who suffered abuse at the hands of an prisoner may no longer consider that person a part of their household, and, indeed, might have difficulty with the concept of the abusive spouse being "assigned" back to the household for counting purposes.

Where Is "Home" for a Released Prisoner?

Lotke and Wagner (2004:592) argue that "if the [prison] doors were open, [prisoners] likely would return to where they came from. That's the place that most people in prison consider their home, and where most will return within a few years." The extent to which this is true is not clear empirically. Moreover, the level of geographic resolution at which prisoner returns to "home" locations—whether to the same specific address as before incarceration, to the same neighborhood, or to the same city or county—is likewise unknown.

Practices vary by state and jurisdiction as to what happens at the moment of a prisoner's release, either into a parole arrangement or the end of a sentence—for example, whether any small sum of money is given to the prisoner or whether any transportation is provided. Some states require that prisoners be released to the county of sentencing, and state prisoners housed in another state's prison may be obligated to return to their home state for release. After release, parole restrictions may be imposed, requiring that the person stay in a particular county or jurisdiction and governing the conditions under which they may travel (Allard et al., 2004:4).

Accordingly, the basic matter of housing is "an often overlooked challenge facing the returning prisoner," argue Travis et al. (2001:35). Criminal histories may bar former prisoners from some housing options; for instance, certain criminal activities preclude persons from being eligible for public or some subsidized housing. "One option for ex-prisoners is to stay with family members following release"; however, Travis et al. (2001) find "some evidence to suggest [that,] among the many who do, these arrangements are often short-lived solutions"—"familial relationships may [be] so severely strained and tenuous that staying with family members or friends is not a viable option."[24]

Travis et al. (2001:40–41) report striking geographic concentration of released and paroled prisoners. In 1996, roughly two-thirds of newly released prisoners returned to "core counties" of metropolitan areas, concentrating the prisoner outflow in a relatively small set of center-city areas. By one estimate,

---

[24]See also Petersilia (1998, 2001) and Roman and Travis (2004) for additional discussion of prisoner reentry.

New York City's resident population count would increase by 36,000 if prisoners housed in upstate New York were counted at home. At a more detailed level, Travis et al. (2001:41) display tight geographic clustering of parolees in Brooklyn, New York, in a small number of census block groups. However, the exact correspondence between these locations and the prisoner's preincarceration address (home of record) is not known.

Is Choice a Factor in Defining Residence?

In Section 2–C.2 and elsewhere in this chapter, we discuss the question of whether intent to remain at a location should be factored into the determination of a "usual" residence. The counting of prisoners raises a related question—should determination of a residence consider whether the choice of residential location is voluntary or not? The treatment of prisoners in the census is frequently compared to the handling of other parts of the broader group quarters population (as it was in the *Bethel Park* case) even though—by the nature of incarceration—prisoners do not choose to be at the facility, and generally do not have discretion in which prison they are held. The involuntariness of confinement is a strong barrier to the idea that prisoners are part of the surrounding community that hosts the prison.

Legal Standards on Residence and Voting

All but two states deny persons the right to vote if they are currently imprisoned; some 32 states further restrict voting by prohibiting persons with felony convictions or who are in active probation arrangements from taking part in elections. Yet several states make clear in their laws that certain conditions—for instance, military service or attendance at college—that take people away from their place of residence do not void their claim to a residence (and voting rights therein), and confinement in prison is one such condition that is included in some state legal text. Article II, Section 4 of the New York State Constitution states:

> For the purpose of voting, no person shall be deemed to have gained or lost a residence, by reason of his or her presence or absence, while employed in the service of the United States; nor while engaged in the navigation of the waters of this state, or of the United States, or of the high seas; nor while a student of any seminary of learning; nor while kept at any almshouse, or other asylum, or institution wholly or partly supported at public expense or by charity; nor while confined in any public prison.

Similar (often nearly word-for-word) clauses are included in other state constitutions and codes.[25] Though other parts of electoral law may deny them the right to vote, these legislative clauses hold that residences are not gained or lost solely due to their confinement; thus, the claim of a legal tie to a preincarceration address can be said to exist.

The two states that do permit prisoners to vote are Vermont and Maine. Under Vermont election code (17 V.S.A. §2121–2122(a)), "a person in a correctional institution must register to vote in the last town in Vermont that the person resided in prior to incarceration"—that is, a prisoner must vote absentee as if he or she still lived at their preincarceration address. Furthermore, the code (28 V.S.A. §807) directly prohibits prisoners from registering to vote in the town where the correctional facility is located. Similarly, Maine stipulates that prisoners vote as though still at their preincarceration location, but adds an element of intent: "a person incarcerated in a correctional facility may apply to register to vote in any municipality where that person has previously established a fixed and principal home to which the person intends to return."

In late 2005, Congress passed the fiscal 2006 appropriations bill including funds for the Census Bureau; the conference report for the bill directed the Census Bureau "to undertake a study on using prisoners' permanent homes of record, as opposed to their incarceration sites, when determining their residences." The Bureau complied with the appropriators' 90-day deadline and submitted its report (U.S. Census Bureau, 2006b), which had been preceded by two reports from proponents of changing the policy (Levingston and Muller, 2006; Wagner et al., 2006). We will discuss the recommendations of these reports in Section 7–E.

3–D.2 Jails

The 2000 census counted approximately 630,000 people at local jails. Though county and municipal jails may house fewer people than prisons at any given time, the flows of prisoners into and out of jails uniquely complicate the problem of counting people within their walls.

As Frase (1998:474) notes, "jails lie at the center of the criminal justice system. They are intimately related to every stage of pretrial and posttrial procedure, and are the detention facility that affects the community most directly and most frequently." As is the case with prisons, jails can vary greatly in their size and characteristics: they may be general population facilities or may be exclusive along gender (e.g., all female) or age (e.g., all juvenile) lines.

[25]These include the Nevada State Constitution, Article II, Section 2, and the Arizona State Constitution, Article VII, Section 3. As shown in Box 2-4, California is one state where "domicile" cannot be gained or lost by a person "while kept in an almshouse, asylum, or prison" (California Elections Code §2025).

However, they can also vary significantly in terms of their designated functions and the mix of services they perform, including pretrial detention and for service of short sentences. In many cases, jails serve some combination of these functions; Frase (1998:474) terms them "the custodial dumping ground of last resort, when no other appropriate holding facility is available."

To be clear, the basic definition of jail used in federal statistical data collection excludes some facilities that are of less interest for purposes of the census but are of keen interest to correctional studies. In particular, the basic definition essentially excludes police, sheriff's, and court lockups—cells that process a large number of persons but for very short time periods. Frase (1998:476) suggests that "there are at least as many police and sheriff's lockups as there are 'jails';" 1993 statistics showed that 3,200 police departments operated such lockups, "which had an average capacity of ten inmates and an average maximum holding authority of twenty-two hours" (Frase, 1998:476).

Even excluding the very-short-term lockups, the variation in functions performed by local jails means that it can be difficult to characterize the typical jail stay. Some states sharply limit the length of time a person may spend in jail—in many cases, the limit is 1 year—but others may permit longer sentences to be served in local facilities. As we discuss in the previous section, state departments of corrections may also make use of capacity in county and municipal jails if there are bed shortages in the prison system; local jails may be contracted to hold prisoners for longer terms. "Considerable variation exists from state to state in the types of sentences that may be served in jail. Thus, although most states limit jail terms to one year, some states (e.g., Pennsylvania) regularly use local jails for longer sentences. Also, some states (*not* necessarily the same as above) make very heavy use of jail for felony sentencing, and consequently have comparatively large jail populations relative to their state prison populations" (Frase, 1998:479).

As described in the introduction to this section, the concept of "prison" as a long-term facility, distinct from short-term "jails," is one that evolved over the course of American history, with "jail" having been the exclusively used term in the early days. The instructions given to enumerators in the 1910–1950 censuses—"in jails, you must enumerate all prisoners, however short their stay"—is thus consistent with historical usage.[26] In 1910 and 1920, census authorities acknowledged the potential for people incarcerated in jails to be duplicated in the census because they were counted with their families at home. Those two censuses included a specific warning:

> To prevent duplication, do not report outside of the institution any person who formerly lived with a family in your district but who at the time of the enumeration is an inmate of such institution as above described,

[26]The specific language quoted is from the 1950 instructions.

even though that person may have entered the institution only the day before the census day.

The standard of counting persons in jail at the jail location continued in 1960–1980. As discussed in Section 3–C, the preliminary residence rules document for the 1990 census suggested that the Bureau intended to draw a distinction between short-term and long-term institutional facilities. Specifically, "county jails" were raised as a specific example of a case where people would be counted "at their usual place of residence if they have one." In practice, however, this was not the case; the technical documentation accompanying 1990 census data releases indicated that persons in custody at "local jails" were counted at the jails.

In the 2000 census, residence rule 20 stipulated that persons "under formally authorized, supervised care or custody" in a "local jail or workhouse" should be counted at the jail, with no provision for specifying a usual home elsewhere.

3–D.3 Juvenile Facilities

Just as boarding schools and colleges present residential ambiguity challenges because parents may be inclined to preserve family structure when reporting who lives in their households, correctional facilities that specialize in housing juveniles may be troublesome for enumeration.

Separate judicial processes for juveniles and adults were first established in 1899 and have evolved over the ensuing century to include separate correctional facilities. Pursuant to the federal Juvenile Justice and Delinquency Prevention Act, "juveniles alleged to be or found to be delinquent . . . will not be detained or confined in any institution in which they have contact with adult inmates." As Sickmund (2004:18) comments, this provision is more commonly known as the "sight and sound separation requirement." A similar provision applies to separation in "any jail or lockup for adults," although exemption is made for juveniles being tried as adults or who have been convicted as criminal felons.[27] Owing to resource constraints, some states have created juvenile detention centers that are colocated within adult facilities, subject to constraints on use of common areas.

The juveniles serving in these facilities are strictly that: juveniles. Most states set an upper age limit for these facilities of 18 years, at which age the juvenile is either released or transferred to an adult facility. However, as of 1999, 10 states set the upper age for juvenile facilities at 16, and 3 states (Connecticut, New York, and North Carolina) are limited to inmates 15 and younger.

[27] Additional temporary hold exemptions apply; alleged juvenile delinquents can be held in secure custody in adult lockups for a 6-hour grace period (including a similar buffer when juvenile offenders must be placed in such lockups for court appearances). The grace period is extended to 48 hours in rural facilities (Sickmund, 2004).

Facilities may be of many different types, ranging from "boot camp" and "boys ranch" (or "girls ranch") settings to highly secure, fenced, and locked facilities (Greenwood, 1995). Generally, "national accreditation standards for juvenile facilities express a preference for relying on staff, rather than on hardware, to provide security. The guiding principle is to house juvenile offenders in the 'least restrictive placement alternative'" (Sickmund, 2004:16). As with other correctional facilities, recent decades have seen the emergence and expansion of privately owned and operated facilities for the treatment and housing of juveniles (Greenwood, 1995).

Time spent in juvenile facilities is difficult to quantify because—much like jails—juvenile facilities play different roles in sentencing and adjudication. Some juveniles may be *committed* to facilities—that is, they are sentenced to the facility for some fixed term by a juvenile or criminal court; others may be *detained* at a juvenile facility while waiting on further action (e.g., pending adjudication or waiting on sentencing); still others may be *diverted* to serve a period at a facility as part of a legal agreement in order to avoid adjudication (in the hopes of deterring the juvenile from further delinquent acts). A National Research Council and Institute of Medicine (2001:156) report finds that no national data on sentence length, time in confinement, and time spent in parole are published. Parent et al. (1994) estimate that juveniles who were released from correctional facilities in 1990 had average stays ranging from 15 days to 32 weeks, depending on the type of facility. They note, however, that their measure gives greater weight to facilities with high turnover, thus understating the lengths of stay of juveniles confined on any given day.

The Office of Juvenile Justice and Delinquency Prevention (OJJDP) operates several survey programs to assess the size and living conditions of the incarcerated juvenile population. Since 1997, OJJDP has contracted with the Census Bureau to conduct the biennial Census of Juveniles in Residential Placement (CJRP).[28] The CJRP is a mail survey sent to all known juvenile residential custody facilities in the United States,[29] excluding facilities that are exclusively for drug or mental health treatment (which may house some juveniles) or for abused or neglected children. The CJRP also, by design, does not cover juveniles that may be housed in predominantly adult facilities. In 2000, OJJDP initiated the separate Juvenile Residential Facility Census (JRFC). Focused on the capacity and quality of the facilities, the JRFC shares the same limitations (excluded facilities) as the CJRP.

The 2000 JRFC found that 110,284 juvenile (under 21 years old) offenders

---

[28]This program is the successor to the Census of Public and Private Juvenile Detention, Correctional, and Shelter Facilities—better known as the Children in Custody Census—that was first conducted in the 1970s.

[29]"Residential" means that they are capable of holding juvenile offenders overnight; some facilities to which the CJRP instrument was mailed turned out to not have this capacity, and so were identified as out-of-scope.

lived in 3,061 facilities; about 70 percent of them were concentrated in the 40 percent of facilities that are government owned (Sickmund, 2002). The 1999 administration of the CJRP (Sickmund, 2004) arrived at a higher number of juvenile offenders—134,011, itself an increase from 125,805 in 1997. About 74 percent of the juveniles encountered in the CJRP have been committed while about 25 percent are detained; only 0–1 percent are under diversion agreements. The total population of juvenile offenders in residential placement is disproportionately composed of racial and ethnic minorities, with blacks (39 percent) and Hispanics (18 percent) being large shares (38 percent are reported as white).

3–E CHILDREN IN FOSTER CARE

Foster children are those who, for a variety of reasons (including abusive or criminal activity by parents, abandonment, or death of parents) are in the care of a state government agency. These children are a unique and challenging population for censustaking purposes because of the variety of their living arrangements. Foster care may be provided in state-funded facilities or through placement of children in the homes of individual families; in the latter case, the state government agency reciprocates by providing regular payments or the foster child's care.[30]

The Pew Commission on Children in Foster Care found that 523,062 children were in some form of foster care arrangement in the United States and Puerto Rico in fiscal 2003; California's 97,261 cases accounted for the largest share of these (18.6 percent), with New York state's 37,067 cases (7.1 percent of the total) ranked second.[31] Demographically, Pérez et al. (2003) found that, for fiscal 2000, approximately 30 percent of foster children are between ages 11 and 15; roughly 25 percent each are in the 1–5 and 6–10 age groups; and about 4 percent are 1 year old or younger. Only 2 percent are aged 19 and older; typically, foster children "age out" of the system on reaching legal adulthood at 18. Pérez et al. (2003) also found that the foster child population is disproportionately African American (41 percent); American Indian and Alaska Native children are also somewhat overrepresented in the foster child population relative to their prevalence in the general population.

[30]Schwede (2003:xiii) summarizes ethnographic observation from the 2000 census that suggests that—for some non-English-speaking households—foster children may be miscounted due to terminological confusion. Specifically, the ethnographers found that the Spanish-language questionnaire used the phrase *hijo de crianza* for "foster child"; however, the Spanish phrase can be interpreted as a child one is raising for a friend or relative—losing the connotation of placement and supervision by the government. Conversely, the Korean-language questionnaire used a term for foster child that translates literally as "child under trusteeship," which confused respondents.

[31]See http://pewfostercare.org/research/docs/Data102705a.pdf [8/1/06].

Individual states vary in the number of children in foster care and in the growth trends of foster care caseloads. The number of "children in substitute care" in Illinois peaked at 51,331 in 1997, according to the state's Department of Children and Family Services, declining to 17,735 by July 2005.[32] The state of New York has experienced a significant decline in the number of foster cases in recent years; the number of children in foster care decreased smoothly from 42,921 in 2000 to 29,680 in 2004 (New York State Office of Children and Family Services, 2004:11). Texas has seen its foster care caseload grow from 18,626 to 26,133 between fiscal 1999 and 2003 (Strayhorn, 2004:3).

For the nation as a whole, Pérez et al. (2003) found that the majority of foster children, 72 percent, are housed in individual families' homes: 47 percent are in foster homes in which they have no relation to the head of the household, and 25 percent live with a relative. Another 4 percent of foster children in fiscal 2000 were living with families while adoption processes were pending, and 3 percent were housed with families on a trial visit basis. By comparison, a significant portion of foster children lived in institutional settings, either designated institutions (10 percent) or group homes (8 percent). The exact mix of in-family placement versus more institutional settings varies by state. New York state reported 74.6 percent of its 2004 foster cases as living in individual homes and 23.9 percent in "congregate care," institutional settings (New York State Office of Children and Family Services, 2004:16). Figures from Texas suggest only about 12.1 percent of foster children living in institutional "residential treatment facilities" in fiscal 2003, but group homes are combined with general foster homes in the tabulations (Strayhorn, 2004:7). January 2006 statistical reports by the Illinois Department of Children and Family Services showed 11.4 percent of the state's foster children living in institutional/group home settings.[33]

The ways in which children can transition *out* of foster care—including "emancipation" (reaching legal adulthood), adoption, or reunion with parents or caretakers—are widely varied, as are the circumstances by which children enter foster care. For census purposes, foster children can be a tricky population group to count accurately because of transitions *within* the foster care system. Although researchers suggest that the notion of "foster care drift"—"children trapped in foster care and constantly moving from one placement to the other" (Usher et al., 1999:22)—can be overstated, national-level records do suggest that foster placements can be transitory. On average, foster children experience three different placement sites during their state supervision (Pew Commission on Children in Foster Care, 2004). The Texas state comptroller (Strayhorn, 2004:7) found that children in the state's foster care

[32]See http://www.state.il.us/dcfs/foster/index.shtml [8/1/06].

[33]The department's monthly tallies are regularly updated at http://www.state.il.us/DCFS/docs/execstat.pdf [8/1/06].

system in fiscal 2003 had averaged 4 different placements; "those remaining in foster care for a decade or more could expect to be moved about once a year."[34] Because of the movement of children in foster care placements, the ties between children and households (when foster children are placed in family homes) could be viewed as tenuous or transitory; this could affect the householder's notion of whether the child is a "usual resident" there. If foster caregivers know that a child's placement in their home is likely to be short term (as opposed to a full adoption), they might believe they should not include the child in a household roster.

Children in "congregate foster care"—group-based settings—may present difficulties for enumeration because of the variety of housing arrangements in which they may be found. These congregate living arrangements run the gamut from "institutional" approaches such as dedicated orphanages and youth centers to group home situations that are indistinguishable from other housing stock. Indeed, a goal of some state-sponsored homes for these youths is to make housing as home-like and nonclinical as possible (e.g., providing access to regular schools and treatment at medical clinics outside the living facility). Address lists generated without full participation of child welfare agencies could lead to operational problems and data oddities that may challenge editing and imputation routines in the census. For instance, the census return from what seems to be a typical suburban house may appear to be a grouping of 6–10 minor children with no adult as a permanent resident (even though staff would be always present at the home, they would not likely reside there full time). Given the age of the children, direct delivery and administration of questionnaires to the children in the facilities would likely not be tenable. As with other group quarters, the accuracy of enumeration in these settings would depend on the willingness and ability of facility staff to provide responses or on the quality of facility or administrative records.

3–F MILITARY AND SEABORNE PERSONNEL

The 2000 census recorded about 282,000 residents of military barracks and other on-base residences, another 30,000 at short-term or transient housing on military bases, and 44,000 men and women on military ships (see Table 3-1). These counts reflect only a part of the broader military population (and their dependents) serving in the United States, in its waters, and overseas. The military population is challenging to count because of the diversity of service locations and the changing nature of assignments. As Hollmann (1987:279) notes, the "usual residence" principle that "is readily applicable to the vast majority of the population" is less suited to the military because they "are by

[34] As extreme cases, 12 Texas foster children in fiscal 2000 were found to have gone through at least 40 different foster placements.

their essential nature more mobile, less uniformly distributed and often occupying less typical forms of shelter. At the same time, all but the youngest, most recently inducted and lowest ranking of service people are also members of more typical households, from which they may often be absent." The military is also uniquely prominent in the context of census residence rules because of the weight that has been put on including members of the military in apportionment totals; along with federal government employees on overseas assignment, military personnel stationed overseas have been included in census apportionment totals in recent censuses (they are only assigned to a state, though, and are not included in redistricting totals). The overseas dimensions of counting the military are a major part of debates on census residence issues in recent decades, and we describe them in greater detail in Appendix C.

In this section we also discuss the special problem of counting people working and stationed on boats, and so not directly tied to any location on land. The treatment of people on ships—whether military, privately owned, or the Merchant Marine fleet that transports goods during peace time but serves as an auxiliary to the navy during war—has been a long-standing concern in the census and practice has varied over the years.

3–F.1 Personnel Stationed at Domestic Bases or Living in Nearby Housing

The 1880, 1890, and 1900 census enumerator instructions were the first—and among the relatively few—sets of census rules to explicitly mention the task of counting people at military stations in the United States (although the direction on how exactly to count them was not clear). The 1880 rule stipulated that "all soldiers of the United States Army, and civilian employees, and other residents at posts or on military reservations will be enumerated in the district in which they reside, equally with other elements of the population." The 1890 rule modified the rule to read that the personnel "will be enumerated in the same manner as has been provided for institutions." That rule seemed to tip the decision toward counting military personnel assigned to domestic bases at the base, but the 1900 census instructions reversed that decision:

> If a soldier, sailor, or marine (officer or enlisted man), or civilian employee in the service of the United States at a station at home or abroad, is a member of a family living in your district, he should be enumerated as a member of that family, even though he may be absent on duty at the time of enumeration.

Specific directions on counting the military population are then absent from enumerator instructions until 1950; in the interim, the general process of counting military stationed at land bases in the United States (by means of specially appointed enumerators) seems to have been the norm.

Beginning in 1950, census practice shifted toward an approach that put emphasis on separate enumeration, using deputized personnel on military bases and vessels to conduct the count, and, ultimately, toward uses of service records. A dual approach took shape in the 1950–1980 censuses: military personnel posted to a domestic base would receive (and be required to return) an individual census report, but off-base households (where the same service members could live with their families) would be covered by the normal household enumeration. The 1950 enumerator instructions for completing the basic population schedule indicated that "soldiers, sailors, marines, and airmen" were only to be enumerated as residents on the standard form if they "are stationed in your vicinity and [live] and sleep off post in your [enumeration district]" (see Box 5-5); instead, military personnel stationed elsewhere were identified as a "special class" to be enumerated by alternate procedures. The question of overlap—of the service members being counted in both places—was left ambiguous; census materials do not suggest any provision for reconciling the individual service reports with household returns from nearby areas. The Bureau's procedural history of the 1970 census, for instance, says only that the "Report for Military and Maritime Personnel" questionnaires collected from on-base service members "were included in the preliminary population counts for the places where they were stationed" (U.S. Bureau of the Census, 1970:8–25); if and when in processing these records were compared with off-base reports is not specified.

The 1990 and 2000 censuses modified these long-standing procedures by permitting personnel living on base to claim a "usual home elsewhere" (see rules 4 and 13 in the 2000 rules, Appendix A). In practice, what resulted in the 1990 and 2000 approach was a multiple-form process. All service personnel stationed to bases were required to fill out individual MCRs, distributed and collected by base personnel. At the same time, housing units located off base received questionnaires in the usual manner (e.g., mostly by mail). Both forms were expected to be returned. If a service member reported an off-base household as his or her usual residence, the MCR record could be linked to the census return from the off-base household to ensure that no duplication had taken place.[35]

This two-form approach suggests one possible source of error in collecting residence information from military personnel and dependents at or near

[35]The MCRs used in the 1980 and 1990 censuses are unusually direct and specific in asking the question about another residence. Question 2b of the 1990 MCR form asks, "What is the address where you usually stay at least 4 nights a week?"—a majority of days in a week as the criterion for usual residence, and not a more generic "most of the time" query. The questionnaire includes "building or barracks number" as a field, so the service member could report on-site housing as applicable; question 2c confirms whether or not the address is on a military base or not. Hollmann (1987) indicates similar text on the 1980 MCR. The 1990 SCR administered to sailors asked the more general question, "Do you have a residence (house, apartment) where you usually stay when off duty?"

domestic bases—confusion in the data collection process. Census materials tried to make clear that completing two forms (the MCR by the service member and, if applicable, the normal census questionnaire at the off-base home) was normal and expected. Instructions on the 1990 MCR advised that:

> Military personnel living away from this installation, but within the census area, also will receive a census form at home. To ensure that such personnel are assigned to the correct jurisdiction, it is important that YOU MAKE SURE YOU ARE INCLUDED ON BOTH FORMS—the report and the census form sent to your home.

Likewise, press releases emphasized the completion of both forms.[36] Still, these efforts may have left some military families confused and claiming that—one form having been filled out—they saw no need to complete another. For instance, public affairs officials at Fort Detrick, Maryland, reported particular confusion following reports in a base newsletter that "the census at Fort Detrick was 100 percent completed." That report referred only to the distribution and collection of MCRs; regular census enumeration and nonresponse follow-up were still in process off base (Duble, 2000).

In addition to possible confusion over two forms, counting personnel at or near domestic military bases is also complicated because bases include a variety of housing arrangements and styles that defy easy categorization.

- *Barracks and dormitories:* A large component of military housing is functionally equivalent to other types of group quarters, with groups of service men and women sharing common facilities, as well as bedroom-bathroom combinations. These quarters range from the highly institutional barracks in boot camps and basic training facilities to the emerging standards of "four plus one" (four people with separate sleeping quarters but shared commons) or "one plus one" (two people with connected, shared living area but possibly separate bedroom and bathroom) configurations.
- *Officer housing:* The military services may offer housing options to officers that lack the "group quarters" aspects of barracks or dormitories; these can be private apartment units that are functionally equivalent to civilian apartment or condominium complexes or stand-alone houses that are likewise indistinguishable from conventional housing stock. Depending on where exactly these units are located (within base confines or on the periphery), there may be ambiguity as to whether the housing units are included on the Master Address File and, hence,

[36] See, e.g., an article from a Fort McCoy, Wisconsin, newsletter (http://www.mccoy.army.mil/vtriad_online/03102000/Military Census.htm [8/1/06]) for a base census project officer's explanation of the two-form approach.

whether they may be picked up by normal census operations in addition to military count operations.

- *On-base and off-base housing for families and dependent children:* Depending on rank and marital status and dependent children, military personnel may live in on-base housing or receive a monetary basic allowance for housing to be applied to off-base housing. As discussed earlier, off-base housing raises the potential of confusion of overlapping operations: if family members live in quarters within the physical confines of the base, they could be missed by the normal household census operations (questionnaire mailout) if address listing is not done carefully.
- *Public partnerships:* In response to long-standing concerns on the quality and quantity of affordable military housing, the Military Housing Privatization Initiative was begun in 1996 in order to facilitate private-sector financing, construction, and maintenance of housing for military servicemembers.[37] Under the initiative, the military services are authorized to enter into agreements with private developers, who may own and operate military family housing under 50-year lease agreements. This private ownership may affect the information that base personnel have on the occupancy and exact structure of housing units during address listing operations.

Like their counterparts serving overseas, domestic-based military personnel have some "home of record" on file in Department of Defense records. Periodically, questions have been raised as to whether military personnel should be counted using administrative records, relying on that "home of record" information. That proposition was raised at the same 1999 congressional hearing that discussed pending legislation on the counting of prisoners. Speaking against the idea, census director Kenneth Prewitt commented that "if we had to match completed census forms for armed forces members to Defense Department administrative records, that would require a massive, costly, and time-consuming operation that we could not undertake without putting the census at risk." Further, Prewitt questioned the nature and consistency of "home of record" coding; in some cases, it could reflect place of birth and not more recent residence information (U.S. House of Representatives, Committee on Government Reform, 2000). One state—Kansas—disagrees with the Bureau's default placement of military personnel as well as college students and fields its own survey in order to adjust census totals so that military personnel stationed in Kansas but whose home of record is elsewhere are excluded from the counts used to redraw legislative districts (we discuss this further in Section 7–E).

[37]See http://www.acq.osd.mil/housing/mhpi.htm [8/1/06].

3–F.2 Shipboard Personnel

The 1850 census enumerator instructions—the first to begin to include detailed residence rules—sought to establish a clear policy for shipboard personnel: "errors necessarily occurred in the last census in enumerating those employed in navigation, because no uniform rule was adopted for the whole United States." Specifically, the rules directed that people should be counted at their homes on land, if possible, and at the ship's "homeport" otherwise:

> The sailors and hands of a revenue cutter which belongs to a particular port should be enumerated as of such port. A similar rule will apply to those employed in the navigation of the lakes, rivers, and canals. All are to be taken at their homes or usual places of abode, whether present or absent; and if any live on board of vessels or boats who are not so enumerated, they are to be taken as of the place where the vessel or boat is owned, licensed, or registered.

Instruction language in 1860 repeated this basic rule, adding the specific instruction that "persons on board any description of ships or vessels accidentally or temporarily in port . . . are not to be enumerated in your district," but rather at their land homes or at "the place where they have been engaged, shipped, or hired." The presumption that sailors were to be counted at their land homes was sharpened further in 1870, with the instruction that "seafaring men are to be reported at their land homes, no matter how long they may have been absent, if they are supposed to be still alive." This latter wording was directly repeated in the instructions for several subsequent censuses, the 1890 instructions being the first to specifically acknowledge that military sailors and marines, as well as navy yard personnel, would be counted by special enumerators.[38]

As it did for counting the domestic military population, the 1950 census marked the arrival of a revised approach for counting naval personnel and the seaborne population generally. The enumerator instructions pointedly told household enumerators not to enumerate "officers and crews of ships" when "they are reported as absent members by their families."[39] Instead, the 1950 census practice signaled a stronger partnership with the Department of Defense and related agencies (U.S. Bureau of the Census, 1955:12):

---

[38]As mentioned in the previous section, enumerator instructions for 1900 indicated that sailors "at a station at home or abroad"—presumably including assignment to a ship—be enumerated at their land home. The 1900 rules added a general provision that "all persons who claim to be residents of the United States" who can be found "in vessels, steamboats, and house boats at wharves and piers or river landing[s]" on the morning of Census Day be enumerated "by the enumerators of the districts contiguous to the water front." Whether those residents could be attributed to land home is not specified.

[39]However, "men employed on vessels on the inland waters (rivers, canals, etc.) of the United States, other than the Great Lakes," were to be "report[ed] in the regular way" (that is, on the household form).

> Naval, merchant marine, and other vessels were enumerated with the cooperation of the Navy Department, Coast Guard, Coast and Geodetic Survey, and the Merchant Marine. Mailing registers were established listing the vessels and the approximate number of crew members aboard each vessel. Packages of the special enumeration forms for crews of vessels (P4) with letters of instructions were mailed directly to the captains of all vessels in the Navy and Coast Guard and to other government-operated vessels. Those for the merchant marine were grouped and mailed to the companies operating the ships and then reshipped by them to each vessel.

Just as the revised military counting procedure brought with it some ambiguity (in the form of uncertainty as to how military person records would be unduplicated), the revised practice for enumerating shipboard personnel left it uncertain as to whether ship crews would be counted at a port or at individual homes. The procedural history of the 1970 census (U.S. Bureau of the Census, 1970) indicates that census forms from military ships and other American flag vessels were routed to the Bureau's processing center and sorted (by reported location on a control card) into "U.S." and "At Sea" groups. The returns from each "U.S." vessel were then "coded to an enumeration district . . . and a census block assigned to the pier area of the port where the vessel happened to be or, in the case of a coastwise vessel, its home port." That is, the physical location of the ship around Census Day appeared to take precedence over the residences reported on individual reports. Accordingly, the physical location of these ships—each of which could house hundreds of service men and women—became a point of concern in communities with seaports. Hollmann (1987:280) recounts that "during the 1960s there was an abundance of stories, most of them no doubt apocryphal, of the efforts of city officials to get as many of the home-ported vessels and as many transients as possible into port to tie up and be counted on Census Day. A major storm which would make life difficult for door-to-door enumerators was supposedly a heaven-sent gift to the local finance officer of chamber of commerce of the port city."

The 1980 census revamped the process for enumerating personnel on ships. Naval ships in the 6th or 7th Fleets were automatically designated part of the overseas population, because deployment to those Mediterranean- and Pacific-based fleets is generally long term. Other naval vessels were "attributed to the municipality that the Department of the Navy designated as its homeport." If the ship had fewer than 1,000 personnel, the crews were counted at the physical homeport location; personnel on ships with crews of 1,000 or more with reported usual residences within 50 miles of the home port were allowed to be counted at those residences, and at the home port location otherwise. Crews of merchant vessels were not allowed to report usual residences on land: they were counted at the U.S. port at which they were berthed on Census Day, at their port of destination if their vessel was not berthed but was still in U.S. territorial waters, and at their home port if they were still in U.S. waters but headed overseas.

The 1990 and 2000 censuses dropped the 1,000-personnel and 50-mile provisos of the rule for counting naval vessels. They replaced the 50-mile radius with the ambiguous "nearby," and permitted reports of household locations on land and counting military shipboard personnel there if possible. The two censuses also permitted crews of merchant vessels to report off-ship residences; those who did not were counted at the berthing port (if berthed on Census Day), the U.S. port of departure (if at sea), the U.S. port of destination (if headed back to the United States from a foreign part), or as part of the overseas population otherwise. Citro (2000:205) relates:

> For the 2000 census, the Census Bureau worked with the U.S. Department of Defense and U.S. Coast Guard to identify both housing units and group quarters on military installations and ships assigned to a home port in the United States. Questionnaires were mailed to housing units on military installations; other methods, such as visiting installations and ships to enumerate people at their work stations, were used to count the military population in group quarters.

The U.S. Maritime Administration assisted in identifying and arranging to mail questionnaires to other maritime vessels in operation around Census Day; these vessels included "factory trawlers, floating processors, tuna boats and National Oceanic and Atmospheric Administration vessels" (U.S. Census Bureau, 1999:1).

Discussion of the military seaborne counting procedures at the 1986 COPAFS conference on residence rules suggested general agreement that the use of home port information rather than current location was generally good. However, accuracy depends on how home ports are recorded and whether they are shifted, for instance, when a ship is detailed to a navy yard for months for extensive repairs or upgrades. The discussion also pointed out remaining ambiguities in the interpretation of what a home port location means. With specific reference to San Diego, Hollmann (1987) describes how different interpretations where a ship is located (e.g., physically tied to a pier within city limits, even though the land below "mean low water" underneath the ship's hull might belong to another jurisdiction) could credit ships whose home port is San Diego to the cities of San Diego, National City, or Coronado. That specific case was argued over most of the 1980s, culminating in a California court ruling that credited the ships to Coronado even though the waters in which they float may be in San Diego's jurisdiction.

– 4 –

Complex and Ambiguous Living Situations

IN THIS CHAPTER we look beyond the residential ambiguities inherent in the group quarters or nonhousehold population to examine other living situations and social trends that complicate the specification of residence. We consider five main categories:

- people with ties to multiple nonpermanent residences, including those who are highly mobile (Section 4–A);
- groups for whom residential ambiguity is created by complex household structures and the changing nature of "family" in American life (4–B);
- homeless people with ties to no fixed residence, living on the streets or making use of shelters and other services (4–C);
- people who may be missed by existing census operations and questions, whether due to gaps in residence rules and questions or to structural features in the census questionnaire (4–D); and
- people whose true residential location—and even their counting in the census—may be affected by the nature of the housing stock in which they live (4–E).

4–A MULTIPLE RESIDENCE AND HIGHLY MOBILE POPULATIONS

This section describes some living situations in which a person's residence lacks a degree of permanence: people in these situations tend to move between more than one residential location, with varying frequency. Gober and Mings (1984:165) argue that "nonpermanent changes in residence have received little attention by social scientists in the western-industrialized world because the process of nonpermanent movement falls between the crack, so to speak, of what we usually define as migration and tourism." For many reasons, even the rough size of the populations in these living situation groups is difficult to ascertain, much less definitive information on trends in their growth as a share of the total population.

Alone among modern U.S. censuses, the 1980 census included in its supplementary reports a separate tabulation of nonpermanent residents, tallying "elderly seasonal migrants, owners of second homes, itinerant farm workers and business people who reside part-time in out-of-town accommodations" (Gober and Mings, 1984:164). These estimates "were never intended as estimates of seasonal or any other form of temporary migration" (McHugh et al., 1995:253), and they excluded nonpermanent residential situations such as those "staying temporarily at the home of a permanent resident or in hotels, motels, campgrounds or other tourist-like settings" (Gober and Mings, 1984:164). Despite their limitations, the 1980 estimates were something rare in the study of multiple-residence situations—national-level information, rather than case-study analysis.

Both Smith (1989) and McHugh et al. (1995) suggest broader conceptual frameworks for understanding multiple residence. These frameworks suggest that the assumption that people are linked to a single usual fixed place of residence may be unrealistic.

4–A.1 "Snowbirds" and "Sunbirds"

"Anyone who has spent spring vacation in Daytona Beach, August at the New Jersey seashore, or February in Sun City, Arizona, knows that many places have large numbers of temporary residents who live there for a few days, weeks, or even months." These temporary residents, Smith (1989:430) notes, "often have a tremendous impact on an area's economic, social, and physical environment, as they increase the demand for housing, shopping centers, health care, water, electricity, transportation, recreational facilities, police protection, and many other types of goods and services." Lowry (1987:15) comments that some of the "seasonal inflow of residents" in such areas are "casual visitors who will not return; but others return year after year, either to rented quarters or to seasonal homes that they own." This latter group is particularly interesting because, by virtue of the frequency with which they return to the area, they develop strong ties and "often become substantially involved in local affairs through their status as property owners." However, "they are not usually able to vote in local elections because they have registered elsewhere."

The term "snowbird" has gained popular usage to describe people—often retirees or the elderly—who leave cold-weather areas during the winter months for warmer climes, such as Florida, Arizona, southern California, and the Rio Grande Valley of Texas. Hogan and Steinnes (1996) and Smith and House (2005) adopt the term "sunbird" to describe those who follow the counterpart trend, fleeing hot summer areas for milder northern climes in places like Cape Cod, the Maine shore, and northern Wisconsin. As discussed in Box 2-1, the existence of such seasonal migration has been known for decades, and tradeoffs in its perceived impact were considered in adjusting the date of the census in the early 1900s. However, the census has not historically asked the questions to make an informed assessment of the size and growth of the snowbird and sunbird communities.[1]

Instead, the research that has been done to date on the demographics of the snowbird population has tended to be local area studies done in areas with high concentrations of seasonal residents:

- Smith and House (2005:4) and colleagues surveyed 500 Florida households each month between September 2000 and December 2003 by telephone. Of the 7,041 respondents aged 55 and older, 83 percent were permanent, full-time Florida residents; 12 percent were permanent Florida residents who indicated that they spend more than 30 consecutive days at some other location (sunbirds), and 5 percent were not permanent Florida residents but were currently in the state on an extended stay (snowbirds). See also Smith and House (2004a,b) and Galvez (1997) for further comment on the Florida snowbird population.
- Research on the Arizona snowbird population has taken several shapes. Happel and Hogan (2002) and colleagues conducted a statewide household survey during 1990–1991, intended to reach 400–500 households a month. That survey showed a variation in the percentage of nonpermanent households, ranging from 0.7 percent in September to 6 percent in February and March (peak season). Though the survey offered other insights, it was terminated in 1991 due to funding cutbacks. However, Arizona State University sponsored smaller-scale, targeted interviewing at mobile home and recreational vehicle parks, beginning in 1984 and continuing through 2000.

  Happel and Hogan (2002) estimated that Arizona may have 273,000 long-term seasonal residents who come each year at the peak of the season. Similarly, Smith and House (2004a) estimate that there may be as many as 920,000 seasonal residents arriving each year in Florida, not counting short-stay tourists.

[1]The 1980 special tabulation on nonpermanent residents covered only those snowbirds whose destination was Arizona or Florida (Hogan and Steinnes, 1996).

- Researchers from the University of Texas–Pan American (UTPA) estimate that 127,000 "winter Texans" took up residence in the Rio Grande Valley (63,500 households), spending a total of $420 million while there; the average length of stay was 3.7 months, and most had made the same seasonal move for 10 years or more (Valley Markets and Tourism Research Center, 2005). The UTPA study also collected survey information from full-time valley residents, as well as Mexican nationals who cross the border for very short (often one day or less) visits, principally to shop. UTPA researchers have conducted a study of "winter Texans" biennially since 1986 (see also Center for Tourism Research, 2003).

Snowbirds were also the focus of an ethnographic study conducted as part of the 2000 census experimental program; Hunter et al. (2003) and de la Puente (2004) summarize the results of the study, performed by Mings (2001).

Though the current levels of snowbird and sunbird migration are not well determined, researchers suggest that seasonal migration will almost certainly continue to increase due to several factors, among them the aging of the "baby boomer" generation, longer life expectancies, and rising household wealth.

The local area surveys that have been done on the snowbird populations focus principally on the nature and impact of snowbirds on their *destination* communities, where they go to for their seasonal trips. Much less is known about the *origin* locations of the snowbird population—the small-area impacts of depopulation in northern, cold-weather communities and consequent lessening of the need for some services (Happel and Hogan, 2002). In much the same manner as in localities housing college students or prisoners, the location at which seasonal residents are counted in the census is important to both communities—arguably more so, since large seasonal swings in population can create large drains on local resources like transportation and health care. The limited work that does exist on snowbirds' origin destinations suggests that some common assumptions may be erroneous. For instance, "it is popular wisdom that the usual residence for nonpermanent households is 'far away,'" noted Mann (1987:11): "for example, 61 percent of all New York State households initially enumerated in a place other than their usual residence were found in Florida, hardly a surprise." However, there are limits to which snowbird movement can be thought of as a cross-state phenomenon; Mann cites a Census Bureau report on nonpermanent households as reporting that "47 percent of California's nonpermanent households lived permanently elsewhere in California; 44 percent of New York State's nonpermanent households lived in New York and 33 percent of Texas' lived in Texas." Relatively little research has been done to date on either the origin or destination of sunbirds; see Hogan and Steinnes (1996) for a contrast between snowbirds "flying from" Minnesota and sunbirds originating in Arizona.

The residence rules for the 1990 census (see Box 5-4) would treat a snowbird or sunbird as "a person who has more than one home and divides time between them," in which case the rules directed that they be counted at "the household where he/she spends the greater part of the calendar year." The problem of counting seasonal residents was singled out for commentary in the Rolark (1995:3) memorandum outlining the 2000 census residence rules:

> [The 1990 rule for "snowbirds"] was never clearly explained on the census questionnaire and we never told respondents which calendar year (i.e., the census calendar year, January 1, 1990 to December 31, 1990; the calendar year preceding the census, January 1, 1989 to December 31, 1989; or the year from April 1, 1990 to April 1, 1991) [to use as a reference]. Using any of these reference periods presents a unique set of problems ranging from the accuracy of the respondent's memory to the respondent's inability to project future behavior. We recommend that persons who live in one residence during the winter months and live in another residence during the remainder of the year be counted at the place where 'they spend most of their time' during the 'yearly cycle'[—]the nonwinter residence. This gets back to the general rule for persons with multiple residences.

Accordingly, snowbirds are referenced in rule 2 of the 2000 census residence rules as being counted "where they spend **most** of the time during the week, month, or year, etc." and are cited as an example of a "yearly cycle" in the attachment to the rules.

Conceptually, the accurate counting of the seasonal snowbird and sunbird populations is complicated by several factors, including the timing of census operations and the cross-national scope of seasonal migration. First, with respect to timing, the 1920 decision on the timing of Census Day (see Box 2-1) alludes to a basic tradeoff that must be weighed: the early months of the year—including the peak census activity months of March and April—are times when the census may find snowbirds at their winter (and likely shorter-term) address. At their nonseasonal addresses, some information on their characteristics may be gleaned from proxy interviews with neighbors or landlords, but the census questionnaire provides no mechanism to suggest a connection between the two addresses. The issue of timing is complicated further when follow-up interviews and coverage measurement operations extend into May, June, and July. Snowbirds may return to their nonseasonal homes, but the later interviews may miss sunbirds who depart at the beginning of the summer season.

Second, snowbird and sunbird migration is not only within the United States. Coates et al. (2002) observe an increasing tendency for Canadians to move to warm-weather locations in the United States for the winter months. "Detailed statistics on the scale of the seasonal migration are sketchy at best" and support widely varying estimates, but they conclude (based on 1999 data) that on the order of 300,000–375,000 Canadians join the snowbird migration. Foreign nationals receiving a census questionnaire at their seasonal home might be less inclined to return it, but that question does not appear to have been studied empirically. In addition, Coates et al. (2002) suggest an increased tendency for snowbirds (both U.S.- and Canada-based) to spend winter months in Mexico, making it impossible to contact them directly at the seasonal residence.

Both Happel and Hogan (2002) and Smith and House (2004a) argue for the need for the Census Bureau to be able to generate national estimates of the seasonal migrant population. Though improvements could be made to the decennial census to acquire such estimates, both sets of authors suggest that the American Community Survey (ACS, the replacement for the census long-form sample) is an ideal vehicle for generating these estimates on a regular basis. In particular, Smith and House (2004a) advocate asking specific questions on the ACS in order to differentiate short- and long-term visitors to an area: "Is the current address your usual place of residence?" If not, "Where is your usual place of residence?" And, "How many months will you be at the current address during your current visit?"

In this section, we have focused on one component of seasonal migration, specifically the movement of older, retired people for stays of several months. It is worth nothing that other significant short-term population movements affect younger age groups and even shorter "seasons." One such case are groups of migrant farm workers, which we discuss in Section 4–A.6; another seasonal population of interest in major hosting cities like Las Vegas, Chicago, and Atlanta is the steady stream of short-term attendees of professional meetings and conferences. Still another major, regular such shift is the influx of college students on spring break that arrives at beach communities, such as South Padre Island, Texas, and Panama City, Florida, or other destinations. Because colleges vary in the exact date of their spring break or semester/quarter break, these communities may experience major population surges—and both the economic benefits and resource drains that accompany that growth—for much of March through April. Studies of the size and nature of the spring break population is limited, but the short-term effect on small areas can be large. Vincent et al. (2000) estimated that 186,000 college students came to the South Padre Island area alone in March 2000, spending on the order of $156 million and resulting in the creation of 4,276 jobs during the spring break season.

4–A.2 Modern Nomads: Recreational Vehicle Users

The "life course" framework for multiple residence suggested by McHugh et al. (1995) identifies snowbird and sunbird movement as a cyclical migration pattern prevalent among retirees and the elderly. Another pattern evident in the same age group—albeit less tied to specific seasons—are those people who essentially travel year-round in recreational vehicles (RVs), touring various parts of the country and visiting family members and friends. Part of this population is also known as "Good Sams," after the Good Sam Club formed so that members could have a signal to know of other motorists and facilities where they could get assistance on the road. In the extreme case of year-round RV users, these people might be considered "homeless snowbirds"—constantly on the move with the seasons (Happel and Hogan, 2002).

Sponsored by the Recreational Vehicle Industry Association, the University of Michigan Survey Research Center has conducted national surveys of RV users every 4 years, as part of its Surveys of Consumers program. The 2001 National Survey of Recreation Vehicle Owners was based on a representative sample of households contacted between January and June 2001. The survey estimated that 6.9 million U.S. households own a RV,[2] representing 7.6 percent of all vehicle-owning households, consistent with rates around 7 percent from previous versions of the survey and suggesting mild growth from a recent low of 6.8 percent in the 1993 data (Curtin, 2001). Owners of full-size motorhomes (1.9 percent of all households) tend to be older (average age of 59) than those with less fully equipped RV types like folding camping trailers (average age of 41). Only 18 percent of motorhome owners reported having dependent children, compared with 61 percent of folding camping trailer owners.

For the 2000 census, a new residence rule 16 was meant to clarify the way in which the RV population should be counted rather than change it. The rule allowed this group to claim a usual home elsewhere; if the respondent could not specify a place where they live most of the time, they were to be counted at the camp where they were found. The rule further noted that if a respondent considers the RV to be his or her usual place of residence, then the RV is considered a housing unit and would be tabulated with the household (not group quarters) population.

Hunter et al. (2003) and de la Puente (2004) summarize ethnographic research by Mings (2001) that focused on a set of RV campgrounds in Arizona. This work suggests some of the conceptual difficulties involved in counting RV users:

- *Definition of the physical place where they are "most of the time":* The place where truly diehard RV users live or stay most of the time may in fact be the RV, if they keep to a steady rotation of visiting friends and relatives for weeks at a time. The 32 RV-using snowbirds interviewed by Mings

[2]For the purposes of the study, "recreational vehicle is defined to include all types of motorhomes, conventional travel trailers, fifth-wheel travel trailers, folding camping trailers, and truck campers" (Curtin, 2001:5).

(2001) constitute a rather small sample; still, of that number, some 20 percent do not maintain a permanent residence.

- *Inaccessibility by common enumeration techniques:* By their nature, intensive RV users can be difficult to reach by physical mail, much less in a timely fashion. Hence, it is possible that a census questionnaire could be sent to a home they maintain, or to the home of friends or relatives where RVers commonly return, but there is no guarantee that the people in question will be there to receive or answer it. Or they may also rely on a rented post office box for the accumulation of their physical mail and thus might not be included in the census mailout population to begin with. The constancy with which they are on the road means that they can be difficult to reach by personal visit; they might be contacted by a one-shot deployment of enumerators to campgrounds (e.g., a T-Night-type operation as described in Section 4–C), but they might not. The best chance of their inclusion in the census might be from proxy reporting by a neighbor or family member, but that approach may impair the accuracy of the data.
- *Incomplete inventory of RV campsites:* Mings (2001), as summarized by Hunter et al. (2003), found that some of the sites where he interviewed dedicated RV-using snowbirds (with no other residence) were not visited by enumerators in any census operation. He urged that future censuses pay more attention to undeveloped and public land campsites rather than commercial RV resorts and privately owned campgrounds. Public land campsites differ from their commercial counterparts in at least two respects that further complicate census enumeration matters. First, they typically do not offer any kind of mail delivery (some commercial RV grounds do), making even a fluke census questionnaire delivery unlikely. Second, public grounds typically impose a limit on the length of stay—for instance, 14 days—that encourages frequent moving by their users (Hunter et al., 2003:5).

4–A.3 Commuter Workers and Commuter Marriage Partners

Work and employment can be a source of ambiguity in residence definitions for a variety of reasons. In the next section we discuss cases where ambiguity stems from the nature of the job itself; here, we discuss ambiguity caused by the location of work. Economic or family circumstances compel some people to live lengthy distances from their work location, in the exurbs and rural expanses surrounding metropolitan areas. These long-distance commuter workers may live one place (near work) during the work week but return to another the rest of the time, making it difficult to identify one usual residence.

From the survey or census context, the problem of trying to obtain information from such a commuter worker—single, without family—is vexing enough. More difficult still are "commuter marriages" where work conditions may split spouses or family members. Each party, reached separately, may not think of the place where he or she lives simply for the purpose of going to their job as their "usual" residence. Commuter marriages are one of many changing and evolving structures that have drawn the attention of family demographers in the past few decades.

Actual data on the extent of commuter workers or commuter marriages are difficult or impossible to obtain. For a group for which so little is known about its basic size and trends, the commuter population has nonetheless drawn a great deal of attention in residence discussions. In summarizing the 1986 Council of Professional Associations on Federal Statistics (COPAFS) residence rules workshop, CEC Associates (1987:23) wondered whether "usual residence" was the appropriate benchmark for this group and whether something more akin to "an 'habitual' residence, perhaps that residence where the person has lived consistently over a longer period of time," would be more meaningful. More fundamentally, the report focused on a critical concern in addressing this group: namely, the tendency for people to connect "household" with family, rather than the census concept of "household" as a collection of people at a housing unit. Choices in treating commuter workers can induce "potential bias of family composition data, creating a class of fictitious single parent families, possibly with incomes greatly underrepresented by excluding that of the absent spouse."[3] CEC Associates (1987:22) concluded that the rule for commuter workers "probably needs to be retained" for the 1990 census, but that "research needs to be undertaken to determine the impact of this rule on family and household statistics, on occupancy statistics, and most importantly on apportionment and redistricting data."

The 2000 census residence rules attempted to add some clarity to the handling of commuter workers, though how well that worked is unclear. Rolark (1995:5) observed that the 1990 residence rules strictly considered commuter workers as "persons with one residence where they stayed on weekends and another residence where they stayed during the week while working"; hence, the rules would count the weekday residence as the usual residence since it would be the place where the greatest amount of time was spent in a given week. "Other patterns of staying (e.g., one month away, one month home, etc.) were not explicitly addressed." With the attachment on weekly, monthly, and yearly cycles included in the 2000 rules, the Bureau opted to leave determination of the appropriate time cycle to the respondent; "persons having

[3]This specific phenomenon regarding family and household data is not a specific concern for the 2010 short-form-only census, since household income is traditionally a long-form (and now ACS) question.

a weekly, monthly, or other cycle of commuting should be counted at the place where they spend the most time in that week, month, or other cycle." However, the decision was not universally accepted by the staff committee developing the rules:

> Some team members strongly disagreed with this decision feeling that this rule does not provide enough guidance to cover the varied patterns of commuter workers. Those who disagreed also felt that, from a respondent's perspective social attachments were stronger determinants of usual residence than the place where the person slept most nights. They felt that although some family members worked and lived away from the family home, they would still be included on the form that his/her family filled out, despite what the rules stated.

Though the Rolark memo implies a more flexible concept of commuter worker arrangements, the phrase "commuter workers" only occurs in the discussion of a "weekly cycle" in the 2000 census residence rules. In addition, the Web-posted version of the residence rules speaks only to "commuter workers living away part of the week while working," advising that they be "counted at the residence where they stay most of the week."

The National Household Travel Survey (NHTS), administered by the Bureau of Transportation Statistics and the Federal Highway Administration, provides some empirical glimpses at long-distance commutes.[4] The NHTS defines "stretch commutes" as those involving trips of at least 50 miles one way: 2001 NHTS data estimated that 3.3 million Americans stretch commute annually, with about 19 percent of those commuters traveling one-way distances of 100 miles or more. In a 4-week period (20 business days), 65 percent of those stretch commuters with a travel distance of 50–99 miles made the commute 16 or more days; that percentage drops to 33 percent for those with one-way trips of 100–199 miles and to 12 percent for drives of 200 or more miles. What the survey does not immediately reflect is changes to household structures that result from these commuter patterns—for example, 65 percent of commuters traveling 200 or more miles make that commute only 1–4 days in a 4-week period, but it is not clear whether that is because the traveler maintains a second residence closer to the work location.

Nested under the broad heading of commuter workers is one very small, highly mobile, and very influential subpopulation—members of Congress.[5] The Bureau's procedural history of the 1970 census reports that "each member of Congress had the option of being enumerated at his Washington-area

---

[4]For additional information on the NHTS, see http://www.bts.gov/programs/national_household_travel_survey [8/1/06].

[5]Similar arguments hold for state legislators, who may need to have residences in the state capital while also maintaining residence in their home districts. In many states, though, legislative sessions are short duration.

address in his home State. For Congressmen and Senators who made the latter choice, the appropriate population and housing data were tabulated for his home address" (U.S. Bureau of the Census, 1970:8-2). Mann (1987) suggests that the same choice was offered in 1980, and the the U.S. Supreme Court's summary of the government's brief in *Franklin v. Massachusetts* (Box 2-5) noted this provision as well. The residence rules for the 1990 census indicated that "the Census Bureau plans to count Members of Congress either at their Washington, D.C. area address or at their home state residence, dependent upon which one they consider to be their usual residence" in following the instructions on the questionnaire. For 2000, it is not clear whether any special program was in place, but members of Congress were listed as an example of a multiple-residence category to be handled by rule 2—no guidance on whether a weekly, monthly, or yearly cycle was most applicable. The counting of members of Congress is one instance for which the difference between different standards of defining residence—and the varying standards used by federal and state agencies—are particularly vivid, as strict adherence to a usual residence standard defined by spending four or more nights a week in a place would seem to make the great majority of them ineligible for reelection from their districts, inasmuch as they would be "residents" of the Washington area (Mann, 1987:7–8).[6]

4–A.4 Residential Ambiguity Due to Occupation

The basic nature of work and employment can serve to complicate residence patterns. As cases for which a strict interpretation of living or staying someplace "most of the time" can be problematic, Sweet (1987:12) suggested:

> Consider the long haul truck driver. Perhaps he (or she) is on the road 200 or more days a year. Yet he has a family and maintains a household at some fixed location. He and his family regard him as a member of this household, and it would seem to be a mistake not to classify this person as a member of his household. Other occupations where a similar pattern might often occur include traveling sales people and installers of large, complex equipment. What distinguishes this group from others that we have discussed is that they normally do not maintain another residence in a fixed location somewhere else.

In such cases, "it would seem desirable to classify these persons as residents of their home"—wherever they might specify it to be—"even though they do not spend a large share of the time there."

[6]Likewise, absolute adherence to a Census Bureau-style "usual residence" standard ("live and stay most of the time") in all settings would typically establish both the President and Vice President of the United States as residents of the District of Columbia. The 12th Amendment's stipulation that the two must be from different states would then seem to make the president-vice president ticket ineligible for reelection.

The idea that people employed in transportation and related services have jobs that require constant movement and extended time away from "home" by their very nature is a long-standing one in the census context. We comment on the rules applying to seaborne personnel (emphasizing the military but applying as well to private-sector crews such as freight ships and cruise ships) in Section 3–F. Enumerator instructions for the 1900 census directed that "persons engaged in internal transportation"—including "canal men, expressmen, [and] railroad men"—should "be reported as of their families, and not where they may be temporarily staying on [Census Day]" provided that "they habitually return to their homes in the intervals of their occupations." The 1950 enumerator instructions added to the list of examples of persons "engaged in transportation services or traveling" to include bus drivers, railway mail clerks, and traveling salesmen. These persons "usually have homes to which they return at intervals and which constitute their usual place of residence"; they were to be counted at that place unless they had no such usual place of residence, in which they were supposed to be counted where they were found. Rule 1 of the 2000 census residence rules covered "[persons who live] in this household but [are] temporarily absent on Census Day on a visit, business trip, vacation, or in connection with a job (e.g., bus driver, traveling salesperson, boat operator)," and ruled that they be counted at the household. Other occupations that fit this general type include airline flight crews and employees of moving companies.

Previous census enumerator instructions and residence rules have also covered a group for whom the opposite situation may apply, in which a person's workplace and residence may coincide: live-in or long-term domestic workers and caregivers. In 2000, such workers were to be counted as part of the household provided that the person "works for and lives in this household and has no other home."

The long shifts—often multiple 24-hour periods—that are part of the work experience of public safety personnel like firefighters or paramedics/emergency medical technicians can also blur the line between "work" and "home" locations. The relative ease of transportation and the increasingly large geographic scope of companies and corporations also creates some jobs where extended travel (e.g., supervising a geographically disperse set of plants or facilities) is the norm. Companies may detail employees to a regional office elsewhere in (or out of) the country for weeks or even months, or a person could be transferred full time to another position elsewhere but elect not to move their family members from their present homes.

4–A.5 Minority Men

Studies of census error since the 1950 census have pointed to high net undercount among adult males of minority racial groups, particularly among young black men. While the 2000 census was successful in reducing the net undercount among minority race groups, minority males are still a group that have historically proven to be complicated to count accurately. Though the undercount for black males decreased in 2000, rates were still elevated for them and other racial and ethnic minority groups, notably Hispanics, relative to whites.

A small-sample but intensive study of households in predominantly black inner-city areas by Valentine and Valentine (1971a,b) suggested one possible contributing source to this "missing men" problem. Later described by Brownrigg and Wobus (1993:156) as "the man under the bed" hypothesis, the argument suggested in this research holds that minority men may be deliberately omitted (concealed) from household counts, either because the men "do not want to be found" or "the men are intentionally not revealed by respondents for ulterior reasons." The Valentine and Valentine (1971a,b) analysis found rates of missing young adult and adult males as high as 61 percent in their small-area study, and they concluded that "practically all the significant inaccurate information came from adult females who [neglected] to mention productive men residing in their domiciles" (quoted in Federal Committee on Statistical Methodology, 1990:43).

Many of the households where males were unreported received welfare assistance; accordingly, a suggested reason for the failure to count the men was a legacy of welfare "man in the house" policies.[7] Under the "man in the house" rules used by some states, welfare payments could be cut off if it were shown that a man was present in the house (ostensibly, filling a "substitute father" role and providing support). In the census context, then, respondents might be unwilling to report to any federal agency the presence of males in the household. "Man in the house" rules were struck down by the U.S. Supreme Court's ruling on Alabama's policy in *King v. Smith* (392 U.S. 309, 1968), but whether people realized that the rules had changed was a different matter; indeed, contemporary researchers still find that some people believe that some variant of a "man in the house" rule is in effect (Edin and Lein, 1997). Fear of jeopardizing eligibility for benefits due to the presence of a male in the household (and, hence, another resource provider) may still hold among people who are residents of public or federally subsidized housing.

The notion of deliberate concealment of household members—particularly men—was later corroborated by Tourangeau et al. (1997), who found that minority males were more likely to be omitted from a household roster where full names were required and more likely to be reported in list-

[7]The Valentines also speculated that another financial consideration might motivate the failure to count men in some households: "many of the unreported men were also engaged in some form of illegal economic activity, e.g., the stolen goods market," so suppressing mention of those men had the added incentive of preserving a household income stream (Federal Committee on Statistical Methodology, 1990:44).

ings when full names were not requested. This finding of sensitivity of the reporting of names was consistent with ethnographic work by Hainer (1987): interviewers found that "any question that was linked to anyone's name was too personal and threatening," and that "informants assumed that any information given to one source is shared by all others" (Federal Committee on Statistical Methodology, 1990:44).[8]

Other analysts suggest that a more benign explanation—confusion, not concealment—may contribute to the census undercount of minority men. Specifically, analysis of data from the 1993 Census Bureau-sponsored Living Situation Survey by Martin (1999) found inconsistencies between people listed on rosters as having any attachment to the household in a 2-month period and the people identified by respondents as "members" of the household. None of the reasons specified for omitting a person as a member of the household hinted at concealment or deception; rather, the explanations spoke to the nonmembers as having tenuous ties to the household, whether for family or economic reasons (e.g., not contributing to rent) or because they were perceived as only being temporary or transitory residents. The confusion hypothesis is consistent with the findings of Iversen et al. (1999); that study administered the 1990 census questionnaire (a mix of both short and long forms) to families in a poor area of inner-city Philadelphia. Using cognitive testing techniques, respondents were encouraged to think aloud as they completed the form and were debriefed afterward, generating a question-by-question description (and explanation) of errors and difficulties. Almost all of the completed forms (104 of 107) included at least one error. The think-aloud comments suggested major problems with literacy and interpretation of even basic questions; a cognitively complex question such as the basic household roster—prefaced, in 1990, by a lengthy presentation of "include" and "exclude" instructions—was particularly troublesome.

An additional reason why some minority men may be missed in the census is because of the lack of a permanent attachment to any particular household. Instead, they may move frequently between the homes of their mothers, their kin, their girlfriends, and their peers, so that a respondent at any one of these locations might not feel obliged to report them in a household listing (Edin and Lein, 1997).

None of the speculated reasons for the undercount of minority males are adequate, in isolation, to explain the full magnitude of the problem. For instance, by the logic of the concealment hypothesis (the fear that reporting men in households might jeopardize eligibility of benefits), the major drop in welfare enrollment in the mid-1990s should have yielded a major increase

[8]An ethnographic study done for the 2000 census focused specifically on a group of young urban male gang members. Those ethnographic interviews suggested sensitivity over confidentiality—coupled with a general "strong aversion to the government" and law enforcement—as possible reasons for concealment from census returns (Hunter et al., 2003:2–3).

in the minority male count in 2000. Though the estimated net undercount rate for black males was reduced relative to 1990, the finding of an undercount among minority men remained intact. The exact combination of causal factors for undercount—concealment, confusion, and high mobility among them—remains a noticeable gap for further research to fill.

4–A.6 Migrant Farm Workers

The highly detailed enumerator instructions of the 1950 census were the first to explicitly mention the counting of worker camps: "persons in railroad, highway, or other construction camps, lumber camps, . . . or places that have shifting populations composed mainly of persons with no fixed places of residence" were to be counted where found. The special case of agricultural worker camps, occupied by migrant farm laborers during harvest periods, was explicitly referenced in the 1980 and 1990 census residence rules (see Box 5-4) and later in rule 14 of the 2000 census residence rules. The 1990 rule took a stronger stance than the 1950 instructions, directing that such agricultural workers be counted at the camp and omitting the clause about the rule applying to workers with no other usual place of residence. The 1980 and 2000 rules permitted migrant agricultural workers to specify a "usual home elsewhere" location; if no such usual home elsewhere was specified, they were to be counted at the camp.

The class of agricultural workers who reside in camps or in owner-provided barracks or dormitory-type housing during harvest periods is only one part—and likely a small part—of the overall class of migrant farm workers. Particular attention has been focused on the temporary farm labor pool composed of foreign migrants, particularly from Mexico but also from other Central American countries, in light of ongoing debates on immigration policy in the United States. As noted in discussing our charge in Chapter 1, our focus is not on whether illegal immigrants should be counted in the census but rather—assuming the goal of generating a resident count—to discuss how residence rules and instructions apply and may be improved.

Just as terms like "usual residence" are complex and can be defined in multiple ways, the meaning of "migrant" farm workers can also vary considerably. Garcia and Gonzales (1995:7) observe that some definitions focus on those workers who are in constant motion, "mov[ing] from one farm area to another in search of agricultural employment. This worker moves about without ever settling down permanently in one location." Other definitions incorporate a seasonal or temporal dimension, "depict[ing] a migrant as a worker who is in an area for a given amount of time" for the harvest of a particular crop; "once the crops are picked, the worker moves on to another area or returns home." A more expansive definition relaxes the specificity on one particular crop or one particular season. Under this definition, "a migrant is a worker who leaves his home community for other areas to work for an indefinite amount of time, that may vary from three months to three years." The key characteristic under this definition is that the worker "does not create a new permanent home in the area" but rather "returns to his home base after he has earned what he set out to make."

The local area and time-specific impacts of migrant farm workers can be massive. For example, Palerm (1994:6) estimated that—on its own—strawberry harvesting in California's Santa Maria Valley consumes 8.5 million person-hours, "represent[ing] nearly 4,000 full-time jobs if employment were distributed evenly throughout the year. In actuality Santa Maria strawberry farms employ as many as 10,000 individual workers, many of them intermittently, during a four-to-five month period and some during even shorter periods of time." While some agricultural products in the valley (such as broccoli and cauliflower) are harvested nearly continuously, other crops (such as grapes and celery) have short, well-defined harvest seasons that—like strawberries—require a significant short-term labor force. Ethnographic researchers have also observed that migrant farm workers from Mexico have a broader geographic effect than some might expect, as they "are no longer limiting themselves to farm areas in the U.S. southwest. Today, they venture to communities and work in agricultural industries found throughout the country, including the U.S. northeast. In Pennsylvania, New York, and New Jersey, for example, vegetable, fruit, and horticultural producers are hiring Mexican laborers in large numbers" for harvest periods (Garcia and Gonzales, 1995:2).

The U.S. Department of Labor's Employment and Training Administration sponsors the National Agricultural Workers Survey (NAWS), which directly interviews farm workers in the continental United States; the survey aims to interview 3,200 workers each fiscal year. Interviews are administered three times a year (based on the seasonality of the industry), in February, June, and October. The 2001–2002 report of survey findings (U.S. Department of Labor, 2005) provides information on a range of characteristics of migrant workers:

- 42 percent of the crop worker force are migrants, defined as traveling at least 75 miles during a 12-month period to do farm and agricultural work. NAWS analysts further differentiate migrant workers as *shuttle* migrants (who work multiple sites in a 75-mile radius of a single U.S. location) and *follow-the-crop* migrants (who travel to multiple U.S. locations to do farm work, depending on the harvest season). *Foreign-born newcomers* are new workers to the survey who have not yet demonstrated a migration pattern. Table 4-1 illustrates the shares of these groups as a percentage of all migrants and of all hired crop labor, as well as the percentages within each group who report that they have

Table 4-1 Classification of Farm Workers

Migrant Type	1993–1994		2001–2002		
	Percent of All Farm Workers	Percent of All Migrants	Percent of All Farm Workers	Percent of All Migrants	Living Apart from Children, as Percent of All Migrants
"Home" Outside the United States					
International shuttle[a]	13	29	13	30	43
International follow-the-crop[b]	5	10	2	5	7
Foreign-born newcomer[c]	10	23	16	38	35
"Home" Inside the United States					
Domestic shuttle[a]	9	18	6	13	6
Domestic follow-the-crop[b]	9	20	6	14	9
Nonmigrant	52	—	58	—	—

NOTES: Percentages do not add to 100 percent due to rounding.

[a] Shuttle migrants may work several farm sites when not at their "home" location, but do all of their farm labor work in a 75-mile radius within the United States.

[b] Follow-the-crop migrants travel to various locations in the United States to do farm labor, depending on the season.

[c] New entrant to the farm labor force who has not yet demonstrated a migration schedule.

SOURCE: U.S. Department of Labor (2005:Table 1.1).

children at their "home location" who do not live with them while they are working.

- 75 percent of the hired crop labor force were born in Mexico and 23 percent in the United States.
- 83 percent identified themselves as members of a Hispanic group.
- Only 21 percent are legal permanent residents of the United States; 53 percent are not legal immigrants and are not authorized to work in the United States.
- 61 percent of the migrant farm workforce reported living in rented property owned by someone other than their employer; 27 percent lived in housing provided for free by the employer, while 6 percent rented from the employer. Only 1 percent reported staying free of charge with family or friend, and 4 percent reported owning their housing property.
- 6 percent of migrant farm workers (and 3 percent of all farm workers) reported living in dormitory or barracks facilities.

The limited glimpse at migrant farm workers' living conditions provided by the NAWS highlights some of the conceptual difficulties involved in contacting them in the census and gathering accurate residence data. The illegal and undocumented status of a large share of the population raises the possibility of undercounting the workers: contacted employers may be apprehensive about providing full reports if they hire undocumented workers, or householders may not readily "volunteer information about the presence of undocumented kin or friends in the home" (Palerm, 1994:26, 28). The NAWS data suggest that only a very small share of farm workers live in dormitory- or barracks-type housing, such as might be reached in group quarters enumeration. Rather, most of the workers contacted in the survey reported living in rental property, possibly sharing the space with other laborers. These short-term rentals pose difficulty for enumeration in several respects—the likelihood that an English-language questionnaire would reach a predominantly Spanish-speaking household, the possible perceived disincentive for foreign citizens to cooperate with the U.S. census or to reply to a government survey (if their immigration status is illegal), the tenuousness of ties to the housing unit, and the possible inaccuracy of proxy reporting by neighbors or landlords. The seasonal nature of crops may also work against complete enumeration due to the timing of Census Day. Returning to the Santa Maria Valley, California example, Palerm (1994:12) comments:

> [On April 1, Census Day,] only one-half or less of the [seasonal] migrants are actually there. By early April the strawberry and lettuce harvest is just beginning to build-up steam but is not yet in full swing. Moreover, having just arrived, most migrants are still in the process of making their

living arrangements for the season, creating with their great numbers havoc in the local housing situation and probably producing the worst possible conditions for the completion of a sound and accurate population count.

4–B COMPLEX HOUSEHOLD STRUCTURES: THE CHANGING NATURE OF FAMILIES

As discussed in Chapter 2, the construct of the family is fundamentally linked to notions of usual residence or household composition. It is also one that varies across cultures and has shifted with time; demographic and sociological research continue to probe "the dimensions of long-run changes in the American family" (Ruggles and Brower, 2003:73):

> The past 150 years have witnessed extraordinary change in American living arrangements. In 1850, for example, 70 percent of the elderly resided with their children, and 11 percent lived alone or with only a spouse; by 1990, only 16 percent resided with children, and 70 percent resided alone or with a spouse only. The changes have been almost as great for the young: since 1910, the percentage of children under age five residing without two parents has increased more than fourfold, to 27 percent in 1990; among blacks, the figure is 67 percent.[9]

Some of the range of contemporary household and family types is expressed in Table 4-2, which describes the number of children (under age 18) in various household compositions using data from the Census Bureau's Survey of Income and Program Participation (SIPP). Single-parent households are increasingly common, accounting for roughly one-quarter of childrens' living situations; about 4 percent of children live with neither parent.

Some of the living arrangements described in this chapter constitute what Census Bureau researchers have termed "complex households"—"those where the web of relationships within the household is other than one nuclear family (i.e., nuclear family being married couple with or without its own biological children)." Examples of these complex households include the presence of nonrelatives in the household, such as unmarried partners and gay partners; more distant relatives such as grandparents, cousins, and aunts and uncles; children who are shared across households; and "people who may be mobile or ambiguous in terms of household membership" (Schwede, 2003:vii). The challenges involved in identifying and labeling such complex households are plentiful and interesting, including the interpretation of the census questionnaire's relationship question depending on which household

[9]Ruggles and Brower (2003) cite a then-forthcoming version of Carter et al. (2006) as the source of these data.

Table 4-2 Children Under Age 18 by Household Composition, 1996 and 2001 (in thousands)

Children Living With	2001	1996
Two Parents	51,113	50,685
Married parents	48,987	49,186
Unmarried parents	2,126	1,499
Biological mother and father	45,103	44,708
Married parents	43,287	43,401
Biological mother and stepfather	4,050	3,723
Biological father and stepmother	815	1,004
Biological mother and adoptive father	445	479
Biological father and adoptive mother	56	37
Adoptive mother and father	605	702
Adoptive mother and stepfather	16	23
Adoptive father and stepmother	19	9
Stepmother and stepfather	4	—
One Parent	18,472	18,165
Mother only	16,297	16,340
Biological	15,980	16,051
Father only	2,175	1,825
Biological	2,082	1,737
Neither Parent	2,917	2,644
Grandparents only	1,407	1,266
Other relatives only	889	688
Nonrelatives only	520	622
Other arrangement	101	69
At Least One Biological Parent	68,531	67,739
At Least One Stepparent	5,081	4,902
At Least One Adoptive Parent	1,372	1,484
At Least One Foster Parent	260	313
Total	72,501	71,494

NOTES: —; rounds to zero.

SOURCE: Kreider and Fields (2005:Table 1), based on tabulations from SIPP, 2001 Panel, Wave 2.

member—as "Person 1"—completes the form.[10] Our review in this section, however, is limited to the effects of these household structures on the derivation of household rosters and counts.

4–B.1 Children in Joint Custody

At the 1986 COPAFS residence rules workshop, Lowry (1987:17–18) observed that "the table of residence rules [for the 1980 census] does not explicitly mention children who divide their time between two or more residences, usually because their parents are divorced or separated." The rule that seemed most applicable to these children is the general guidance of only counting nonpermanent "visitors" at the household if they have no other place where they live and sleep most of the time. However, he concluded, "I doubt that a divorced parent who completes a regular household census form would be likely to [consider] his or her offspring as a 'visitor.' "

No residence rule in the 1990 census specifically mentioned children in joint custody, but the 2000 census residence rules added joint custody as an example case in rule 2, which covered cases where residence follows a weekly, monthly, yearly, or other cycle. Though the rule text does not specify which of these is most applicable, the attachment defining the cycle types implies that joint custody children are supposed to be counted based on where they spend the majority of the month:

> Some children live with one parent for one week out of the month and the other parent the remaining three weeks during the month. We consider these persons to be on a "monthly cycle" and they should be counted at the place where they spend most of their time during the month (e.g., children in joint custody situations, etc.).

If time is split equally between multiple residences, the rule implies, the children should be counted where they are on Census Day.

The Panel to Review the 2000 Census (National Research Council, 2004c:47) speculated that double counting of children in joint custody arrangements—as part of the broader population with ties to more than one residence—was an important source of duplication in the 2000 census. Evidence on this point was provided by two Census Bureau coverage evaluation

---

[10]"Interrelationships among other persons in the household can be masked and not be identifiable," and family structures within "households" can be difficult or impossible to recover, depending on who fills out the census form (Schwede, 2003:viii). For example, consider a case where a man and woman live together, unmarried, along with the woman's child from a previous relationship. If the man is the census respondent, the woman may be reported as an unmarried partner or an "other nonrelative," while the child would likely be "other nonrelative"; the biological link between woman and child is obscured. If the woman is the respondent, the biological link between her and her child would be preserved, but it would be ambiguous to family and household researchers whether the male is the child's biological father or not.

studies. A detailed reinterview of 17,500 people and a clerical review of their census records identified children in joint custody as a source of error: they were counted at a census household when proper application of the residence rules would place them at another location (Adams and Krejsa, 2001).[11] In addition, methodology for matching census records based on name and date of birth to detect duplicate census records (discussed further in Section 8–B) suggested spikes in the age distribution of estimated duplicates that were consistent with children in joint custody. This finding was first noted when the detailed matching was done within the coverage evaluation samples (Fay, 2002a) and corroborated by matches of the entire set of census records against itself (Fay, 2004).

The general problem of defining residence for children in joint custody arrangements is complicated by the number of ways that these arrangements come about. Custody awards may be made by judicial ruling as part of the formal dissolution of a marriage; in describing the nature of shared custody arrangements and trends in their development, much of our discussion is focused on this group because of the formality of the arrangement and its presence in legal records. However, it is very important to note that shared custody arrangements can arise without a formal, legal divorce: they can develop in cases where a divorce is not fully executed, such as a separation or estrangement. They may also arise in the absence of a marriage; in Section 4–B.2 we discuss cohabiting couples and the problems inherent in counting children in such households, but it is also essential to consider the question of how children are counted when cohabiting couples split up. Much of what is known about the magnitude of and trends in custody arrangements focuses on custody awards after divorce; detailed empirical information on more informal shared custody arrangements is scarce.

The institution of joint custody arrangements in divorce proceedings is a relatively recent phenomenon. By the 1920s, a strong preference among divorce courts to award custody to mothers "became as firmly fixed as the earlier paternal preference" that had dominated for decades. As divorce rates escalated in the 1960s, fathers in divorce cases began to argue that maternal-dominated custody decisions violated equal protection and constituted sex discrimination, resulting—following passage of the federal Uniform Marriage and Divorce Act in 1970—in the adoption of gender-neutral custody laws based solely on the "best interests" of the child (Kelly, 1994:122). Over the course of the 1970s, the notion of "joint custody" emerged as a viable option in divorce proceedings. California passed the first law formally recognizing

[11]The 17,500 people in this evaluation follow-up study were drawn from the "E-sample" whose census records has been matched to a "P-sample" (independent Accuracy and Coverage Evaluation interviews) to estimate net census error.

Box 4-1 Types of Child Custody Arrangements

Sole Custody

- **Sole legal custody**—*Custodial parent* is assigned all legal rights, duties, and powers as parent; all decisions affecting welfare of child. *Noncustodial parent* has limited rights and powers; has access to child's medical and school records.
- **Sole physical custody**—*Custodial parent* has primary physical custody of child. *Noncustodial parent* is usually awarded visitation rights.

Joint Custody

- **Joint legal custody**—Both parents retain rights and powers to make decisions regarding child's health and welfare. Decisions must be specified in some states to preserve the authority.
- **Joint physical custody**—Both parents retain right to share in day-to-day residential care of child; generally not defined as equal sharing, but intended to grant substantial periods of time to each parent.

Divided Custody—Each parent has child for a portion of the year or in alternating years. Each parent has legal rights for decision making when child is in that parent's care.

Split Custody—Each parent has sole legal and physical custody of one or more children; noncustodial parent has visitation rights.

SOURCE: Kelly (1994:Table 1).

joint custody in 1979; "by 1991, more than 40 states had statues in which joint custody was either an option or a preference" (Kelly, 1994:123).

Kelly (1994) summarizes various types of custody arrangements; these are listed in Box 4-1. Within this framework, an important point related to the residence determination of affected children is that "joint" physical custody is not necessarily (and perhaps not often) a strict fifty-fifty divide. Instead, variation in custody awards and state laws means that the "actual resident time [that a child may spend with a custodial parent] may range along a continuum from somewhat expanded visiting to equal time in each household" (Kelly, 1994:23). Cancian and Meyer (1998:150) suggest differentiating between equal shared (physical) custody and unequal shared custody, defining the latter as an arrangement where the child lives with one parent 30–49 percent of the time and with the other parent the remainder of the time. The cutoff of 30 percent is relevant for their analysis of Wisconsin divorces because that figure defines a threshold in child-support formula calculations.

The itemization of types of custody arrangements suggests a rigidity in classification; this is not always the case. In considering children in joint custody, it is important to bear in mind that even families with legal custody arrangements do not always follow them: the legal decision about which parent(s) a child will live with does not always correspond to where the child actually lives. This may be even more likely as children grow up and as the time since the initial custody decree increases. Even families who do not go to court for a legal settlement may still have children who spend part of the time in each parent's household. Actual living arrangements among these children may be sufficiently fluid to resist easy categorization and are certainly more dynamic than is documented in administrative records of custody awards.

A major gap in understanding the size and dynamics of children in joint and other custody arrangements is that there is no uniform national-level source of data on child custody arrangements. At least two major federal data collection programs provide some information on shared custody situations: vital statistics and the Current Population Survey (CPS). Both have limitations, though, and are generally focused on legally ordered custody arrangements (as in divorce proceedings).

The first of these data sources is the vital statistics program of the National Center for Health Statistics (NCHS). Until 1995 NCHS gathered information on (physical) custody arrangements as part of the divorce portion of the vital statistics, compiling counts from records of registered divorces maintained by state government departments. However, budget constraints led NCHS to abandon its detailed marriage and divorce data files in favor of basic counts and rates. Yet even when the detailed data were being gathered, relatively few states reported the full data; the current basic count data continue to suffer from inconsistent or nonexistent reporting, to the extent that NCHS no longer publishes a national count of divorces.[12] (A supplementary source of detailed divorce data—a marital and fertility history supplement to the Current Population Survey, separate from the child support supplement discussed below—was also terminated in 1995. As a result, there exists no national source of information on divorce trends in the late 1990s and early 2000s.)

The most recent surviving data from the vital statistics program to be analyzed by NCHS staff date from 1989–1990 and include reports of physical custody arrangements from 15 and 19 states in 1989 and 1990, respectively. Those data, shown in Table 4-3, suggest that about 40 percent of divorces include a determination of custody. Among those cases where custody is awarded, roughly 80 percent are a grant of sole physical custody to one parent, 15 percent are joint custody, and the remainder either split custody or placement of children with another person entirely.

Short of detailed information on the number of children involved in shared custody agreements, the vital statistics data do shed light on trends in divorce in the United States. Divorces need not involve custody awards—there may be no children in the family, or the children may have reached legal adulthood—but trends in divorce are at least a rough benchmark to bear in

[12]NCHS attributes this latter decision to lack of reports from California, Georgia, Hawaii, Indiana, Louisiana, and Oklahoma (National Center for Health Statistics, 2005).

Table 4-3 Divorces by Whether and to Whom Physical Custody of Children was Awarded, Selected States, 1989 and 1990

	1989[a]		1990[a]	
	Count	Percent	Count	Percent
Custody Awarded	128,507	40.2	110,787	38.2
Sole, to husband	11,186	3.5	9,046	3.1
Sole, to wife	92,330	28.9	79,001	27.2
Joint	20,183	6.3	18,980	6.5
Split[b]	3,104	1.0	2,579	0.9
Other person	1,704	0.5	1,181	0.4
No Custody Awarded[c]	149,674	46.8	131,561	45.3
Not Stated[d]	41,514	13.0	47,818	16.5
Total	319,695	100.0	290,166	100.0

[a] The 1989 data are based on reports from 15 states (Alabama, Connecticut, Idaho, Illinois, Kansas, Michigan, Missouri, Montana, New Hampshire, Oregon, Pennsylvania, Tennessee, Utah, Wisconsin, and Wyoming), and the 1990 data on reports from 19 states (adding Alaska, Nebraska, Rhode Island, and Virginia).

[b] Some children of the divorcing family were awarded to one parent and some to the other parent.

[c] Includes divorces in which the number of children under 18 years of age was reported as "none," and therefore the custody item was not applicable; also includes divorces in which children were reported, but no custody was awarded.

[d] Includes divorces where custody was not reported, except those in which the number of children was reported as none. Such divorces were assigned to "no custody awarded."

SOURCE: Clarke (1995:Table 17).

mind when considering trends in shared custody. Based on data from reporting states, the vital statistics on divorce suggest a major growth from a steady state of roughly 400,000 divorces per year from 1950 through 1960 to a peak of 1.213 million divorces in 1981 (Clarke, 1995). In 1990, an estimated 4.7 million children (7.3 percent) lived with an unmarried parent, 5.9 million (9.1 percent) with a divorced parent, and 4.8 million (7.4 percent) with a separated or divorced parent (Shiono and Quinn, 1994). Since the 1981 peak, "marital dissolution rates have been constant for almost two decades" (Bumpass and Lu, 2000:29).

Although national-level compilation of court and government records on custody arrangements is lacking, researchers have studied custody issues using detailed extracts of court records. Cancian and Meyer (1998) use Wisconsin court record data from 21 counties, compiled over several years, and summarize findings from other studies. Additional court record compilations, for sets of counties, have been compiled for California (Maccoby and Mnookin, 1992), Michigan (Fox and Kelly, 1995), and Minnesota (Christensen et al., 1990).

A second national source of information on custody arrangements is the CPS conducted by the Census Bureau. Every two years from 1994 to 2002, a child support supplement to the CPS (conducted in March/April) asked survey respondents whether they had ever been awarded joint legal or physical custody of a child. Specifically, the questionnaire asked whether "a court or judge" gave the respondent custody (U.S. Census Bureau, 1998:Items S503, S504). Child Trends (2002) analyze data from the 1994, 1996, and 1998 supplements, and some of their calculations are shown in Table 4-4. Generally, the CPS data suggest a relatively stable distribution of custody awards in aggregate; joint custody awards account for 8–9 percent of settlements, with the most frequent result (68 percent of cases) being the mother being awarded sole legal and physical custody. The more recent sets of CPS child support data corroborate this result, as summarized by Grall (2003:1):

> In the spring of 2002, an estimated 13.4 million parents had custody of 21.5 million children under 21 years of age whose other parent lived somewhere else. About 5 of every 6 custodial parents were mothers (84.4 percent) and 1 in 6 were fathers (15.6 percent), proportions statistically unchanged since 1994.

The CPS data capture some part of the resolution of custody in out-of-wedlock cases, and that marital status has an effect on custody outcomes: unwed mothers are most likely to be awarded full legal and physical custody, while dissolved marriages are more likely to have shared custody of some sort.

For purposes of determining residence in the census and in surveys, the critical question is how each of the parents in a divorce understand and interpret questions about where their children live. A body of research summarized by Lin et al. (2004) and Schaeffer et al. (1998) suggests important differences in how divorced mothers and fathers report their childrens' living arrangements; the two parents may report things differently (such as the amount of child support money paid), and each parent "tended to report that their children spent a greater number of nights with them than with the other parent" (Lin et al., 2004:386).

Lin et al. (2004) report the results of a survey of about 1,400 divorced parents in Wisconsin that speaks directly to responses to residence questions.[13] Their work included both qualitative and quantitative assessments, allowing respondents to describe arrangements in their own words in open-ended responses as well as to provide a direct count of the number of nights spent in

[13]The Lin et al. (2004) study was not limited to pairs of divorced parents; 440 of the 1,392 interviews were with only one of the parents in a proceedings.

Table 4-4 Type of Child Custody per Most Recent Agreement, 1994–1998 (in percent)

	Mother Legal and Physical	Mother Physical, Joint Legal	Father Physical; Legal Sole or Joint	Joint Physical and Legal	Other
1994	68	8	12	8	4
Marital Status					
Never married	85	3	6	2	4
Single, previously married	63	12	13	11	2
Currently married	62	7	15	9	7
1996	65	10	11	9	5
Marital Status					
Never married	82	4	8	2	4
Single, previously married	56	14	14	14	2
Currently married	62	9	10	10	9
1998	68	7	10	9	6
Marital Status					
Never married	83	2	7	3	4
Single, previously married	58	10	15	14	4
Currently married	65	7	9	10	9

NOTES: Estimates only calculated for households with a child (under age 21) who lives with one biological parent and whose other parent is absent. "Other" includes split custody arrangements.

SOURCE: Child Trends (2002:Tables P19.1–P.19.3), based on Child Support Supplement of the CPS, 1994, 1996, and 1998.

a location. Their results suggest important sensitivity to particular words—divorced fathers were less likely to use the explicit phrase "live with" in describing where their child lives, although they were more likely to explicitly say that the child "live[s] with" both parents. Asked to count how many nights the child spent living with their mother in a year, mothers placed the count about 44 days longer than did fathers. Their analysis concluded that "parents are more apt to say that their children live with one of the parents when children spend at least 58% of the time in the full year with one parent" (Lin et al., 2004:391). Divorced parents' qualitative assessments of where their children live may vary from a count of nights stayed; even "when parents report the same number of nights that children spend with the mother, mothers are more likely than fathers to say that the child lives with the mother, and fathers are more likely than mothers to say that the children live with the father or both parents" (Lin et al., 2004:395).

4–B.2 Cohabiting Couples

An increasingly important group in the studies of family dynamics are cohabiting couples who live together—perhaps for long periods of time—but do not marry.[14] Though debates on the nature and effects of cohabitation frequently force cohabitation into a dichotomous framework, either as "a stage in courtship that leads to marriage or an alternative to marriage," the reasons for living together (and for separating) are more complex than that framework allows (Seltzer, 2004:58).

Cohabitation rates have increased dramatically over the past three decades—the 1970 census found 523,000 households maintained by unmarried couples, compared with 4.9 million in 2000—and this growth has been shared among all demographic groups (see, e.g., Bumpass and Lu, 2000; Bumpass and Sweet, 1989; Simmons and O'Neill, 2001; Waite, 1995). The 1995 National Survey of Family Growth found that 45 percent of white non-Hispanic and black women aged 19–44 reported that they had ever lived in a cohabiting relationship, as had 39 percent of Hispanic women; those figures represented increases from the 1987–1988 wave of the survey, from 32 (white), 36 (black), and 30 (Hispanic) percent, respectively. "Marriage and remarriage rates have declined markedly, though these declines have been largely offset" as "cohabitation has grown from a rare and deviant behavior to the majority experience among cohorts of marriageable age" (Bumpass and Lu, 2000:29).

Perhaps the most basic conceptual challenge that cohabitation raises for collection of residence and household composition data is the length and nature of cohabiting relationships. Though cohabiting relationships can last years or decades, it is commonly a short-term pairing, "with about half lasting a year or less, only one-sixth lasting three years, and about a tenth lasting five years or longer" (Bumpass and Lu, 2000:33). The tenuousness and fragility of relationships in early stages can complicate questions about residence—at what point does the cohabiting partner switch from a person who "stays" here versus one who "lives" here? As Sweet (1987:11) describes, "the [cohabiting] process may begin with an occasional night together. Gradually one partner may spend more and more time in the household of the other, while maintaining their own apartment. Eventually one partner gives up his or her own place and they are unambiguously living together." Short of that "unambiguous" level of commitment, "one partner may feel that they 'live together,' while the other partner may feel that he or she lives somewhere else, and merely stays here most, or all, of the time" (Sweet, 1987:11).

A critical question involved with the responses of cohabiting couples is how children in this situation are counted. Children of cohabiting couples

[14] As with most research on cohabitation, our focus here is on heterosexual couples, though we comment on the special case of gay couples at the end of the section.

may be the biological child of both partners or of just one partner (long-term cohabiting couples may have children in the household who are the biological child of neither partner—adopted or foster children). Just as the adult partners in a couple may have discrepant ideas as to whether each person "lives" or "stays" in the household, the same ambiguity may carry over to the counting of children that are "his" or "hers" rather than "ours."

In addition, cohabitation can be a transitory phenomenon, with partners moving in and out of cohabiting relationships. Just as partners in the relationship might not know whether the current cohabitation will last, parents or other family members might also view the arrangement as temporary or a passing thing; particularly for younger people living together, "enduring ties" might compel parents or others to count them at home.

Though attitudes toward cohabitation have relaxed over recent decades, it is not universally embraced either. Some sense of a lingering stigma could also be a mechanism for the "concealment" or denial of a coresident partner. This is particularly so for a special case within the general class of cohabiting couples: same-sex couples, for whom an enduring stigma and fear of discrimination may contribute to underreporting (Smith and Gates, 2001). Following Black et al. (2000), Smith and Gates (2001) compared the 601,209 same-sex households containing two people identified as "unmarried partners" by the 2000 census' relationship question with other survey measures. They argue that the 2000 census undercounted gay and lesbians living in coresident relationships by 62 percent.

4–B.3 Recent Immigrants

The discussion in Section 4–A.6 focused on migrant farm workers, many of whom are undocumented persons who cross the U.S. border in search of employment; they were considered as a special case where employment is a cause of residential mobility. More generally, new migrants and immigrants to the United States present challenges for census enumeration due to their family structures and the living situations they create.

Analysis of the 2000 census public use microdata sample by Hernandez (2004) found that 47 percent of children in immigrant families live in crowded housing (defined as having more than one person per room), compared with 11 percent among children in native-born families. Immigrant children are also about twice as likely to have other people living in the household than native-born families (grandparents, other relatives besides parents and siblings, or nonrelatives). Several of the ethnographic studies conducted as part of the 1990 census focused on immigrant families, and concurred that "the most common living arrangement among the immigrants is an extended household. Nuclear families share their homes with kinsmen; they take in and shelter close relatives, like a parent, sibling, uncle or aunt, cousin, or *compadre* or *comadre* [fictive kin/co-parents]. In many cases, the kinsmen are migrants, who work in the area without their families" (Garcia and Gonzales, 1995:18). A body of research (see, e.g., Massey, 1986; Singer and Massey, 1998; Curran and Rivero-Fuentes, 2003) examines the general settlement process of migrants arriving in the United States.

Immigrant households are also highly likely to face a considerable potential barrier to accurate enumeration: language. The Census Bureau defines a household as "linguistically isolated" if "all household members age 14 years or older speak a language other than English and have no English proficiency" (Lestina, 2003:iii). An evaluation conducted by the Census Bureau based on the 2000 census estimated that 4.1 percent of households are linguistically isolated, an increase from 3.2 percent in 1990. These households can comprise a major share of the total population of either counties or census tracts; eight counties in Texas consist of at least 25 percent linguistically isolated households, and 11 tracts nationwide are at least 75 percent linguistically isolated households (Lestina, 2003:iii).

Operationally, linguistically isolated households present complications because they are less likely to avail themselves of self-response options to the census. Presented with an English-language questionnaire, they may fail to return the form simply because they do not understand it; they may also elect not to call telephone numbers on the questionnaire to request a foreign language questionnaire. In 2000, 57.7 percent of linguistically isolated households were enumerated in the census by mailing back a census form (this includes specially requested foreign language questionnaires and interviews completed when a respondent called the telephone questionnaire assistance line); by comparison, 71.2 percent of non-linguistically isolated households responded by mailback. Thus, linguistically isolated households are more likely to require the more labor-intensive follow-up operations.

In addition, new immigrant households, like migrant households, may be unwilling or unable to provide complete household listings due to social and cultural concerns. They may harbor great fear and mistrust in the government and, with it, the census, or they may fear jeopardizing their immigration status, particularly if they perceive that the data could reveal income streams from other kin or housing violations. Depending on the norms of their origin countries, they may also have differing views as to whether it is culturally appropriate to report to census takers and other officials about infants and children in the home, as well as older household members.

Massey and Capoferro (2004:1079) also observe that "undocumented migrants often live in irregular housing and frequently sublet rooms or even floor space in homes and apartments registered in the names of others. Often these arrangements are illegal, which virtually guarantees that the unauthorized residents will be unreported." Particularly vivid examples of the irregular and crowded residences held by migrant and immigrant families are found in the

Box 4-2 *Colonias*

Las colonias—using the Spanish term for neighborhood or community—are predominantly low-income residential areas on or near the U.S.-Mexico border that are notable for their lack of basic services such as electricity and water/sewer systems. All of the border states—California, Arizona, New Mexico, and Texas—have *colonias*, but they are most commonly associated with Texas, which has the highest number of these settlements. The Federal Reserve Bank of Dallas (1996) estimated that there are more than 1,400 *colonias* in Texas, housing over 400,000 people, and projects that an additional 700,000 people will need affordable housing along the Texas border by 2010—many of whom will gravitate toward *colonia* developments.

Colonias date back to at least the 1950s, when developers created unincorporated subdivisions in agriculturally barren land that—despite the lack of services—draw buyers due to the low cost. The population of the *colonias* is predominantly Hispanic, and unemployment rates range widely (from 20 to 80 percent, in one study) but are generally very high (Federal Reserve Bank of Dallas, 1996:7). Housing stock includes "many homes, built without regard for indoor bathrooms or plumbing, [that] are rated as substandard or dilapidated by housing inspectors" (Federal Reserve Bank of Dallas, 1996:11).

As part of the 2000 census research program, ethnographers conducted interviews in four *colonias*: two in Dona Ana County, New Mexico, one in El Paso County, Texas, and one in Riverside County, California (de la Puente and Stemper, 2003). (Riverside County does not directly border Mexico; under U.S. Department of Housing and Urban Development standards, *colonias* can be defined in a 150-mile border region.) Their research suggested that the *colonias* concentrate and combine four major barriers to accurate enumeration:

- *Irregular housing stock:* Ethnographers noted situations where owners of one-acre lots informally "subdivide" that lot and sell parts to family members who put trailers or campers on their spot. Only one address is associated with the lot, regardless of how many trailers may be there. They also observed small housing units nested within others, as well as housing unit "shelters" behind bushes or other physical obstacles.
- *Limited education and limited English knowledge:* Three of the observed colonias were in update/leave areas (see Box 2-2), in which questionnaires were left at households by enumerators. These English-language forms were thus left with predominantly Spanish-speaking households. In their interviews, the researchers found that "none of the interviewees noticed the Spanish language message [to call an 800 telephone number] for assistance at the bottom of the first page" of the questionnaire; "because it lies at the end of a page written in English, we suspect most interviewees . . .overlook the message" (de la Puente and Stemper, 2003:9).
- *Concerns regarding confidentiality:* Many *colonia* residents are undocumented or illegal immigrants, and hence may not be eager to participate in a government census or survey. Yet the ethnographers also noted a strong undercurrent of appreciation for the census; residents seemed to see participation in the count as a validation or empowerment process.
- *Complex and fluid households:* Several of the *colonias* have large proportions of seasonal migrant workers.

colonias along the U.S.-Mexico border; see Box 4-2. (We return to the problems associated with unusual housing stock in Section 4–E.)

4–B.4 Issues Unique to Native Americans

Unique among racial and ethnic minority groups in the United States, American Indians and Alaska Natives have a distinct legal and political standing.[15] Due to treaty obligations, tribal governments are essentially sovereign but dependent nations within the United States. In addition to the challenges associated with counting a population with demographic and family structures that differ from other groups in the nation, the legal standing of tribal governments creates basic logistical concerns for the census. Cooperation with tribal governments is necessary to generate accurate address listings and coordinate outreach activities and follow-up enumeration.

The accurate collection of information on native American ancestry is itself a complex topic—how well does the Census Bureau's self-identification system square with legal status as American Indian (listing on tribal rolls) or cultural standards (e.g., participants in ongoing Indian society, whether living on American Indian reservations or not)? However, it is not one that is directly related to our charge; a previous National Research Council (2004a) report discusses race and ethnicity data collection in the census more generally.

Based on ethnographic observations from the 1990 and 2000 censuses, and extant demographic research, some of the issues that can complicate the meaning and measurement of residence among native Americans include the following:

- *Fundamental difference in "household" concept:* Lujan (1990:10) observes that "the census schedule is founded on a western European image of how society is oriented" and its notions of residence are "based on the nuclear family household"; however, "most American Indian tribes and Alaska Native villages are based on the extended family concept and current residence patterns reflect this lifestyle." American Indian "households" are likely to include more multigenerational kin and, consequently, larger and more extensive household rosters, as well as frequent mobility among the housing units on a reservation. Schwede (2003:xiv)

[15]Indeed, though the historical review in Chapter 2 did not examine the topic in depth, American Indians have had a distinct standing in the census through most of American history. The Constitution directed that the population counts used for apportionment be calculated "excluding Indians not taxed." As a result, American Indians were excluded in the 1790–1850 censuses. The 1860 census included coverage of "civilized Indians," a definition based on land ownership. American Indians have been consistently counted in the census since 1890, though they were not included in apportionment counts from 1890 through 1940 (Lujan, 1990).

summarized ethnographic observation of Navajo and Inupiaq Indian respondents by noting that they tend to "not identify households in terms of shared physical structure [as the Census Bureau does], but rather on the basis of sharing of domestic functions such as earning and pooling income, cooperating in subsistence activities, cooking, child care, child raising, and other domestic tasks."

Ethnographic observation of American Indians living in the San Francisco Bay area confirmed that household structures can be fluid in the community, with family members or fellow tribe members staying for times when they do not have any other place to go. Even though the off-reservation American Indian population, such as that in San Francisco, is geographically disperse, "certain 'key' households in the community are known for taking in those who need a place to stay. Oftentimes, these are the homes of female members of the community who serve maternal roles by providing shelter." Hunter et al. (2003:7) concluded that these house guests—even if related—are highly unlikely to be reported as household members.

- *Resistance to or reluctance to federal government questioning:* Given the long and difficult history of American Indian relations with the federal government, it is perhaps understandable that a reluctance to cooperate in federal activities like the census might linger. Extensive ethnographic observation of the urban Indians in St. Louis prior to the 1990 census (cited in Lujan, 1990) suggested that 23 percent of the Indian population interviewed signaled that they would not be inclined to answer the census, citing confidentiality concerns and general distrust of the government. Similar ethnographic study of the Colville Indians (cited in Lujan, 1990) had the same finding, suggesting more general reasons that the Indians would be reluctant to answer, including fear of exposure of housing regulation violations in federally subsidized tribal homes and reluctance to discuss cohabitation. As a result, American Indians were a major focus of the extensive outreach and advertising campaign of the 2000 census, including custom-themed posters and targeted activities carried out in partnership with tribal governments.
- *Language differences:* Lack of fluency in English can hinder a declining but still-existent small group of people whose primary language is a native tongue from participating in the census.
- *High mobility rates:* Summarizing past work, Lujan (1990) divides reasons that have been associated with high mobility in the American Indian population into traditional and contemporary influences: Traditional influences include active participation in native American culture, which can involve attendance at tribal celebrations and participation in

ceremonies. Contemporary influences include subsistence activities, reflecting stark economic conditions on many Indian reservations. In the absence of employment opportunities on the reservation, many residents seek jobs elsewhere. Since their efforts may still be contributing to the economic well-being of the family at the reservation, "enduring ties" may drive both the worker and family members to consider the family home as the "usual residence" even if the worker is away most of the time. Bonvillain (1989), in an ethnographic study of the St. Regis Mohawk tribe for the 1990 census, found that men of the tribe are in high demand by construction companies for their worksmanship. Hence, groups of men from the Mohawk tribe regularly set up temporary group households at their employment sites, returning to St. Regis when they could.

4–C THE HOMELESS POPULATION

An examination of complex ties between people and residences is incomplete if it does not consider people with ties to no fixed residence in addition to those with ties to two or more residences. As a panel, we have not weighed recommendations on the exact manner by which the street homeless population should be counted, a group that also includes those making use of shelters and relief facilities like soup kitchens. However, we do briefly comment on the nature of homelessness and the operations by which the census has tried to include them in the counts.

A basic question we have tried to answer for the various groups and living situations profiled in this report is the size of the population in question. But the basic question—"how many people are homeless in the United States?"—is very difficult to answer. Homelessness is not necessarily a permanent state; instead, people and families living in impoverished conditions experience homelessness on an episodic basis. Thus, a definition that considers homelessness strictly as a point-in-time phenomenon—portraying only a snapshot of "persons literally without a roof over their head, or forced to sleep in public or private shelters" at a single instance—is unduly restrictive and "seriously underestimates the level of homelessness in society" (Kusmer, 2002:4).[16]

Wright and Devine (1992b:212) summarized the measurement difficulties:

[16]Indeed, some advocacy groups like the National Coalition for the Homeless oppose "the release of a separate 'count' of people enumerated in homeless situations (at selected service sites and identified outdoor locations) because such a number would be, by its very nature, both inaccurate and misleading, and therefore lead to uninformed decision-making by policymakers." See "NCH's Position on the Census and Homelessness" at http://www.nationalhomeless.org/publications/reports.html [8/1/06].

> A moment's reflection will make plain that 'the total number of the homeless' is of necessity a 'soft,' ambiguous number that probably cannot be known with a high degree of precision. . . . There [is] no shared or widely agreed-upon definition of just what constitutes a 'homeless' condition, and so different investigators are free to define the phenomenon in different ways. There is a large and obvious difference between the number literally homeless on any given night (a point prevalence rate), the number homeless at least once in the course of, say, a year (a period prevalence rate), or the number who become homeless during a given year (an annual incidence rate). No matter how inclusive the definition and how systematic the search, it is obvious that the homeless are a mobile, even nomadic, and certainly hard-to-locate group, and so the possibility is always open that large numbers of them have been missed in the counting effort. The above and a range of related factors imply that no study can provide a *definitive* count of the size of the homeless population. The best one can hope for is a more or less plausible count with known and small uncertainties attached to it.

"Public awareness of the 'new homeless' can be traced to the late 1970s," notes Kusmer (2002:239), "when beggars and 'street people' became increasingly noticeable in the downtowns of many cities." Homelessness continued to grow in prominence as a problem through the 1980s. In 1984 the U.S. Department of Housing and Urban Development estimated that, on any given night, at least 250,000 people lived on the streets or in shelters. "By 1990, that figure had doubled. A 1996 Urban Institute survey estimated that on an average night 470,000 persons in the United States were sleeping in shelters but that a much larger number, close to 2 million, had experienced homelessness at some point during the previous year" (Kusmer, 2002:239). Contemporary estimates put the point prevalence rate of homelessness (the number literally homeless on a particular day) at on the order of 840,000 people; over the course of a year, it is suggested that at least 2.3 million, and perhaps as many as 3.5 million, experience a spell of homelessness (National Law Center on Homelessness and Poverty, 2004; Burt and Aron, 2000). Limited survey data suggest that episodes of homelessness average 5 months in length (National Law Center on Homelessness and Poverty, 2004:7).[17]

Very few survey measures have attempted to give a nationwide examination of homelessness. The most recent national-level study was conducted in 1996, when the Census Bureau was the contracted data collector for the National Survey of Homeless Assistance Providers and Clients (NSHAPC).[18]

---

[17]See also "How Many People Experience Homelessness?," Fact Sheet #2 published by the National Coalition for the Homeless, June 2005 version, posted at http://www.nationalhomeless.org [8/1/06].

[18]See Burt and Aron (2000) and Burt (2001) for additional discussion of NSHAPC findings and Burt et al. (1999) for a full description of the survey's design. Prior to the NSHAPC, the last

The NSHAPC included two separate surveys, one focused on administrators or representatives of programs providing services to the homeless and the other on the users or clients of those surveys. The surveys reached 6,300 program representatives and 4,200 users, respectively. Specific programs eligible for inclusion in the survey were emergency shelters, transitional housing, voucher distribution centers, food pantries, soup kitchens, mobile food programs, targeted outreach programs (including mental health care, alcohol/drug, and HIV/AIDS programs), and drop-in centers. The sample was constructed in a way that emphasized services in about 70 urban areas but also included a component of smaller cities and rural areas. However, given the survey's focus on formal programs and agencies, the survey may have understated the level of rural homelessness since local governments and organizations may not have formally designated specific programs for the homeless. The completeness of the survey's coverage of service-providing programs has also been criticized because the programs were surveyed in October 1996 while clients and users were surveyed in February; some services that may be open during winter months (February) may not be open in the fall (October).

The 1940 census was the first to use rules for nontypical residences, designating a specific Transient Night (T-Night) for the enumeration of the population that could be found at hotels, tourist or trailer camps, missions, flophouses, and other such places. More recently, the 1980 census did not include any program to count people at street locations. Instead, it mounted a "Mission Night" operation to canvass people at shelters, low-cost hotels and motels, and local jails, followed in the summer of 1980 by a daytime "casual count" operation in some cities around employment and welfare offices and select other locations.

The 1990 census mounted a more prominent effort to count segments of the homeless population through the institution of the S-Night (Shelter and Street Night) operation; see Box 4-3. "The intent (or perhaps the hope) [of S-Night] was that the shelter, street, and abandoned building enumerations would yield a fairly complete, accurate, and reliable point prevalence estimate of the number of homeless in the U.S." (Wright and Devine, 1992b:213). However, the Census Bureau avoided casting S-Night tallies as a comprehensive count of the homelessness; moreover, Cordray and Pion (1991:595) argued that the Bureau "studiously avoided providing any definition of homelessness" and "skirted the issue by simply designating locations where and time when persons would be counted."[19]

---

nationwide survey directly focused on the homeless was commissioned by the Urban Institute in 1987 (Burt and Cohen, 1989), the source of the estimate of roughly 500,000 persons experiencing homeless on any given day that gained some general acceptance over the 1990s.

[19]Cordray and Pion (1991:595) cited Taeuber and Siegel (1991) as saying that the 1990 census did not impose a definition, but noted in fairness that no agreed-upon definition exists and that "some would suggest that there is little merit in arriving at a definition of homelessness."

Box 4-3 S-Night

Methods for 1990 Census Shelter and Street Night (S-Night)

- Local governments were asked to identify places where homeless people live and congregate, including street locations, shelters, abandoned buildings, inexpensive hotels and motels (e.g., "flophouses"), and bus and train stations.
- *Phase 1:* Enumerators canvassed shelters and inexpensive hotels and motels between 6:00 p.m. and midnight on March 20, 1990.
- *Phase 2:* Enumerators canvassed street locations, parks, all-night restaurants, and other locations between 2 a.m. and 4 a.m. on March 21, 1990; abandoned buildings were observed between 4 a.m. and 8 a.m. in some cities.
- Supplemental sites were visited on March 21 and 22, 1990.
- *Results:* 34,000 sites were canvassed, tallying 178,000 people at shelters and 49,000 at street locations.

SOURCES: Citro (2000); McNally (2002); U.S. General Accounting Office (2003).

Despite the Census Bureau's efforts, "the S-Night operation attracted considerable media attention and was controversial because the results were viewed as an undercount of the homeless" (Citro, 2000:206). S-Night was the focus of a special issue of *Evaluation Review* (Wright and Devine, 1992a; Edin, 1992; Hopper, 1992; Cousineau and Ward, 1992; Stark, 1992; Devine and Wright, 1992; Martin, 1992). Researchers conducted evaluations of S-Night procedures for five U.S. cities, using observers to watch the S-Night enumerators at work and, in at least one case, deliberately "planting" several ostensibly homeless people at eligible sites to see if they were captured in the enumeration or not.

The 1990 S-Night also occasioned a lawsuit by the National Law Center on Homelessness and Poverty, the U.S. Conference of Mayors, the cities of Baltimore and San Francisco, and 11 individuals. "On September 15, 1994, the District Court rules in favor of the Census Bureau, stating that the agency's alleged failure to count large numbers of homeless persons did not constitute a violation of its constitutional duty to conduct the decennial census, since individuals do not have a 'right' to be counted" (Bureau of the Census, 1995:1–43). On August 9, 1996, the U.S. Court of Appeals for the District of Columbia upheld the lower court's ruling, albeit for different reasons; the appeals court concluded that the plaintiffs had not demonstrated direct harm as a result of the Census Bureau's procedures for counting the homeless, and hence rejected the case because the plaintiffs lacked standing (*National Law Center on Homelessness and Poverty v. Kantor*, 94–5312).[20]

[20]"In short, we cannot determine—indeed we have no idea—what effect any methodology for counting the homeless would have on the federal funding of any particular appellant recipient,"

Box 4-4 Service-Based Enumeration

Methods for 2000 Service-Based Enumeration (SBE) Program

- During 1999, local governments and community-based organizations were asked to identify service locations anticipated to be open at census time.
- Eligible locations were shelters (including emergency and short-term shelters, as well as transitional shelters where clients can stay up to 2 years); some hotels and motels; soup kitchens; mobile food van stops; and selected outdoor locations.
- Census field staff visited SBE locations several weeks prior to enumeration to verify and update information.
- *Phase 1:* Enumerators canvassed shelters on evening of March 27, 2000; respondents were asked to complete Individual Census Reports (see Box 3-1); the long form was administered to every sixth person.
- *Phase 2:* Enumerators canvassed soup kitchens and mobile food van stops during daytime hours on March 28, 2000; soup kitchens were visited during the meal serving the largest number of clients; interviews were conducted using a modified census questionnaire, including a request for usual home elsewhere (UHE) information; only soup kitchen clients were eligible for the long-form questionnaire.
- *Phase 3:* Enumerators visited selected outdoor locations during early morning hours of March 29, 2000; enumerators were accompanied by contact person/"gatekeeper" identified through local partnerships; interviews were conducted using Individual Census Reports and treated as group quarters.
- People who returned "Be Counted" forms who specified "no address on April 1, 2000" were counted as part of SBE population.
- Enumerator failure to record a group quarters identification code on returned questionnaires resulted in some SBE questionnaires being excluded from data capture and processing; the number of questionnaires so affected has not been specified by the Census Bureau.
- *Results:* 14,817 sites were visited (51 percent of these were shelters), tallying 258,728 people. Soup kitchen and mobile food van respondents were permitted to claim a UHE; of the 71,632 person records collected from those sites, 24,846 were successfully geocoded to a housing unit, and 9,618 were subsequently chosen by the Bureau's Primary Selection Unit as the actual household location.
- The Bureau intended to use facility usage data from the preenumeration visit to adjust data for emergency and transitional shelters to account for those people who regularly use the facility but not specifically on the night of March 27. However, concerns with the resulting data quality led the Bureau to abandon separate tabulations of shelters as a category in initial summary files; the shelter data were later released in tract-level summaries and analyzed by Smith and Smith (2001).

SOURCES: Citro (2000); McNally (2002); U.S. General Accounting Office (2003).

For 2000, components of the former S-Night program were reformed into a new multistage operation termed service-based enumeration (SBE); see Box 4-4. Various components were covered through large deployment of enumerators on different nights around Census Day—shelters first, then soup kitchens and selected outdoor locations identified by local officials and community groups. Rule 17 of the 2000 census residence rules—consistent with 1990 practice—directed that people "at a soup kitchen or outreach program (e.g., mobile food van)" be permitted to indicate a usual home elsewhere. If no such home could be specified for a person, they were to be counted at the facility as if they reside there—even if the facility offers no beds or, by definition, people are legally barred from actually lodging in the facility (as is true for soup kitchens). The Bureau also mounted a T-Night operation on March 31, 2000, interviewing people at migrant work camps, campgrounds, fairs and carnivals, and marinas; people counted at these locations were permitted to report a usual home elsewhere.

4–D PEOPLE MISSED BY CENSUS QUESTIONS AND OPERATIONS

4–D.1 Census Day Movers

Throughout history, Americans have never been known for their propensity to stay put in one place for a long time. Rather, the nation is one that is constantly on the move; as of the mid-1990s, it was estimated that 17 percent of the population move their residence at least once a year (Hunter et al., 2003:1). People move for many reasons, including the loss of a job or a transfer within the same company, the formation or dissolution of marriages or partnerships, or simply the search for new opportunities and new settings.

Conceptual problems—for both the person answering the questionnaire and for the Census Bureau in processing the results—arise for people who move residences on or around Census Day. This includes people reached at a current address for whom a move is imminent, people who have just moved to a new location, and people who are in the midst of a move (perhaps temporarily staying in a hotel or with friends or family). The major conceptual question posed by movers is whether a "usual residence" can be defined prospectively (as the place where the person *expects* to spend most of the time) or must be defined retrospectively. The 2000 census marketing campaign emphasized the impact of census data on the placement of such services as fire stations and health care facilities; to the extent that these themes resonate with the public, census respondents may be more inclined to think of themselves as counting at the new location even if they have not yet actually lived there.

---

noted the court; nor could the court find evidence that possible undercount of the homeless diluted the vote of areas with large homeless populations.

The conceptual challenges of defining residence among movers overlap with operational concerns:

- Could the timing of the census mailing—or, for that matter, the placement of Census Day itself on the first day of a calendar month—exacerbate difficulties in response (e.g., if there is a strong tendency for new apartment leases or the like to start at the beginning of a month)?
- What becomes of census questionnaires that arrive at recently vacated units, and how good will proxy data from neighbors or landlords (collected in follow-up activities) be? Should forwarding or other information about the departed household residents be gathered?
- How different are the households that occupy a unit before and after the move? In particular, the accuracy of matching makes the treatment of movers a major concern when comparing census returns with independent measures from postenumeration surveys, such as the 2000 Accuracy and Coverage Evaluation.

Work that has been done on the general timing of moves suggests that the March–May time frame of peak census activity is the beginning—but not the peak—of the basic "moving season." Schachter and Kuenzi (2002) analyzed data from a migration history module conducted as part of the 1996 panel of the SIPP. The module (administered between June and September 1996), included questions on the month and year respondents moved to both their current residence and their immediately previous residence; by subtraction, the data also provide insight on duration of stay at the previous residence. Peak moving activity occurs in the summer months of June, July, and August: those 3 months accounted for one-third of all recorded moves in the data set. The months central to decennial census questionnaire mailout—March and April—are periods where moving activity begins to escalate from the low-move-activity winter months.

Neither the 2000 census nor its past few predecessors defined formal residence rules for the handling of movers; technical documentation for data files for the 2000 census note only that "people who moved around Census Day were counted at the place they considered to be their usual residence" (U.S. Census Bureau, 2001:C-2). However, some census operations have explicitly tried to determine how well movers are handled. Lowry (1987:7–9) describes a limited check on movers that was conducted during the 1970 census; Census Bureau staff were allowed to examine change-of-address cards filed with local post offices in parts of 17 metropolitan areas. If the two addresses were in the same census district, enumerators were deployed to see if the person or family had been counted at one of the addresses. The operation suggested that many movers "escaped enumeration; presumably both their previous and prospective dwellings were reported as vacant." Bureau analysts concluded that if the operation were "conducted nationally and (carefully) it would have added 0.2 to 0.7 percent to the 1970 population count."

In 2000, enumerator questionnaires (those used in nonresponse follow-up and coverage improvement follow-up) asked whether the respondent lived at the housing unit on Census Day or whether he or she moved in after April 1. If they moved in after Census Day, a card instructed the enumerator to complete up to two questionnaires at the unit:

- The current resident was asked, "Did you, or anyone else in your household, complete a census questionnaire from your previous address?" If yes, the enumerator was told to "thank the person and do nothing further." If no, or the person did not know, the enumerator was to conduct a census interview on a fresh questionnaire, writing in the person's previous address.
- The enumerator was told to determine the housing unit's status as of Census Day. If it was occupied by a different household, a "knowledgeable respondent," such as a neighbor, was to be contacted, and "as much information . . . as possible for the Census Day residents" was to be entered on the questionnaire for the unit.

The Census Bureau's evaluation of this "mover probe" procedure (Keathley, 2003) indicated that 22,850 households would not have been included in the census if not for the probe—a small number relative to the 105 million occupied housing units measured in the census, but also a small part (0.05 percent) of the total follow-up load. Operational evidence suggested that the probe might not have been well explained to enumerators, which could be improved by computerizing the follow-up interviews in 2010. The operation suffered from the severe limitation that information on both the pre- and postmove addresses were not gathered or linked in any sense, sharply limiting analysis that could be done on the characteristics of movers.

4–D.2 Census Day Births and Deaths

Just as moves of residence on and around Census Day can create counting complications, so too can much more drastic changes in status: births and deaths. Table 4-5 illustrates vital statistics data on births and deaths by month for the year 2004. Deaths historically show some seasonal effect, increasing somewhat during the winter months; the peak census months of March and April are at the end of the seasonal escalation.

No formal residence rule was defined for births or deaths on Census Day, nor were they explicitly included in the questionnaire or enumerator instructions for the most recent censuses. Early census enumerator instructions treated the question matter-of-factly; the first set of instructions, accompanying the 1820 census, directed that (Gauthier, 2002:6):

Table 4-5 Births and Deaths in the United States by Month, 2004, Provisional Vital Statistics Data

	Births		Deaths	
Month	Number	Rate per 1,000 population	Number	Rate per 1,000 population
January	334,000	13.5	236,000	9.5
February	317,000	13.7	203,000	8.8
March	347,000	14.0	212,000	8.5
April	334,000	13.9	197,000	8.2
May	339,000	13.6	195,000	7.9
June	346,000	14.4	187,000	7.8
July	360,000	14.5	191,000	7.7
August	357,000	14.3	189,000	7.6
September	357,000	14.8	185,000	7.7
October	350,000	14.0	195,000	7.8
November	337,000	13.9	192,000	7.9
December	345,000	13.8	211,000	8.4

SOURCE: U.S. Centers for Disease Control and Prevention (2005). "Births, Marriages, Divorces, and Deaths: Provisional Data for 2004." *National Vital Statistics Reports* 53(21):Table B. Originally published June 28; updated October 18.

> [marshals] are to insert in their returns all the persons belonging to the family on [Census Day], even those who may be deceased at the time when they take the account; and, on the other hand, that they will not include in it, infants born after that day. This, though not prescribed in express terms by the act, is the undoubted intention of the legislature, as manifested by the clause, providing that every person shall be recorded as of the family in which he or she shall reside on [Census Day].

By 1940, instructions strove to be more specific—enumerators were to count people living as of 12:01 a.m. on Census Day. Babies born after 12:01 a.m. should not be counted, but people dying after 12:01 a.m. should. This instruction was repeated in 1960 (1950 had a simpler "count people living on Census Day" instruction). The 1970–1990 censuses did not explicitly give directions, but the Bureau's implied suggestion was to count anyone alive at any time on Census Day.

The three phenomena described thus far in this section—moves, births, and deaths around Census Day—can create complications for enumeration and household rostering based on the timing of the interviews. In particular, the more time passes between the administration of the census questionnaire and follow-up operations (e.g., enumerator contact in the case of large households with more than seven members or the postenumeration survey used for coverage evaluation), the more such changes may create problems in matching. In Section 6–F, we discuss a particular flaw in current census operations where moves, births, and deaths may cause problems. The flaw is the designation of April 1 as Census Day at the same time that the letter accompanying the census forms (and other publicity) encourages respondents to complete and return the questionnaire before April 1.

4–D.3 Babies and Young Children

The accuracy with which the census counts children, and infants in particular, has been a point of concern for at least a century. Early census ledgers requested that enumerators list infants' ages as fractions of a year, so that a 3-month-old would be listed as 3/12 and an 18-month-old as 1 6/12.[21] However, the 1850 census volume (DeBow, 1853:iv) chided that, "in regard to ages the assistant marshals are often remiss with infants. They omit fractions, and show all to be of one year of age, instead of noting the parts of the year, etc. On this account some counties include no births within the year." Young (1901:254) summarizes findings from the 1880 and 1890 censuses, in which "apparent deficiencies in census reports of children [were] caused by omissions in the enumeration and by overstatement of age." The latter was judged to be "by far the more important"' infants under 1 year old might be seem to be undercounted because parents might report them as being 1 year old, a practice that may have been encouraged by the 1890 census schedule's use of "age at nearest birthday" rather than "age at last birthday" as the question heading. The 1940 and 1950 censuses tried to increase attention to this long-standing problem by requiring the completion of a separate infant card for "each child born during the 4 months from 12:01 a.m. December 1, 1939, to 12:01 a.m., April 1, 1940" (e.g., infants of less than 4 months).

More recently, West and Robinson (1999) analyzed 1990 census data, including results from independent demographic analysis estimates of the population, to confirm the undercount of children. They found that the undercount of children ages 0–17 was particularly concentrated among American Indian, Hispanic, and black households, and among households where the property is rented rather than owned.

West and Robinson (1999) argue that residence rule problems "may disproportionately affect children" because they are central to several household and living situations that make accurate rostering difficult. We have already discussed several of these situations—children living at boarding schools or colleges for part of the year, and children in foster care settings—in Chapter 3, and issues surrounding children in joint custody arrangements and

[21]The 1860, 1880, and 1890 enumerator instructions directed that fractions only be used for children under 1 year of age and that children under 1 month be reported as 0/12. The 1900 and 1910 instructions directed that fractions be used for "child[ren] not 2 years old," while the 1920 and 1930 census asked for fractional ages up until children "not 5 years old."

cohabiting-couple households in this chapter. Generally, children may be raised in households that do not include either of their biological parents for a variety of reasons. Children may be raised in households maintained by their grandparents; West and Robinson (1999) cite research suggesting that as many as 2 million children lived in grandparent households in 1990. Depending on circumstances—such as the loss of parents or removal by incarceration—children may also be raised by other relatives or by family friends. To the extent that such arrangements are not "formalized"—e.g., the grandparents do not actually take legal custody of the children—it is possible that some respondents to a census or other questionnaire asking them to roster their households might not think it proper to include the children.

Census coverage errors among children may arise from respondent misunderstanding of (or resistance to) stated census residence rules, but there is reason to suspect that a more basic, structural feature of the census questionnaire itself plays some part. In 2000, the census questionnaire allowed for detailed entry of data for six people, with space to list only the names of six additional people. These large households were intended to be automatic cases for the coverage edit follow-up operation, but if the household could not be contacted during follow-up, the exact nature of persons 7–12 in the household—and certainly any information in households with 13 or more—would be unknown.

The reason that this structural feature may affect the counting of children is that listing people in descending age order may be a natural approach for the adult responding to the census form. Indeed, this type of ordering was explicitly instructed during the enumerator-contact era from 1850 through 1960.[22] Likewise, a respondent may intuitively reverse-sort people in the roster by the degree to which they "belong" in the household; in cases where extended families and multiple generations live in the same structure, children of kin may simply fall lower in the respondent's mental ordering of people in the household. Hence, they may fall in the section where only names are recorded, or not at all if no space is left. Using 1998 data from the Current Population Survey, O'Hare (1999:8) suggested that 5.6 million children lived in households of seven or more people; to the extent that age ordering is used in answering the census form, these children would be at risk of being missed in the census.

4–E AMBIGUITY DUE TO HOUSING STOCK ISSUES

The "Be Counted" program of the 2000 census placed blank questionnaires in public places so that people who believed that they had not been

[22]Per the 1850 enumerator instructions, "the names are to be written, beginning with the father and mother [or other adult head of family]; to be followed, as far as practicable, with the name of the oldest child residing at home, then the next oldest, and so on to the youngest, then the other inmates, lodgers and borders, laborers, domestics, and servants."

reached by the census could return a questionnaire; the program expanded on the similar "Were You Counted?" program of the 1960–1990 censuses.[23] These programs resulted in the addition of thousands of people to the census—71,000 in 1980 (from 62,000 returned forms) and 260,000 in 1990 (on 352,800 forms). For the 2000 census, the Census Bureau printed over 16 million "Be Counted" forms, about 74 percent in English and the remainder in five different languages; 804,939 were returned, eventually resulting in the addition of 560,880 people (Carter, 2002:iii, 1).[24]

Though the raw counts of persons added to the census by these self-report programs is small relative to the population as a whole, the programs are important because they are the exception to an otherwise strict rule: the only way in which any information—good or bad—can be obtained from respondents is if they are reached by a census questionnaire, delivered in the mail or by an enumerator. In a mail- and address-based census, this can only happen if the housing stock occupied by people is known and reachable; this is not always the case.

As America's urban centers grew in size, and a larger share of the populace lived in cities rather than in rural areas, census officials became aware of an increasing variety of places and shelters where people and families settled. The Census Bureau's official definition of what constitutes a "housing unit" has shifted with time, as shown in Table 4-6, as has guidance on hard-to-determine housing locations. Enumerator instructions for the 1860 census warned that "very many persons, especially in cities, have no other place of abode than stores, shops, etc."—that is, "places which are not primarily intended for habitation." Upon contact by the enumerator, "such buildings will be reckoned as dwelling houses within the intention of the census law" and persons who actually live in them were to be counted there.[25] Officials of the 1880 census considered an even broader array of unusual housing types:

> By individuals living out of families is meant all persons occupying lofts in public buildings, above stores, warehouses, factories, and stables, having no other usual place of abode; persons living solitary in cabins, huts, or tents; persons sleeping on river boats, canal boats, barges, etc., having no other usual place of abode, and persons in police stations having no homes. Of the classes just mentioned, the most important, numerically, is the first, viz.: those persons, chiefly in cities, who occupy rooms in

[23]In addition to circulating forms to mayors' offices, the 1970 census took the step of having the "Were You Counted?" form printed in several large newspapers; "the reader was urged to fill in this form and send it to the census district office if he believed that he or members of his household had been missed in the enumeration" (U.S. Bureau of the Census, 1970:5-32).

[24]Carter (2002:28) concludes that about 15,410 of the returned "Be Counted" forms "were determined to be persons with no usual residence."

[25]However, watchmen or other staff who sometimes "sleep in such store or shop merely for purposes of security" but maintain a home elsewhere in the city were to be counted with their families.

Table 4-6 Criteria for Distinguishing Separate Units in Multi-Unit Dwellings, 1850–2000

Census	Housing Unit Criteria
1850	Living together in a house, or part of a house, upon one common means of support and separately from others in similar circumstances
1860	Living together in a house, or part of a house, upon one common means of support and separately from others in similar circumstances; institutions may be broken into multiple units if there are several tenements or distinct households
1870	Living together under one roof and provided for at a common table
1880–1890	Common roof and table; in "tenement houses and the so-called 'flats' of the great cities," households distinguished by separate tables
1900	"Best test" is number of separate tables; each unit "usually, though not always, has its own meals"
1910–1930	Separate portions of the dwellings house and housekeeping entirely separate
1940	Separate portion of house and separate cooking or housekeeping facilities
1950	Room with separate cooking equipment or two or more rooms with direct access to a common hallway
1960	Live and eat separately from others and direct access to a common hall or cooking equipment
1970	Live and eat separately from others and direct access to a common hall or complete kitchen facilities (the rules were not strictly enforced)
1980	Live and eat separately from others and direct access to common hall (the rules were not strictly enforced)
1990	Live and eat separately from others and direct access to common hall
2000	Live separately from others and direct access to common hall

SOURCE: Ruggles and Brower (2003:Table 2); Ruggles (2003:Note 2).

> public buildings, or above stores, warehouses, factories and stables. In order to reach such persons, the enumerator will need not only to keep his eyes open to all indications of such casual residence in his enumeration district, but to make inquiry both of the parties occupying the business portion of such buildings and also of the police.

The 1880 administrators announced that they would issue letters "to the mayor[s] of every large city of the United States, requesting the cooperation of the police, so far as it may be necessary to prevent the omission of the classes of persons herein indicated." The 1890 enumerator instructions used similar language, though no letters to mayors or police were offered; however, new residence patterns—"tenement houses and the so-called 'flats' of the great cities"—were added to the roster of special cases.[26]

[26]Interpreting eating as a key sign of usual residence, the 1890 instructions directed that "as many families are to be recorded" in these places "as there are separate tables."

As Ericksen (2000:207) summarizes, "whole-household omissions" are easy to understand conceptually; they occur when an entire housing unit is left off the Master Address File and, hence, does not receive a questionnaire. However:

> Some housing units are missed because no one finds the building. Others are missed because they are located in structures that appear to contain only one residence but actually contain several. For instance, a three-story house, originally built for one family, is converted into apartments. From the outside, the census-taker sees only one door and one mailbox, and he or she does not search for nor find the extra housing units to update the mailing list. . . . Housing units for the very poor can be be hard to find. Old apartment buildings, often located in high-crime areas, may not have separate mailboxes or identifiable numbers for the individual apartments. Many very poor people, frequently undocumented aliens, live in places like garages and tents located in the backyards of friends or relatives. Their addresses are not listed and they are not counted.

In the balance of this chapter we briefly consider two special cases where the nature of housing stock precludes an easy determination of residence.

4–E.1 Hotels and Motels

The earliest census enumerator instructions recognized the need to account for the population found in inns and other public houses. Through the 19th century, public houses and hotels tended to be places where people spent relatively long stretches of time; the concept of the hotel as a primarily short-term and transient residential location is a more modern one, particularly spurred on by the emergence of motor lodges or motels in the mid-20th century. The enumerator instructions for the 1900 census were the first to acknowledge the temporary, "transient guests of a hotel"; these "are not to be enumerated as of the hotel, unless they are likely otherwise to be omitted from the enumeration." Instead, only boarders or employees "who regularly sleep" at the hotel (including the owner) were to be counted at the hotel.

The 1930 enumerator instructions acknowledged the basic problem of hotel housing stock: "the distinction between an apartment house and an apartment hotel, and in turn between an apartment hotel and a hotel devoted mainly to transients, will often be difficult to establish." Having laid out the basic challenge, though, the instructions prescribed a rather confusing rule:

> All of the persons returned from a hotel should likewise be counted as a single "family," except that where a family of two or more members (as a husband and wife, or a mother and daughter) occupies permanent quarters in a hotel (or an apartment hotel), it should be returned separately, leaving the "hotel family" made up principally of individuals having no other family relations.

The 1960 census—the last "traditional" personal-visit census—included hotels as one of the "special places" to be addressed by special enumeration techniques. Under the definition of "housing unit" established for that census (see Table 4-6), persons residing in an apartment hotel would be counted as separate units, so long as they have access to a common hallway (Ruggles and Brower, 2003:80).

Since that time, the exact manner in which hotels and motels have been included in (or excluded from) the census is uncertain. Among the most recent censuses, the 1980 census was the only one to attempt direct contact with temporary hotel guests; guests "on the night of March 31, 1980, were requested to fill out a census form for assignment of their census information back to their homes if they indicated that no one was at home to report them in the census" (Bureau of the Census, 1982:C-1). Instead, the 1990 and 2000 censuses left hotels and motels as an ambiguous case—not explicitly covered in the household population, not directly considered part of the group quarters population, and only partly included in the shelter-based population. In 1990, the Bureau did include occupied hotel and motel rooms in its definition of "housing unit," when occupied by permanent residents: that is, persons who consider the hotel as their usual place of residence and have no usual place of residence elsewhere. In addition, some hotels and motels were counted as part of the "emergency shelters for homeless persons" category to be counted in the S-Night operation, specifically, those charging less than $12 a night and those used entirely or dedicated in part to homeless people).[27] The 2000 census included no attempt to count persons living in hotels or motels, and no mention of hotels is made in the 31 residence rules for the 2000 census.

Several challenges are posed by hotels and motels for definition of residence:

- *Proliferation of extended-stay hotels:* Over the past two decades, extended-stay suite hotels have become a larger share of the hotel market. They have become options for people whose employment sends them to another site for weeks or months, and the hotel-type arrangement may be more attractive and less burdensome than solutions like short-term or month-to-month apartment leases.
- *Poor fit with either household or group quarters listing, and with household or group quarters enumeration procedures:* Hotel rooms are a gray area in

[27]These single room occupancy (SRO) hotels, also described as "flophouses," tend to "offer individual, sparsely furnished rooms, with limited cooking facilities and communal bathrooms"; they "have traditionally offered housing to the poor elderly, the low-income single working population, the mentally handicapped, and alcohol and narcotic addicts" (Rollinson, 1990:47). Though the overall stock of SRO housing in the United States is believed to be declining, sizable pockets remain in some large cities; a canvass of such facilities in San Francisco found over 450 families and 760 children living in SROs (Citywide Families in SROs Collaborative, 2001).

terms of potential coverage on Census Bureau address lists. Given that some hotel rooms (particularly extended-stay options) can effectively serve as short- to long-term leased apartments, an argument could be made for their inclusion as potential housing units. That said, hotel stays are much more often short term in nature. Given the highly transitory nature of most of the hotel population, and that the U.S. census remains a *de jure* rather than *de facto* count, it could likewise be argued that special, broadly aimed efforts at enumerating the hotel population would be wasteful.

- *Embedded and associated housing units:* As hinted in the 1900 enumerator instructions, hotels and motels can be deceptive in that actual full-time living situations may be overshadowed by the short-term transitory nature of guest stays. For instance, even the smallest motels may have owners or managers who live on-site; they should rightly be counted in the household population, but they may be missed entirely, depending on the quality of address updating systems. Likewise, it is certainly not unprecedented for hotels to become a full-time living arrangement (e.g., family members of owner/managers or conversion of hotels—in part or in full—to apartments or condominiums).[28]

4–E.2 People Dislocated by Disasters

The destruction caused by natural disasters—hurricanes, tornadoes, earthquakes, fires—can disrupt the lives of thousands of people, forcing them to find shelter or alternate housing for weeks or months following the disaster. The decennial census has had to deal with the short-term dislocation impacts caused by major disasters: Hurricane Floyd (North Carolina, August 1999) in the 2000 census, the Loma Prieta earthquake (San Francisco Bay area, October 1989) in the 1990 census, and Hurricane Camille (Gulf Coast, August 1969) in the 1970 census among them. But the impact of Hurricane Katrina in August 2005—causing not only massive damage but the effective depopulation of New Orleans—has raised particular concern over major conceptual and logistical challenges, for the definition of residence and for basic data collection, that must be addressed by the decennial census (and the entire federal statistical system). This set of challenges is also set against the backdrop of the terrorist strikes of September 11, 2001, and the realization that major dislocation effects may arise from human-caused as well as natural disasters.

The 2000 census residence rules included no specific provision for counting people displaced by disasters. Under a general heading of "people away

[28]A recent high-profile example is the landmark Plaza Hotel on Fifth Avenue in Manhattan. The hotel closed in April 2005 for conversion for mixed use; a smaller hotel component will still operate, with 150 rooms, but 200 condominium units will be carved out for occupancy.

from their usual residence on Census Day," technical documentation for 2000 census data indicates that, "in some areas, natural disasters . . .displaced households from their usual place of residence. If these people reported a destroyed or damaged residence as their usual residence, they were counted at that location" (U.S. Census Bureau, 2001:C-3). How exactly this count was to be implemented was not made clear:

- The closest analogue to the short-term facilities where disaster evacuees might be housed is the "emergency shelter" of rule 24 of the 2000 census residence rules—one for which reporting of a usual home elsewhere is disallowed.
- The documentation suggests that the mechanism for counting people away from their usual residence was "by means of interviews with other members of their families, resident managers, or neighbors" (U.S. Census Bureau, 2001:C-3). However, the standard 2000 census questionnaire included no provision for reporting other address information for household members, and the instructions on the form direct respondents to include "people staying here on [Census Day] who have no other permanent place to stay" at this household (seemingly including disaster evacuees temporarily staying with family members). Reliance on neighbor reports in cases where whole neighborhoods are destroyed or displaced is also problematic.

Conceptually, the case of people dislocated by disasters raises the fundamental question of how much weight should be placed on *intent* in determining usual residence—in this case, on the intent to return to buildings that no longer exist or to housing units that are uninhabitable for the foreseeable future. In Chapter 2, we discussed various conflicting legal and social standards that underlie the meaning of residence. Displaced people face a particularly difficult set of tradeoffs because they must create some ties to their new community that otherwise could be considered evidence of permanent change—such as renting or leasing housing and enrolling children in local schools. At the same time, they are obliged to maintain ties to the disaster area—seeking federal disaster benefits and insurance settlements and paying taxes and land use fees. The Katrina example is particularly telling as the 2006 election season nears; Louisiana election officials plan to permit hurricane evacuees to vote and to allow votes specific to New Orleans offices and issues to be cast at multiple stations around the state. Proposed federal legislation would extend absentee voting to Katrina evacuees in both 2006 and 2008, on the same basis as military personnel stationed abroad, and would further require all states to publicize and promote this absentee voting.[29]

[29]The legislation, the Displaced Voter Protection Act of 2005, would require displaced per-

Logistically, people displaced by disasters raises yet another set of basic questions. Through partnership with federal authorities and private charities, can displaced people be contacted—either in the census, if the disaster occurs close to Census Day, or in ongoing collections like the ACS or CPS? Barring that, can an up-to-date inventory of the destroyed (and rebuilding) housing stock be maintained, and the population in temporary shelters be accurately logged and followed up as they move to more permanent homes?

sons to "submit an affidavit stating that the individual intends to return to the place of residence where the individual is otherwise qualified to vote." The act was introduced separately in the House and Senate as H.R. 3734 and S. 1867, respectively, and has also been incorporated into an omnibus relief bill, H.R. 4197. As of May 2006, all of the bills had not advanced beyond committee consideration (the omnibus H.R. 4197 having been referred to 9 separate committees).

– 5 –

Mirroring America: Living Situations and the Census

IN A PAPER originally prepared for the 1986 Council of Professional Associations on Federal Statistics residence rules workshop, Anderson (1988:50) observed that

> for some groups with ambiguous residential statuses the [Census Bureau] has accumulated a body of information on where they are and has developed mechanisms to count them. I would include students, military personnel, and persons in institutions in this category. For others, less is known. Here I would include undocumented immigrants, the homeless, and migrant workers. I would also include those groups that display newly emerging residential and family patterns: children whose parents share custody; families with earners in different cities or regions and thus two residences; individuals with seasonal residences.

Today, the state of affairs is not much changed. The Census Bureau's ethnographic research program in the 1990 and 2000 censuses yielded small glimpses at the dynamics of some ambiguous residence groups, such as new immigrants and migrant farm workers; see Box 5-1. However, other key groups of interest remain mysteries: the full nature and extent of commuter marriage arrangements is unknown, and information on the snowbird and sunbird populations is limited to the local-area studies conducted by some particularly affected areas. Empirical information on ambiguous residence groups and their response to different question styles and enumeration approaches is rare; the stand-alone 1993 Living Situation Survey (see Box 5-2) and the Alternative Questionnaire Experiment conducted as part of the 2000 census (see Box 5-3) provided some information on general response to residence questions and cues, but not with so large a sample that inference can be made about key groups like households with college or boarding school students or children in joint custody arrangements.

5–A LESSONS FROM A REVIEW OF LIVING SITUATIONS

Our review of living situations in Chapters 3 and 4—situations in which ties to a single, "usual" residence are ambiguous—is meant to be comprehensive but not exhaustive. Moreover, even for the groups we profile in these chapters, not every detail or conceptual wrinkle that apply are described in these pages. However, we believe that the discussion in these two chapters does give rise to several fundamental baseline messages.

The first is a reaffirmation of a basic notion that is all too easy to forget. A recitation of the various hard-to-count population subgroups is a reminder that the American populace has never been easy to count. *Conducting the decennial census is a tremendously difficult and complex task.* It is useful to bear in mind that any process that tries to count approximately 300 million individuals, deploying hundreds of thousands of temporary staff and processing millions of questionnaires of various sorts, in roughly 6–8 months will necessarily have some shortcomings.

Second, the array of challenging living situations should dispel the notion that there are any quick or easy fixes. *No set of residence rules or set of questions, however carefully developed and articulated, can completely eliminate error in the reporting of residence.* Individual living situations, as well as respondents' ability, willingness, and incentive to answer questions, are too varied to be addressed by any single procedure or "one size fits all" approach to data collection. What can be done is to solidify core concepts and adjust enumeration procedures to try for greatest improvement; this is the motivation for our call for a delineation of core residence principles in Chapter 6.

Third, though individuals' living situations can vary greatly, our review does suggest some commonalities, in general structure and motives, across various hard-to-count groups. *Single residence rules cannot be considered in isolation from each other; changes in the way one group is counted may have implications for the enumeration of others.* For instance, the intent to return to a particular location, even when one is staying at a different place for some period of time, plays into the definition of "usual" residence for some people. The fact that prisoners may eventually return to a "home" community is suggested as an argument for counting prisoners there rather than at the prison. But in assessing that change, it is important to consider how much weight to put on intent to return: for prisoners, even though property may have changed hands or family members at "home" may have severed ties with

Box 5-1 Ethnographic Research in the Census

Ethnography is a social science research methodology that provides intensive study of particular communities or groups through observational analysis and extensive interviews. The Census Bureau has used ethnography to study coverage in the census since the 1970s, with major efforts associated with the 1990 and 2000 censuses.

The 1990 Ethnographic Evaluation of the Behavioral Causes of Census Undercount studied 29 sites and groups (de la Puente, 1993). By comparison, the 2000 census program of formal experimentation included six ethnographic research projects, focused on: attitudes toward protecting privacy (Gerber, 2003); participation by "Generation X," persons born between 1968 and 1979 (Crowley, 2003); mobile populations (Hunter et al., 2003); complex households—in particular, on-reservation Navajo Indians, Inupiaq Eskimos, Korean immigrants, Latino immigrants, rural non-Hispanic whites, and blacks in southeastern Virginia (Schwede, 2003); social network tracing of highly mobile people (Brownrigg, 2003); and *colonias*—poor rural communities near the U.S.-Mexican border (see Box 4-2) (de la Puente and Stemper, 2003). The results of each of the six projects were summarized in one of the Bureau's evaluation "topic report" series (de la Puente, 2004), and several of the papers are published in revised or extended form in Schwede et al. (2005).

Among the 2000 census ethnographic studies, Schwede (2003) found that examination of "complex households" in six socioeconomic groups highlighted key differences between the Census Bureau's definition of "household" and the definition envisioned by respondents. For instance, the Census Bureau's conception of a household as the set of all persons living in one housing unit runs contrary to the experience of Navajo Indians and Inupiaq Eskimos, for whom family ties are more central to the notion of a household than physical location. Members of the family who live at a further distance, in sheep camps, in another housing unit on the reservation, or off the reservation may be considered part of the household if they are contributing to the family's income or subsistence; conversely, people living in the same housing unit—even if they are related—may not consider themselves part of the same household if they are not pooling incomes or sharing food.

Ethnographic research also suggested a rich set of high-mobility patterns, for economic and other reasons, that complicate the definition of usual residence (Schwede, 2003; Lobo, 2001; Fleisher, 2001):

- long-distance cyclical mobility to and from Navajo and Inupiaq households for temporary jobs;
- cross-national cyclical mobility between households in Latin America and Latino households in Virginia for jobs;
- seasonal cyclical mobility for subsistence activities among the Inupiat or by snowbirds to escape cold winters;
- mobility for purposes of higher education, found in most of the samples;
- frequent movement of children for schooling and other purposes (among Navajo and Inupiat households) and for joint custody arrangements (among rural non-Hispanic whites);
- cyclical movement of elderly persons between their own houses and their relatives' houses and among households of adult children;
- sporadic movements of tenuously attached persons; and
- temporary ad hoc moves of indeterminate length into the houses of sick or elderly relatives who can no longer manage for themselves.

Box 5-2 Living Situation Survey

The Living Situation Survey (LSS) was a special survey conducted in May–October 1993 by Research Triangle Institute, sponsored and designed by the Census Bureau. The basic objective of the survey was "to test experimental roster probes designed to improve coverage of tenuously attached people. The LSS also collected a great deal of detailed data about movements in and out of sample households" to provide basic information on fundamental residence issues, such as the assumption that each person in the United States has a single, uniquely defined usual residence (Martin, 2004:1).

LSS data were collected from a national multistage probability sample of housing units, designed so that it oversampled minority and renter areas. Interviewers performed personal visits, contacting a household respondent for each sampled housing unit and asking that respondent to list all the persons associated with that unit within the previous 2–3 months. The interviewer then guided the respondent through a series of 13 roster questions; though some emulated the decennial census usual residence approach, the LSS questionnaires did not provide residence rule cues such as lists of specific include/exclude suggestions. After the battery of rostering questions, the respondent was then asked questions about each individual they had identified in rostering. Upon completion of the interview with the designated household respondent, the multistage nature of the sampling design came into play, as selected individuals identified in the rosters were asked to answer questions about themselves. The survey had a 79 percent response rate and was completed in 999 housing units in 1993; those housing unit interviews yielded 3,537 identified persons, 2,825 of which had "more than a casual connection" to the sampled housing units (Sweet and Alberti, 1994:319).

The LSS asked the household respondent to identify whether each person they identified was a usual resident at that address, based on the census definition (where they live and sleep most of the time). The detailed queries also permitted a determination as to whether each person was a usual resident based on a time-based criterion, based on whether they had moved in or out of the residence during the 2- to 3-month reference period used in the survey. Sweet and Alberti (1994:320) report that the two standards agreed most of the time but not exclusively; 94.89 percent of the respondent-labeled and time-based assessments of usual residence agreed with each other; 3.71 percent were dubbed usual residents but did not live in the unit for more than half of the reference window, and 1.41 percent had lived at the unit more than half the time but were not correctly labeled by the respondent. Inconsistency was most prevalent for 18–29-year-olds (Sweet and Alberti, 1994:320).

An important finding of the LSS concerned the reporting of black and Hispanic males aged 18–29. Nearly a quarter of these minority males found by the survey were misclassified in terms of their status as usual residents: 17.5 percent were dubbed usual residents by the survey respondent but had not actually spent most of the past 2–3 months at the residence, and 5.6 percent had spent most of the past several months at the household but were left off the list of usual residents by the respondent (Sweet and Alberti, 1994:321). Related issues are discussed in more detail in Section 4–A.5.

Box 5-3 Alternative Questionnaire Experiments

The past three decennial censuses have all included some form of Alternative Questionnaire Experiment (AQE), designed to test responses to different versions of either the census short-form or long-form sample questionnaire. The 1980 AQE tested two "matrix"-type forms more conducive to the optical sensing technology then used to extract electronic data from microfilmed questionnaire images against a more respondent-friendly design; the 1990 AQE tested five different variations of the census long-form questionnaire (DeMaio and Bates, 1992; Martin et al., 2003).

The 2000 incarnation of the AQE was broader in scope than its predecessors, incorporating three separate subexperiments. One of these focused on the census long form and the branching and "skip" instructions that are meant to guide respondents through the questionnaire (Redline et al., 2002). Another tested different versions of the race and Hispanic origin questions and changes in reporting induced by those questionnaire features (Martin, 2002). It is the third component of the 2000 AQE—"An Experiment to Improve Coverage Through Revised Roster Instructions" (Gerber et al., 2002)—that is germane to this panel's study; hence, in this report, we will often use "2000 AQE" to refer specifically to the portion of the experiment that studied roster and residence concepts.

The full 2000 AQE consisted of eight panels of 5,000 households (10,000 were included in one of the panels); five of these focused on changes to the cues used to help respondents navigate through the census long form, and one presented the 2000 census short-form questions but used the design conventions of the 1990 census. The two panels of key interest from the perspective of census residence rules were a modified short form with revised residence and roster questions and a control group consisting of the standard 2000 census short form. Both of these panels were mailed to approximately 5,200 selected households, and both experienced return rates on the order or 73 percent. The sample of households was stratified by "high coverage areas" (low proportions of minorities and renters in the 1990 census) and "low coverage areas" (high proportions), and were drawn from the Master Address File for mailout/mailback areas of the country (Martin et al., 2003).

Respondent households that returned the revised-roster questionnaire and the 2000-census control questionnaire by mail and included phone numbers were sampled so that the respondents could be reinterviewed (4 months later) and the results used to study the effects of design changes. In total, 4,218 households were selected for reinterview; 2,958 interviews were completed.

The specific changes to the residence question and its related instructions that were tested by the 2000 AQE are described in Box 6-2.

the prisoner; for persons displaced by a natural or manmade disaster, when the structure they occupied at "home" may no longer exist and may not for the foreseeable future; or for long-term hospital patients, for whom there may be the intent to return but not the physical capability.

For another instance, if one considers changing rules or instructions to strengthen some peoples' possibly tenuous ties to a household and remind the person filling in the form that they should be included, how would the changes affect other groups? If such a cue is targeted at partners in cohabiting couples, should similar cues be used to signal the inclusion of foster children (no matter how short the placement) or extended kin and family members staying in American Indian households? As a final example, a consistent approach would seem to be desirable for handling housing units that are inherently mobile and not tied to a particular land location, whether they have small "crews" (long-haul trucks, trains, or recreational vehicles) or large ones (military and commercial sea vessels).

The sheer number of living situations for which residence can be difficult to define fosters our fourth message. *Trying to craft a rule for every eventuality and circumstance is a losing proposition.* The 31 formal residence rules of the 2000 census were an attempt to be more comprehensive and inclusive than the 1990 census (see Box 5-4), and in the history of the American census may be bested only by the 1950 decennial for length and intricacy (see Box 5-5). The 1950 enumerator instructions—like the 31 rules of 2000—were ambitious, and reading through them does give a sense of the magnitude of census operations. Moreover, as we argue in Chapter 6, there remains a need for some detailed explanation of how specific residence scenarios should be handled in the census (e.g., for reference by field enumerators or by census staff manning telephone questionnaire assistance centers). But maintaining a long list of residence rules, built by aggregating experience with difficult problems in previous censuses, results in a flawed product, the internal logic of which is difficult to immediately discern.

Our final message from the review in the past two chapters is simple. *Not enough is known about hard-to-count populations.* Key questions concern the size of the populations in question and the social or demographic trends that are making them more (or less) problematic for the census. As Anderson (1988) noted in the quote at the beginning of this chapter, we do not know enough about the extent and real impact (demographic and economic) of seasonal migration, about degrees of attachment of foster children to their households or group living sites (and vice versa) and the accuracy of gathering data on such in censuses and surveys, or about the blending of short-term and long-term resident situations in the same structure (such as the housing of long-term prisoners in local jails or the establishment of semipermanent living quarters in contemporary hotels and motels). Addressing this deficiency is the focus of the remainder of this chapter.

5–B NEEDED RESEARCH ON LIVING SITUATIONS

Conduct of the decennial census requires a major allocation of money and staff. A large share of these resources must necessarily be dedicated to the basic mechanics of the census—the hiring of staff, the development and processing of questionnaires, and the maintenance of support systems like the Master Address File among them. The resources available for research and refinement of techniques to improve the count are relatively scarce, particularly in the years between censuses. Yet the types of living situations described in this report require continued research to inform improvements in enumeration methods. The need to allocate scarce resources motivates the need for serious attention to quantitative information on groups that the Bureau has more typically addressed through qualitative studies.

Finding 5.1: There is a serious need for additional quantitative information on the magnitude of emerging social trends for groups, as well as a need for further qualitative assessment and better definitions of concepts. Important hypotheses can emerge from qualitative techniques such as ethnographic research, but these need to be tested quantitatively.

5–B.1 Fuller Use of Internal Data

We note first the Census Bureau's unique access to its own microdata. Two of our predecessor panels (National Research Council, 2004b,c) have urged the Census Bureau to continue to mine and analyze the microdata from the 2000 census and its follow-up and coverage evaluation programs, in ways that get beyond the mainly operational focus of the formal evaluation program of the 2000 census. We concur, and add that the Bureau's major survey programs also provide the basis for further investigation of residence-related questions, whether the surveys are self-initiated (e.g., the Current Population Survey or the American Community Survey) or conducted on behalf of other agencies (e.g., its surveys of the jail and prison population).

Finding 5.2: Through its own resources as well as contacts with outside researchers, the Census Bureau has data on diverse residence situations that could be used to inform residence-related decisions.

Recommendation 5.1: **The Census Bureau should conduct and facilitate further research using its detailed census and survey results.**

Box 5-5 Include and Exclude Instructions in the 1950 Census

A section labeled "Persons to Enumerate in Census of Population" in the 1950 census enumerator instructions can be parsed as at least 40 separate instructions on how to address living situations, whether on the standard population schedule or on Individual Census Reports (then used for nonresidents and visitors, such as persons in hotels). Excerpted below are the portions of the instructions specifically formatted as persons to enumerate (on the standard schedule) and to exclude:

Include as members of the household:

a. Members of the household living at home at the time of the enumeration.
b. Members of the household temporarily absent at the time of the enumeration, on vacation, visiting, or on business.
c. Members of the household who are in a hospital but who are expected to return in a short period of time.
d. New-born babies who have not yet left the hospital.
e. Members of the household attending a school below the college level and residing in other [enumeration districts (EDs)]. (Student nurses and students at the college level will be enumerated in the ED in which they are living while attending school.)
f. Domestic or other employees who live with the household, sleeping in the same dwelling unit.
g. Boarders or lodgers who regularly sleep in the dwelling unit.

Exclude as members of the household:

a. Do not enumerate persons temporarily visiting with the household, if they have a usual place of residence where they will be enumerated. . . .
b. Do not enumerate citizens of foreign countries temporarily visiting or traveling in the U.S. or living on the premises of an Embassy, Ministry, Legation, Chancellory, or Consulate. . . .
c. Do not enumerate students or children living or boarding with a household in your ED while attending some regular school below the college level in the locality, and having a usual place of residence elsewhere from which they will be reported.
d. Do not enumerate members of the household who are living elsewhere while attending college, even though they may be at home in your ED on vacation. . . .
e. Do not enumerate persons who take their meals with the household but usually lodge or sleep elsewhere.
f. Do not enumerate domestic employees or other persons employed by the household but not sleeping in the same dwelling unit. . . .
g. Do not enumerate persons who were formerly members of the household but have since become inmates of correctional or penal institutions (including jails—no matter how short the stay), mental institutions, homes for the aged or needy, homes or hospitals for the chronically ill or handicapped, nurses' homes, convents or monasteries, or other places in which residents may remain for long periods of time. . . .
h. Do not enumerate officers and crews of ships and persons living in lighthouses. . . .
i. Do not enumerate persons working abroad for the U.S. Government if their regular place of duty is abroad. Such persons will be enumerated under special procedures. However, you must enumerate as a resident of your ED any person who usually lives there if he is temporarily abroad on a vacation or in connection with his work. . . .
j. Do not enumerate soldiers, sailors, marines, and airmen not now living in your ED. . . . If, however, servicemen are stationed in your vicinity and [live] and sleep off post in your ED, enumerate them as residents.

The Committee on National Statistics' *Principles and Practices for a Federal Statistical Agency* argues that statistical agencies have a responsibility for in-house analysis of their own data. "There are strong arguments for a statistical agency to have staff whose responsibility is to conduct objective substantive analyses of the data that the agency compiles, such as analyses that assess trends over time or compare population groups"; among these are the ability to use the results of those studies to inform changes to the agency's data collection program (National Research Council, 2005:40). Historically, the Census Bureau's focus has been on the production of data and not internal analysis. We urge the Bureau to facilitate ties with external researchers to carry out research objectives, but—pursuant with the practice of in-house analysis—the Census Bureau needs to take a stronger interest in analyzing and interpreting their own data to guide operational improvements. Given their unique access to their own microdata and operational results, there are some problems that the Census Bureau itself must analyze simply because there is no one else who can.

5–B.2 Monitoring Social Trends

In an organizational hierarchy like the Census Bureau's, methodological changes and developments such as those described above tend not to happen without an internal advocate and an established base within the Bureau. There is also a need for additional breadth in subject-matter specialists whose insight on census and survey data can be marshaled. Emerging social trends and their impact on the accuracy of basic residence information is sufficiently central to the purposes of the census that the topic deserves a visible, active research effort and institutional support by the Census Bureau. Accordingly, we believe that the research effort we suggest should be a standing activity, including input from all parts of the Bureau but coordinated in a central place:

> *Recommendation 5.2:* **People's attachment to households and group quarters has changed significantly over several decades and is likely to continue to change in ways that cannot now be predicted with confidence. The Census Bureau should establish a standing research office whose task it is to continually monitor changes in factors influencing people's attachments to locations where they are counted, and the connectedness of changes among them, using such information to generate appropriate research and recommendations for changes in how people can be more accurately enumerated in the decennial census.**

The office we envision in this recommendation would perform a variety of important roles:

- *Analyze Census Bureau data and facilitate cross-divisional ties:* In addition to planning evaluations and analyzing experiments connected to the decennial census, the office would analyze data collected from topic modules on the Current Population Survey and the American Community Survey (discussed below in Section 5–B.3). This research office should also tap the field experiences and operational data from surveys conducted by the Bureau on behalf of other agencies, several of which are noted in the previous two chapters: these include the National Nursing Home Surveys, the National Hospital Discharge Survey, and the Census of Juveniles in Residential Placement.
- *Design new experiments:* Based on perceived gaps in knowledge, the office could design new surveys and experiments of the type we describe further in Chapter 8. Over time, the experience of this office in spurring new research would lay a solid foundation for improving census processes on the basis of sound scientific evaluations.
- *Build and strengthen ties to external research:* An important function of the office would be to monitor and cull from the work of the broader research community. External researchers have developed a number of data sources that focus on living arrangements and difficulties in counting people with ties to one or more places: these include the National Survey of Families and Households (Sweet et al., 1988; Sweet and Bumpass, 1996, 2002), the New Immigrant Survey, the Los Angeles Family and Neighborhood Study, and the site-specific surveys of snowbird and sunbird populations (discussed in Section 4–A.1). Synthesizing the results of researchers' analyses of these data sources would be an important function of the trends office; more significantly, the office should study and learn from the approaches used in these studies to reach targeted populations and collect data from them. This work could suggest improvements that could be used to improve the accuracy of the census. We revisit the need for Census Bureau ties to external research in Chapter 8.

5–B.3 Basic Research on Living Situations

In this closing section, we suggest selected topics that are particularly ripe for basic research, several of which could be done using extant data resources internal to the Census Bureau. This is a selected list, and should not be interpreted as either a comprehensive list or as a specification of the highest priority research topics.

First, existing 2000 census data should be able to help address remaining questions on the characteristics and reporting patterns of large households, those with seven or more members; the 2000 and other recent censuses only collected detailed information for six people in a household and asked only for the name of persons 7–12. In some cases, people beyond the six described on the form were found during census follow-up operations, but in others their characteristics had to be filled by imputation. If there is some instinctive tendency to list household members in rough age order, this structural feature of the census form could help explain undercounting of children and babies; it may also be the case that persons not listed in positions 1–6 are those with the weakest ties to the household, and they may be potential duplicate cases. This potential phenomenon is important to understand for the purposes of an accurate count, but is particularly important with regard to the long-form replacement American Community Survey (ACS). Detailed person-level characteristics measured by the social and demographic ACS questions could be distorted by higher levels of imputation required to fill in persons missing from large households.

> *Recommendation 5.3:* **Extant data from the 2000 census on large households of seven or more members should be reanalyzed for better understanding of the nature of the households and to inform better practices to collect data for large households.**

In particular, the Census Bureau should study trends in age reporting by the person number on the census form, the geographic and demographic concentration of these large households, and the characteristics of persons in positions 7 or higher relative to those for which full information is available in positions 1–6.

Other potentially interesting avenues of research connected to the undercount of babies and young children is to examine the actual reported age of babies in the census in comparison with vital statistics data. It would be interesting to see if birth months are missing equally, or whether a lack of March/April births may speak to structural problems in census reporting (see Section 6–F for further discussion). Ultimately, this work could lead to revisions to the structure of the questionnaire, such as the number of persons for whom full information is requested and the number for which only names are listed. Such revision is consistent with an idea we explore in Chapter 6, that the census short form may be too short.

The choice of large households and the potential effect on undercount of the young as a research topic is a useful one because it focuses on a potential source of census omissions. As general guidance to ongoing research on social trends and enumeration methods, it may approach being a truism but we think it important to stress that sources of both duplication (overcount) and omission (undercount) be considered. As we have noted, duplication is often the more tractable problem to analyze, but appreciating the extent to which social trends and existing census approaches might influence census omissions is vital to overall accuracy.

> *Recommendation 5.4:* **Census Bureau research on living situations that do not easily fit census residence rules should strive to gather data on the sources of omissions in the census, as well as sources of duplication.**

Not all of the important contextual work on residence trends can be achieved by reanalysis of existing data and strengthening ties with external researchers and other survey programs. There is a need for original work as well: the trends office we recommend could design and implement original quantitative work on the basis of identified gaps in existing knowledge. It is surprising that it has been over a decade since the Census Bureau sponsored its LSS; a return to collection of hard data on ambiguous residence situations and questionnaire strategies is long overdue.

> *Recommendation 5.5:* **Data similar to those collected by the 1993 LSS should be conducted on a regular basis. A convenient form for a more regular study could be inclusion of a supplement to the ACS or a stand-alone survey.**

By this recommendation, we emphasize that we do not literally mean that the LSS be replicated and conducted exactly as it was in 1993. The content of the survey should reflect major data needs and, to the extent possible, the survey should be targeted and designed in order to achieve adequate sample coverage in populations of interest. It is also essential that the data from this work be done in a well-documented and publicly accessible way. While we do not advocate an exact replication of the LSS, neither do we think it wise to be overly prescriptive of the shape the work should take; it is for this reason that we suggest a module or supplement to the ACS as a possible vehicle. Current plans for the ACS already include a "methods panel" to test new approaches and questions that would be well suited for additional residence questions; we discuss this further in Section 8–C. A supplemental module to the Current Population Survey (CPS) may also be a useful means for collecting these data, though design differences between the CPS and the census or ACS might make this option less useful for suggesting specific improvements to the census. Regardless of the mechanism by which the data are collected, the crucial thing is that the data be produced, so that future discussions of ambiguous residence situations and possible corrective strategies can be supported by formal quantitative research rather than anecdote.

Part III

Improvements for the Future

– 6 –

Residence Principles for the Decennial Census

CHANGE DOES NOT COME EASILY to the decennial census; it is too large and intricate an operation for massive overhauls of operations or procedures to be feasible in a short amount of time. The quality of the resultant data is paramount, and so implementing procedures that have not been tested is inadvisable. A discussant at the 1986 Council of Professional Associations on Federal Statistics (COPAFS) conference on residence rules noted similar comments on the nature of change in the census in one of the presented papers and wondered if the statement was mildly euphemistic, whether "this is perhaps another way of saying that we are unlikely to do much to make any needed improvements in time for the 1990 census, since we are beginning to think seriously about the problem in December of 1986" (Sweet, 1987:6). Twenty years later, with the 2010 census looming in the not-too-distant future and even with the benefit of somewhat better lead time, we face something of the same problem.

In our case, there are very promising signs of improvements to the collection of residence information in 2010. The advent of the American Community Survey and, with it, the narrowing of focus on a short-form-only decennial census is a considerable simplification and has permitted earlier attention to residence considerations than in the past. As we have noted, residence concepts were a primary focus of the 2005 census test and will be the topic of a further mailout experiment in 2006. Also, as witnessed in our panel's public meetings, the Census Bureau has made good strides in redrafting and revising both the census residence rules and the definitions of group quarters.

We conclude that the Census Bureau would be best served by some radical revamping of its basic approaches to collecting residence information. It is unrealistic to expect large-scale changes in the 2010 census. Rather, our intent in this report is to provide a mix of short-term and long-term guidance: short-term proposals that can be implemented in 2010 and long-term research topics (to be informed by tests conducted in 2010) to provide a basis for decisions in 2020 and beyond.

In this chapter we discuss the basic nature of residence rules and argue for a principle-based approach to defining residence. A core set of residence principles should be used to develop related products and operations (Section 6–A). The second half of this chapter discusses the implications for the most important such product, the census questionnaire itself and the specific residence instructions in it (Sections 6–C and 6–D). In discussing the general presentation of residence concepts to census respondents, we suggest the need for a major change in the way residence information is gathered, switching from an instruction-based to a question-based approach (6–E). In Section 6–F, we discuss a particular problem in communicating with census respondents, encouraging prompt replies to the census form while still preserving the meaning of Census Day. We close in Section 6–G with recommendations for research on presentation issues.

6–A A CORE SET OF PRINCIPLES

As discussed throughout this report, the residence rules for the decennial census must satisfy several needs simultaneously. They have primarily been regarded as an internal Census Bureau reference, but also must be adaptable to the construction and phrasing of items on the census form and must govern the design of related census operations. They must also be a resource for the training of temporary census enumerators and a reference to census staff who must field questions from census respondents.

As we noted in Chapter 2 and Recommendation 2.1, the Census Bureau's current approach of trying to serve all these needs with a single document is seriously flawed. The 2000 census rules were so long and intricate that it was difficult to discern their meaning; with effort, one can intuit some of the logic that guided the construction of the rules, but the task is difficult. The situation needs to be reversed: a concise core set of residence principles should be developed, and all the related products and extracts—question wording and structure, enumerator training materials, and so forth—should be built using the principles as a base.

At the 1986 COPAFS residence rules conference, Lowry (1987:30–31) made an early call for the reduction of formal census residence rules to a set of basic principles. He argued that the Bureau would be better served by having the residence rules (for the 1990 census) "express general principles rather than merely treating the most common examples of residential ambiguity." Specifically, he proposed six principles:

- If respondent lived for more than half of the preceding year at an address that he occupies on [Census Day] or expects to return to in the following year, that address is his usual residence.
- If respondent lived for more than half of the preceding year at an address that he does not expect to return to in the coming year, his usual residence is the address he now occupies or expects to occupy for most of the coming year.
- If respondent has several homes, no one of which he occupied for more than half of last year, he should select one as his usual residence. If an editor must choose, choose the current residence.
- If the respondent has a home to which he returns in the intervals between traveling, that home is his usual residence even though he may have been elsewhere for more than half of the preceding year.
- If the respondent says he has no usual residence or gives no address for his usual residence, he should be enumerated as a transient at the place where he was staying when contacted.
- If the respondent says his usual residence is elsewhere but gives an incomplete, wrong, or invalid address for that residence, he should be enumerated as a resident of the smallest geographic unit that is unambiguously codable from his response—census block, enumeration district, tract, city, county, or state.

Discussants at the workshop found these principles to be generally reasonable; the workshop summary concluded that "a system that relies on well-defined principles, yet can incorporate responses from people in a wide variety of living [situations,] should accommodate appropriate decisions about just where to count each individual as of April 1st" (CEC Associates, 1987:6).

The Lowry (1987) principles have attractive features: the third principle subtly suggests an approach of collecting data on multiple residences and putting the onus of determining the usual residence on the Census Bureau, not the respondent. The sixth principle is intriguing, with its bold approach of associating people only with large geographic areas (e.g., cities, counties, or states) if that is all that can be validated from their provided residence information; how these larger-area counts would be distributed in block-level redistricting counts is not specified and would surely be contentious. A strength of the Lowry (1987) principles is their anchoring to a reference period of 6 months, interpreting either a retrospective (first principle) or prospective (second principle) stay of the majority of a year as the basic definition of usual residence. But that specificity is also a limitation, since 6 months is a coarse time interval with respect to some living situations; the third and fourth principles weaken the hard-line 6-months criterion, imparting some ambiguity in the determination of residence.

We conclude that the census would benefit from the specification of a core set of residence principles and retain some of the desirable features of the Lowry (1987) exemplars, but take a somewhat different approach in our recommendation.

A key component of a set of residence principles for a decennial census is an indication of the general residence standard to be pursued. As we state in Section 2–F.2, we know of no pressing reason why the U.S. census should shift away from the *de jure*/"usual residence" approach as its basic standard (though we also believe that it is useful that this decision be periodically revisited). So, in suggesting a candidate set of residence principles, we support a system that seeks a *de jure* count; the *de facto*/current residence location should be used as a tie-breaker in cases where no usual residence can be determined.

Assuming a general *de jure* model, and taking as given the objective of the census as a resident count (rather than a citizen count), we recommend adoption of a set of principles based on the following:

> *Recommendation 6.1: Suggested Statement of Residence Principles:* **The fundamental purpose of the census is to count all persons whose usual residence is in the United States and its territories on Census Day.**
>
> 1. **All persons living in the United States, including non-U.S. citizens, should be counted at their usual residence. Usual residence is the place where they live or sleep more than any other place.**
> 2. **Determination of usual residence should be made at the level of the individual person, and not by virtue of family relationship or type of residence.**
> 3. **If a person has strong ties to more than one residence, the Census Bureau should collect that information on the census form and subsequently attempt to resolve what constitutes the "usual residence."**
> 4. **If a usual place of residence cannot be determined, persons should be counted where they are on Census Day.**

Unlike the Lowry (1987) principles, we do not specify a fixed reference period in our first principle, the basic definition of usual residence. In our deliberations, we found that the varied living situations that must be accounted for in the census do not lend themselves easily to any choice of a time window we could construct. Ultimately, we find that a definition of usual residence must necessarily be somewhat ambiguous in order to be most broadly applicable. In our first principle, we use the wording "more than any other place" rather than other language (like "most of the time") because we favor use of a plurality rule and because we find it less ambiguous than "most of the time."

Even so, ambiguity remains about the relevant time frame against which "more than any other place" is to be assessed; in addition, the issue of whether any element of *prospective* living arrangements should be considered remains unspoken (e.g., new movers, who intend to live at this location more than any other place but have not been living there for very long). Some of the ongoing research we recommend in the remainder of this chapter should be brought to bear on refining this principle—it should, for instance, consider whether including a specific reference period has a significant impact on response or respondent confusion. In the interim, our definition of usual residence strives to reduce some ambiguity but necessarily leaves some issues open.

The second principle asserts that usual residence should be viewed as an individual-level attribute, not one that is tied to family relationships or type of residence. This principle is included to be consistent with changes that we recommend in enumerating what has traditionally been treated as the group quarters population. As we argue in more detail in Chapter 7 (and as follows from descriptions of some group quarters types like health care facilities and jails—Sections 3–C and 3–D.2), the concept of group quarters enumeration requires comprehensive reexamination. The expected length of stay and actual living situations in some group quarters types is such that it is inappropriate to characterize residence status for all facility residents based solely on the facility's name or type. Including the provision that usual residence does not depend on family relationship speaks to situations such as college students living away or children in joint custody. This is a principle that some respondents may continue to violate, given the strength of enduring ties, but we believe that it is useful to have this concept expressed as a principle rather than something to be inferred from a long list of examples.

The third principle foreshadows arguments that we will make later in this chapter. We favor an approach in which the census form asks a sufficient number of questions to get a sense of each person's residential situation. Significantly, we believe that it is important that the census move toward collection of information on any other residence that a person may be affiliated with. Our ultimate vision is of a census in which residence information is collected *without* burdening respondents with the problem of deciding who is usually resident at their address; those kinds of determinations should be made by the Census Bureau during processing of the forms, based on the data provided by respondents to a series of residence questions.

Finally, the fourth principle is a "tie-breaker" rule. There are some living situations that do not lend themselves to an unequivocal determination of a place where a person lives or sleeps more than any other; examples include dedicated recreational vehicle users and children in joint custody situations who spend equal time with both parents. When a usual residence cannot be determined, the person's location on Census Day—the *de facto* residence location—should be used.

Of course, these are not the only principles that could be developed for the census, but we believe them to be an adequate set. Other possibilities include something akin to Lowry's sixth principle, which raises the question of the level of geographic resolution needed for tabulation; that principle would assign people to the smallest possible level of geography (but not necessarily to a specific geographic coordinate). We revisit the question in Chapter 7 and the counting of the group quarters/nonhousehold population. Another possible principle—perhaps useful for handling some ambiguous residence situations but that may be prohibitively difficult operationally—is to specify that children under a certain age must be counted at the home of their parent or guardian. Though it would provide an alternate solution to the current mismatch between the counting of boarding school and college students and could be more consistent with a "family" interpretation of household, the specification of the age cutoff could make enumeration of college students even more difficult.[1]

Collectively, our four suggested principles imply the exclusion of American citizens (nonmilitary and nongovernment employee) living overseas, consistent with practice in recent censuses. To satisfy the requirements of current law and court precedent, military personnel and federal civilian personnel stationed overseas, and their dependents, would continue to be assigned to their home states of record for purposes of apportionment only.

6–B PRODUCTS FOR IMPLEMENTATION OF THE PRINCIPLES

Articulation of a core set of principles provides a basis from which to work in developing other products, one of which is an explanation of how the principles apply to a variety of living situations. The 2000 census model of trying to craft rules to match all possible living situations is flawed because it lacks a unifying conceptual basis. Yet there is still a definite need for something analogous to the intent of the 31 rules—a list of examples of how ambiguous living situations should be resolved—but one that is grounded in concept and better structured for comprehension.

Such a listing is needed for several reasons and audiences. Though field enumerators should be made familiar with the basic residence principles, it is also useful for them to be aware of concrete examples of the application of the principles to living situations that they are likely to encounter. Similarly, staff who deliver questionnaire assistance by telephone should also have common situations and their treatment as a reference. We suggest in Section 2–F.2 that census residence rules be made more transparent to the public and to decisionmakers, and public posting of examples of the application of chosen

[1]Such an age-based rule would also misplace minors who have won legal emancipation from their parents.

residence principles to various living situations is an important part of that effort.

A good way to meet this need would be with a "frequently asked questions" (FAQ) list, as has become common on Web sites. Residence situations and their treatment under the residence principles should be grouped thematically for easier use. Box 6-1 illustrates a possible mapping of our recommended principles to residence situations. This structure can certainly be improved—particularly for a version to be posted on the Internet—but it provides a basic starting point for development of an FAQ list and other census documents.

The location at which prisoners should be counted has become a major point of contention. Our suggested principles—and our interpretation of them, in Box 6-1—lead us to side with the Census Bureau's current general procedure of counting them at the location of incarceration. At this time, not enough is known about the exact nature of the *alternative* of counting prisoners at a place other than the prison, nor about the accuracy and consistency of facility data on inmate residence, to recommend change in counting prisoners in the 2010 census. However, we strongly urge that the 2010 census include a major test on the collection of additional residence information from prisoners and further assessment of the quality of administrative records that could inform future reconsideration of the prisoner counting issue. We discuss these initiatives below and in Chapter 7.

The residence principles should also be used to generate residence-based products in addition to the FAQ list of applications. Key among these are any specific instructions or other cues included on the census questionnaire itself; we discuss this in greater detail in the balance of this chapter.

The residence principles should be thought of as an integral part of the entire census process, not a small, side component. They should be used as a template for the development of related census operations. Chapters 7 and 8 discuss three major operations—techniques for group quarters/nonhousehold enumeration, programs to update the Master Address File, and routines to unduplicate census records—that should be designed with residence principles as a guiding concept.

Other census operations for which residence principles should be kept in the forefront include:

- development and implementation of unduplication algorithms, including any revisions to the primary selection algorithm (used to screen and combine duplicate census questionnaires) and plans for "real-time" unduplication during the census process (we discuss this briefly in Section 8–B);
- development of the advance letter that precedes the main questionnaire mailout, including instructions for requesting a foreign language ques-

Box 6-1 Illustration of Application of Residence Principles as the Basis for "Frequently Asked Questions"

1. PEOPLE WITH ONLY ONE RESIDENCE

Count at the location where they live or sleep more than any other place. Examples include:

- College students living at their parental home while attending college
- Unmarried partners
- Housemates, roommates, or boarders
- Foster children or foster adults
- Live-in employees (e.g., caregivers, domestic workers)
- Military personnel in the United States
- People who happen to be away
- Resident staff of embassies, legations, and other diplomatic facilities in the United States[a]

2. PEOPLE WITH MORE THAN ONE RESIDENCE

Collect information on "any residence elsewhere." Count at the location where they live or sleep more than any other place. If time is equally divided between locations, count at the location where they are found on Census Day. Examples include:

- College students living away from the parental home, in on-campus or off-campus housing
- Boarding school students
- People living away most of the time while working
- People who split time between two or more residences (during the week, month, or year), such as snowbirds, unmarried partners with separate residences, commuter workers, etc.
- Children in shared custody arrangements

3. PEOPLE WHOSE LIVING SITUATION CHANGES ON CENSUS DAY

Count at the place where they live or sleep—or will live or sleep, in the case of babies—more than any other place.

- Births—babies born on or before midnight of Census Day
- Deaths—people who were alive at any time on Census Day
- People who move into a residence on Census Day, where they will live or sleep more than any other place
- People who move out of a residence on Census Day, but who have not completed the move to a new location on Census Day (the former residence should be considered the place where they live or sleep more than any other place)

4. U.S. CITIZENS LIVING OUTSIDE THE UNITED STATES

Count on the basis of U.S. law and court precedent. Currently, this means including the following groups in apportionment counts only; all other U.S. citizens residing overseas are not included in the count.

- Military personnel stationed abroad
- Merchant Marines
- Other U.S. citizens employed by the U.S. government

5. PEOPLE LIVING IN SPECIAL PLACES ON CENSUS DAY

Collect information on "any residence elsewhere," along with information on duration of stay (tailored institutional forms). Based on duration of stay, availability of residence information, and whether census record for that residence lists that person, count at the residence. If no other information is available, count at the facility. Examples include:

- People in transient quarters, such as hostels, recreational camps, public or commercial campgrounds, racetracks, parks, or carnivals

Box 6-1 (continued)

- People in federal and state prisons
- People in local jails or other municipal confinement facilities
- People in correctional residential facilities
- People in group homes for juveniles, including Job Corps Centers living quarters
- People in residential treatment centers
- People in health care facilities, including nursing facilities and hospitals
- People in emergency and transitional shelters on Census Day, for people experiencing homelessness
- People at soup kitchens, regularly scheduled mobile food vans, and targeted nonsheltered outdoor locations
- People in group homes and other residential treatment centers for adults

[a] Under current law and practice, foreign citizens resident at embassies and legations in the United States may voluntarily decline to be included in the census count.

tionnaire, and any follow-up mailings (e.g., reminder cards or the proposed second/replacement questionnaire mailout);

- refinement of the Bureau's routines for editing census data and imputing for nonresponse;
- development of experiments to be performed during a decennial census and of formal evaluations of census operations; and
- design of public outreach programs.

6–C PRESENTATION OF RESIDENCE CONCEPTS TO RESPONDENTS AND ENUMERATORS

The Census Bureau took a first step toward self-enumeration—that is, questionnaires filled out by respondents themselves—in the 1960 census. Households were mailed an "Advance Census Report" that enumerators later picked up in person and copied onto computer-readable schedules. On the strength of that experience, mailout and mailback of census questionnaires became the dominant form of census conduct in 1970, when approximately 60 percent of the nation's housing units were mailed questionnaires. By 2000, approximately 82 percent of housing units were reached by mail.

Mail administration of the census has obvious operational efficiencies in comparison with deploying enumerators to interview every household. However, self-enumeration profoundly affects the process of a census, shifting the burden of comprehending and interpreting the meaning of questionnaire concepts from a trained enumerator to an untrained respondent. As a result, the modern census questionnaire has to satisfy several, sometimes conflicting, constraints:

- The questionnaire (and any other material in the "package" mailed to households, such as a cover letter) must be self-contained and comprehensive enough to guide the respondent through the process of providing complete, accurate answers. (Of course, capacity should still exist for assistance by phone or on the Internet and for requests of census questionnaires in languages other than English, if necessary, but the hope of mailout/mailback techniques is collection of data from the largest number of respondents possible without any direct intervention.)
- The questionnaire must be designed in such a way that the human reader can easily follow the flow of the questionnaire, but also so that the questionnaire can be scanned electronically for editing and tabulation.
- The questionnaire should be visually appealing (or, at the very least, not off-putting) in order to maximize respondent interest and willingness.
- The questionnaire cannot be unduly long in terms of the number of questions. It must satisfy U.S. Office of Management and Budget limitations on respondent burden under the Paperwork Reduction Act; in recent censuses, the length and content of the census long form has drawn particular concern.[2]
- For maximum efficiency, the questionnaire must also meet physical size and shape limitations imposed by computer scanning technology. Even though the 2010 census is oriented as a short-form-only census—with the long-form content shifted to the American Community Survey—questionnaire content must also be conducive to the development of computer-assisted versions for follow-up by telephone or (in 2010) hand-held computing device.
- Census questionnaires must be printed relatively quickly and cheaply, and in massive quantities.
- Census questionnaires must include space for technical features, such as a block for the mailing address and Master Address File identification number or spaces for enumerators or census clerks to code operational information as needed.

Self-response questionnaires and their properties are a topic of vital research in statistics and survey methodology; indeed, the study of their properties has grown in importance with the availability of new technologies for survey administration such as automated telephone interviews and data collection through questionnaires on the Internet. Methodological work on self-

[2]Most notably, as cited in National Research Council (2004c), the long-form questionnaire was literally thrown back at a Census Bureau director by a congressional appropriator, with the directive to "make it shorter."

response surveys does not suggest a single "best" way to obtain residence data from respondents to the massive-scale survey that is the decennial census. In simplest terms, what modern survey methodology tells us about census questionnaire design and structure is that presentation matters.

> **Finding 6.1:** Responses to self-administered census forms depend upon the visual layout and design of questionnaires as well as the actual wording of questions and residence cues.

The Census Bureau's approach in recent decades has been to produce designs that try to find an elusive, optimal level of instructions and cues at the beginning of the questionnaire in order to try to get best compliance with Bureau residence standards. However, modern survey research suggests a set of basic findings on the impact of questionnaire layout and the interpretation of visual and graphical cues in survey questionnaires. These findings (based, e.g., on Christian and Dillman, 2004; Jenkins and Dillman, 1995; Redline and Dillman, 2002; Schwarz et al., 1998; Tourangeau et al., 2004; Conrad et al., 2006; Tourangeau et al., 2006) should drive the collection of residence and other information in the decennial census.

> **Finding 6.2:** Evidence suggests that people often ignore instructions on questionnaires. In addition, they may disregard instructions with which they disagree, even if they do read them.

"Ignore" and "disregard" are admittedly strong words; respondents' failure to read and follow instructions is not necessarily a hostile act. People may assume the questions are designed to be interpretable and, as a result, may feel that they do not need direction. People who do read the instructions may not completely understand them and, if they disagree with some points (e.g., where to count their college student child), may decline to follow them. Such assumptions and disagreements are particular problems for the basic census residence question, since people are likely to assume that they know where they live and who lives with them. They may also be impatient with lengthy instructions and scan only enough to get the gist (as they see it) of what the question is asking.

Modern survey research also supports three key ideas that should guide questionnaire design:

- When instructions are needed, they should be placed where they are most needed.
- When multiple instructions are needed, the most important one—or the one that is most likely to apply—should be placed first. Respondents are less likely to read an instruction the further down it is in the list.

- Visual cues should be used to convey what respondents are supposed to do. For example, white space (against a light background color) or some other graphical feature should be used to indicate when a response is required. Other graphical cues should indicate what form the response should take (e.g., boxes invite respondents to check one or more items).

6–D INSTRUCTIONS AND RESIDENCE QUESTIONS IN RECENT CENSUSES AND TESTS

As prologue to suggestions on how best to present residence concepts to respondents, it is instructive to consider the ways in which recent U.S. censuses have presented residence instructions and cues. We also discuss the approaches taken in census tests and follow-up census operations, as well as strategies used in foreign censuses.

6–D.1 Previous U.S. Censuses

As noted in Box 5-5, the 1950 census enumerator instructions were the first to include a detailed itemization of persons to include or exclude from the census. That approach clearly carried over to the 1960 instructions to respondents on the "Advance Census Report" mailed to households (Figure 6-1). The "PLEASE BE SURE TO LIST" and "DO NOT LIST" categories dominate the instructions at the top of the form, and overwhelm the basic "usual residence" statements at far left under "PLEASE LIST." Other features of note in the 1960 instructions for respondents include the unconditional plea about "including babies," the emphasis on people staying at the household but "who have no other home" (mentioned twice), and the emphasized assurance that college students, military personnel, and persons in institutions would be counted at their other location.

In preparation for the 1970 census—the first to be conducted principally by mailout/mailback—the Census Bureau developed "approximately 700 different questionnaires, field and administrative forms[,] address registers, handbooks, and manuals" for testing between 1961 and 1967. Experiments covered "type styles and sizes, paper and ink colors, as well as [the] formats in which various items would be printed" (U.S. Bureau of the Census, 1970:4-8). The final version of the 1970 household questionnaire included its residence instructions (Figure 6-2) on a separate page immediately preceding the listing of household members in Question 1. The number of specific include/exclude categories is higher than the number used in 1960,[3] though at least one "addition" comes from listing college students twice—to list those living at home

[3]The Bureau's precedural history of the 1970 census suggests that, "if the enumerator found these instructions insufficient, he referred to a table of residence rules" for additional detail (U.S. Bureau of the Census, 1970:15-4).

PLEASE LIST:

1. Everyone who usually lives here, whether related to you or not.

2. All persons staying here who have no other home.

PLEASE BE SURE TO LIST—

- All members of your family living with you, including babies.
- All other relatives living here.
- Lodgers and boarders living here.
- Servants, hired hands, others not related to you who are living here.
- Any one else staying here but who has no other home.

ALSO LIST—

Persons who usually live here but who are away temporarily on business, on vacation, or in a general hospital.

DO NOT LIST—

- College students who are away at college (or who are here only on vacation).
- Persons stationed away from here in the Armed Forces.
- Persons away in institutions, such as a sanitarium, nursing home, home for the aged, mental hospital.

They will be counted there.

NAMES OF PERSONS LIVING HERE ON APRIL 1, 1960, AND THOSE STAYING HERE WHO HAVE NO OTHER HOME Write names in this order: Head of household on first line; Wife of head; Unmarried children, oldest first; Married children and their families; Other relatives; Others not related to head of household (If you need more space, please use additional sheets of paper) (P2)			What is the relationship of each person to the head of this household? (For example, wife, son, daughter, grandson, mother-in-law, lodger, lodger's wife)	Male or Female (M or F)	Is this person— White, Negro, American Indian, Japanese, Chinese, Filipino, Hawaiian, Part Hawaiian, Aleut, Eskimo (etc.)?	When was this person born? (P6)		Is this person— Married, Widowed, Divorced, Separated, Single (never married)? (Leave blank for children born after March 31, 1946)
Last name	First name	Middle initial	(P3)	(P4)	(P5)	Month	Year	(P7)
			Head					

Figure 6-1 Basic residence question, advance materials distributed prior to enumerator visits, 1960 census

NOTE: The specific formatting shown here is from the "Notice of Required Information for the 1960 Census of Population and Housing" form. The header instructions on this version are identical in wording (but slightly different in spacing) on the two variants of Advance Census Reports mailed to households (one for large cities and the other for smaller places). A third variant of the Advance Census Report included a question on citizenship in the tabular array under the instructions; this form—unique to the 1960 census—was used only in New York State, pursuant to a requirement in the state constitution. The same instructions and basic format were also used on the "Were You Counted?" form that persons who believed they had been missed could return to the Census Bureau.

6. In Question 1 on page 2, please list each person who was living here on Wednesday, April 1, 1970, or who was staying or visiting here and had no other home.

LIST IN QUESTION 1

Family members living here, including babies still in the hospital

Relatives living here

Lodgers or boarders living here

Servants or hired hands living here

Other persons living here

College students who stay here while attending college, even if their parents live elsewhere

Persons who usually live here but are temporarily away (including children in boarding school below the college level)

Persons with a home elsewhere but who stay here most of the week while working

DO NOT LIST IN QUESTION 1

Any person away from here in the Armed Forces

Any college student who stays somewhere else while attending college

Any person who usually stays somewhere else most of the week while working there

Any person away from here in an institution such as a home for the aged or mental hospital

Any person staying or visiting here who has a usual home elsewhere

Note: If everyone here is staying only temporarily and has a usual home elsewhere, please fill this circle → ○ and give their names on page 4 in the space for question 12. Do not answer any other questions. Mail back the form on Wednesday, April 1.

— PAGE BREAK —

1. **WHAT IS THE NAME OF EACH PERSON** who was living here on Wednesday, April 1, 1970 or who was staying or visiting here and had no other home?

Line No.

Print names in this order
Head of the household
Wife of head
Unmarried children, oldest first
Married children and their families
Other relatives of the head
Persons not related to the head

(1) Last name

First name Middle initial

(2) Last name

First name Middle initial

Figure 6-2 Basic residence instructions and Question 1, 1970 census questionnaire

NOTE: Space was provided for entering 8 names; only 2 are shown here.

and to exclude those living at school. The choices made in revising the instructions between 1960 and 1970 are interesting:

- Save for the mention of "persons who usually live here but are temporarily away" and the introduction of the term "usual home elsewhere," literal mention of "usual" residence is absent; the central instruction is to include "each person who was living here," not each person who usually lives here.
- The 1960 instructions on persons staying in the household but who are away temporarily or who have no other place to live have been altered to target two specific groups: boarding school students and commuter workers (persons "who stay here most of the week while working").
- The unconditional instruction that babies should be included in the count is now modified to "babies still in the hospital" (and was further modified to "newborn" babies in the 1980 and 1990 censuses).
- The guidance that college students, military personnel, and institutionalized persons would be counted somewhere else is omitted.

Though awkwardly formatted (tacked on to the end of the "do not list" section), the 1970 questionnaire also introduced a checkbox to indicate that all people at the address in question are only there temporarily.

The include/exclude directions on the 1980 census form (see Figure 6-3) are identical to those used in 1970, save for the dropping of "servants or hired hands living here" as a category of persons to include and a somewhat clearer handling of the "everyone here is staying only temporarily" item. The key difference between the 1980 form and its predecessors is that the earlier forms had respondents jump from a list of instructions directly into rostering and questions. The 1980 form attempted to make respondents work through a preprocessing step—itemizing all the persons belonging to the household in workspace immediately adjacent to the residence instructions (before copying these names to column headings on the next pages of the questionnaire). This layout had the advantage of placing the main residence question in proximity to the instructions associated with it, in contrast with the 1960 and 1970 versions. However, the list of include/exclude instructions that respondents were asked to process could still be interpreted as long and duplicative: for example, college students and commuter workers are referenced in both the include and exclude lists. Sweet (1987:15) criticized the 1980 Question 1:

> [It is] very poorly structured. Census respondents have no reason to read all the details regarding who to include and who to exclude. It would seem to be better survey questionnaire practice to force respondents to consider explicitly whether or not each of the various conditions applies to their household.

Question 1

List in Question 1

- Family members living here, including babies still in the hospital
- Relatives living here
- Lodgers or boarders living here
- Other persons living here
- College students who stay here while attending college, even if their parents live elsewhere
- Persons who usually live here but are temporarily away (including children in boarding school below the college level)
- Persons with a home elsewhere but who stay here most of the week while working

Do Not List in Question 1

- Any person away from here in the Armed Forces.
- Any college student who stays somewhere else while attending college.
- Any person who usually stays somewhere else most of the week while working there.
- Any person away from here in an institution such as a home for the aged or mental hospital.
- Any person staying or visiting here who has a usual home elsewhere.

1. What is the name of each person who was living here on Tuesday, April 1, 1980, or who was staying or visiting here and had no other home?

Note

If everyone here is staying only temporarily and has a usual home elsewhere, please mark this box ☐.

Then please:
- answer the questions on pages 2 through 5 only, and
- enter the address of your usual home on page 20.

Please continue ↗

Figure 6-3 Basic residence question (Question 1), 1980 census questionnaire

Of the past several censuses, the 1990 census form (see Figure 6-4) devotes the most physical space to the basic household membership question and associated instructions, filling an entire page in the questionnaire booklet. "Usual residence" is prominently described as the census standard at the top of the page—although a reader's eye is arguably drawn more to the multiple lines of bold-face type introducing question 1a. The wording of nearly all of the include/exclude instructions was revised between 1980 and 1990. Although no new population subgroups are raised as new categories in either list, the new wording of the bullet points revives the specific example of "live-in employees" (to be included), adds "housemates and roommates" in addition to the lodgers and boarders referenced in previous censuses, and makes specific reference to including foster children. Significantly, the 1990 census was the first of the self-enumeration censuses to instruct respondents not to include prisoners in household counts. The item flagging cases in which all residents are only staying at the address temporarily is retained, though space to write in (a single) usual address has been added on the same page. Previous censuses effectively cut off the census questioning if the all-temporary condition applied, but the 1990 census had respondents continue with the full questionnaire.

The 2000 census form (which was displayed in Figure 2-1 and is also illustrated as part of Box 6-2) was the product of an extensive design overhaul based on several objectives. Principal among them were the user friendliness of the form (with attention to color and graphical cues to aid navigation) and easier automated data capture and optical character recognition. Like the 1990 form, the 2000 form asks a respondent to go through a preprocessing step, providing a count of people in the household in Question 1 before providing information on them on the subsequent pages. The list of include/exclude instructions is streamlined: college students are no longer mentioned in both categories, and the provision to include "foster children, roomers, or housemates" consolidates multiple points used in previous censuses. The word "usual" does not appear anywhere in the instructions, though the working definition of usual residence as the place where one lives or stays most of the time is embodied in the last bullet point of both the include and exclude lists. Rather than "usual," one of the bulleted instructions introduces a different concept—persons staying at the home on Census Day are to be counted there if they do not have another "permanent place to stay." In terms of physical layout, the 2000 form is different from its predecessors in that the largest part of the instructions for Question 1 (the include/exclude lists) are placed *after* the answer space, not side by side (as in 1990) or before the answer space (1970 and 1980).

Though the "verbal and visual changes" improved the 2000 questionnaire, Iversen et al. (1999:121) criticize the 2000 census questionnaire development, arguing that "far less research attention was devoted to errors in responses provided by those who answer the census, to the factors that contribute to such errors, and to changes that could contribute to the accuracy and completeness of the information provided by respondents."

6–D.2 Coverage Probes

In addition to the basic residence question (asking for a count or listing of household members), past decennial censuses have made different use of coverage probe questions. These questions, placed slightly later in the questionnaire, serve to jog respondents' memories and prompt them to reconsider additions or deletions to the list of household members. Even before the advent of self-enumeration in the census, the 1950 census schedule included such a coverage probe as Housing Item 8: "We have listed (number) persons

Page 1

The 1990 census must count every person at his or her "usual residence." This means the place where the person lives and sleeps most of the time.

1a. List on the numbered lines below the name of each person living here on Sunday, April 1, including all persons staying here who have no other home. If EVERYONE at this address is staying here temporarily and usually lives somewhere else, follow the instructions given in question 1b below.

Include	**Do NOT include**
• Everyone who usually lives here such as family members, housemates and roommates, foster children, roomers, boarders, and live-in employees	• Persons who usually live somewhere else
• Persons who are temporarily away on a business trip, on vacation, or in a general hospital	• Persons who are away in an institution such as a prison, mental hospital, or a nursing home
• College students who stay here while attending college	• College students who live somewhere else while attending college
• Persons in the Armed Forces who live here	• Persons in the Armed Forces who live somewhere else
• Newborn babies still in the hospital	
• Children in boarding schools below the college level	
• Persons who stay here most of the week while working even if they have a home somewhere else	• Persons who stay somewhere else most of the week while working
• Persons with no other home who are staying here on April 1	

Print last name, first name, and middle initial for each person. Begin on line 1 with the household member (or one of the household members) in whose name this house or apartment is owned, being bought, or rented. If there is no such person, start on line 1 with any adult household member.

	LAST	FIRST	INITIAL		LAST	FIRST	INITIAL
1				7			
2				8			
3				9			
4				10			
5				11			
6				12			

1b. If EVERYONE is staying here only temporarily and usually lives somewhere else, list the name of each person on the numbered lines above, fill this circle ⟶ ○ and print their usual address below. DO NOT PRINT THE ADDRESS LISTED ON THE FRONT COVER.

House number	Street or road/Rural route and box number	Apartment number
City	State	ZIP Code
County or foreign country	Names of nearest intersecting streets or roads	

NOW PLEASE OPEN THE FLAP TO PAGE 2 AND ANSWER ALL QUESTIONS FOR THE FIRST 7 PEOPLE LISTED. USE A BLACK LEAD PENCIL ONLY.

Figure 6-4 Basic residence question (Item 1), 1990 census questionnaire

who live here. Have we missed anyone away traveling? Babies? Lodgers? Other persons staying here who have no home anywhere else?"

The 1970–1990 censuses included a set of probes on the self-enumeration questionnaire. Figures 6-5 and 6-6 illustrate the probes used in 1970 and 1980, respectively. The 1990 form used two probe questions, omitting the "did anyone stay/visit" question akin to 1980's H3. An unspoken goal of the questions was to make people reconsider (and revise, as appropriate) the main household roster. Those respondents who wanted to cite specific reasons for including or excluding certain people were directed to a different page in the 1970 and 1980 questionnaires; the 1990 form gave two short lines after the coverage probe questions to write in responses. Mailed questionnaires that had "yes" or blank answers to these questions were flagged for follow-up by district office enumerators.

No such probe was included on the 2000 census questionnaire used in the main mailout/mailback component of the census, but two probes were included on the enumerator questionnaires and used in areas where enumerators conducted direct interviews (see Box 2-2). The same enumerator questionnaires were also used in the nonresponse follow-up and coverage improvement follow-up operations. Question C1 on the 2000 census enumerator form probed for possible *undercount*, asking, "I need to make sure I have counted everyone who lived or stayed here on April 1, 2000. Did I miss—

- any children, including foster children?
- anyone away on business or vacation?
- any roomers or housemates?
- anyone else who had no other home?"

Question C2—aimed at possible *overcount* or duplication—asked, "The Census Bureau has already counted certain people so I don't want to count them again here. On April 1, 2000, were any of the people you told me about—

- away at college?
- away in the Armed Forces?
- in a nursing home?
- in a correctional facility?"

In both cases, a "yes" answer meant that the enumerator was supposed to check "yes" and then make alterations in the roster in Question 1, indicating in spaces there whether the entry was an "add" or a "cancel" due to an answer to the probes. However, Nguyen and Zelenak (2003) found that the completed questionnaires show that the enumerators did not follow the procedures fully; only 21.8 percent of forms answering "yes" to C1 has someone listed on the roster with the "add" box checked. Likewise, only 43.4 percent of "yes" returns to C2 showed someone with "cancel" marked. Accordingly,

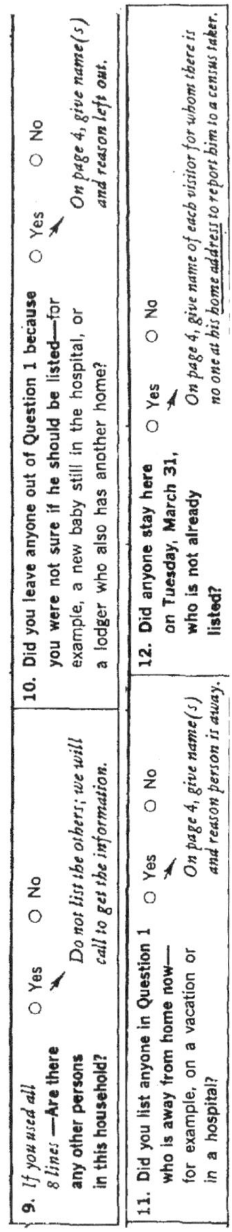
9. *If you used all 8 lines:* **—Are there any other persons in this household?** ○ Yes ○ No
Do not list the others; we will call to get the information.

10. **Did you leave anyone out of Question 1 because you were not sure if he should be listed**—for example, a new baby still in the hospital, or a lodger who also has another home? ○ Yes ○ No
On page 4, give name(s) and reason left out.

11. **Did you list anyone in Question 1 who is away from home now**—for example, on a vacation or in a hospital? ○ Yes ○ No
On page 4, give name(s) and reason person is away.

12. **Did anyone stay here on Tuesday, March 31, who is not already listed?** ○ Yes ○ No
On page 4, give name of each visitor for whom there is no one at his home address to report him to a census taker.

Figure 6-5 Coverage probe questions, 1970 census questionnaire

H1. Did you leave anyone out of Question 1 because you were not sure if the person should be listed — *for example, a new baby still in the hospital, a lodger who also has another home, or a person who stays here once in a while and has no other home?*

O Yes — *On page 20 give name(s) and reason left out.*
O No

H2. Did you list anyone in Question 1 who is away from home now — *for example, on a vacation or in a hospital?*

O Yes — *On page 20 give name(s) and reason person is away.*
O No

H3. Is anyone visiting here who is not already listed?

O Yes — *On page 20 give name of each visitor for whom there is no one at the home address to report the person to a census taker.*
O No

Figure 6-6 Coverage probes (Questions H1–H3), 1980 census questionnaire

Nguyen and Zelenak (2003) caution about making inference about the nature of people added or deleted based on the questions. Overall, based on checked "add" and "cancel" boxes, the coverage probes resulted in 77,050 people being added to the census and 83,160 deleted. Some impressions from the data are of interest even if they are merely suggestive—nearly 60 percent of the people added were young (ages 0 to 24) and nearly 70 percent of the deletes were aged 15–24, possibly college students who had been listed in the initial roster.

6–D.3 Foreign Census Questionnaires

Appendix B summarizes the approaches taken by censuses in selected foreign countries, including the specific residence instructions and questions on the questionnaires. In terms of the presentation of concepts to respondents, some basic impressions from a review of these other censuses include the following:

- Some national censuses have been able to develop layout and question wording in order to effectively guide respondents through a series of questions while economizing the amount of space devoted to instructions and the overall length of the questionnaire. The Swiss (person-level, not household-level) and New Zealand census forms are largely instruction free, with the bilingual (English and Maori) New Zealand form promoting effective flow through a streamlined questionnaire.
- There is considerable variation in the space dedicated to instructions. The Canadian census instrument for 2001 is comparable to the instruction-heavy 1990 U.S. census form, with bulleted include/exclude

instructions taking up a full page. (Those listings do suggest different choices and priorities between the two countries, such as the prominent treatment of children in joint custody arrangements and refugees on the Canadian form.) The 2001 United Kingdom census form devotes a full page to instructions and rostering of household members (and, separately, short-term visitors). Though it is still a rather large presentation, the United Kingdom form is a remarkably concise distillation of formal residence definitions and rules that are—if anything—more elaborate than the U.S. residence rules (see Appendix B).

- It is not unusual for multiple residence questions to be asked and additional address information collected on the census form. The Australia and New Zealand *de facto* censuses also ask for usual residence information, and the Swiss form asks for two addresses (and includes a check box to indicate which is the place where "you usually reside (4 or more days a week)" in a small amount of physical space on the page. (Again, the Swiss form is completed by each person, not each household.)
- Some of the forms include reminders or cues to respondents, in addition to the standard instructions. The most prominent example of these is the New Zealand census form that—despite being light on formal instructions—makes repeated entreaties to respondents to be sure to include babies in the report. The Canadian census form for 2001 includes the reminder that the respondent be sure to include himself/herself in the count.

6–D.4 Alternative Questionnaire Tests and Approaches

The 1980, 1990, and 2000 decennial censuses each included an Alternative Questionnaire Experiment (AQE) as part of their programs of testing and experimentation. Samples of census households (between 42,000 and 50,000 households in 1990 and 2000) received forms that differed in various ways from those received in all other households. The 1980 experiment focused mainly on variations on matrix-style forms most conducive to the electronic data-capture technology used at the time. The 1990 AQE was the most ambitious of the three; its experimental panels included completely redesigned versions of the census long form as well as radical structures, such as a kit of personal-response census forms rather than a single household form (DeMaio and Bates, 1992).

The 2000 AQE included experimental groups that assessed the impact of graphical and narrative instructions to guide the flow through the census long form; it also contained a group that repeated the census race and Hispanic origin questions in their 1990 form (for contrast with the 2000 version, which introduced the option of multiple-race reporting and restructured the questions generally). For the purposes of this report, the component of the 2000 AQE that is of most interest is one that varied the initial residence/roster instructions.

Box 6-2 illustrates the specific roster instructions tested by the 2000 AQE, side by side with the residence question asked on the standard census form. The revised questionnaire constituted a bundle of at least 10 changes, from the bluntly worded "master" instruction statement (to count people "using our guidelines") to finer variations (placing the broader category of "people who live here most of the time" rather than a more specific group as the first item in the include list). Bureau analysis ultimately concluded that the 2000 AQE form worked at least as well as the standard form, yielding a slight improvement in overall response to question 1 and negligible differences in mail return rates. A small reinterview program suggested that the AQE form may have improved coverage (fewer omissions) in some areas, but effects on duplication were negligible (Gerber et al., 2002; Martin et al., 2003). However, as a consequence of its design, the 2000 AQE was not a controlled experiment that could lead to clear answers to such general design questions as:

- What happens when the instructions come before or after the question, or when they come before or after the answer space?
- Will people skip the instructions when they come first and instead go right to an item labeled with the number 1?
- Does a general summary or instruction statement improve understanding of the inclusion and exclusion instructions?
- Does it matter which inclusion or exclusion criteria come first?

As an operational trial, the 2000 AQE indicated that the alternative form worked at least as well as the regular form, but it did not provide information on exactly why it worked.

6–D.5 Toward 2010: Mid-Decade Census Tests

In preparation for the 2010 census, the Census Bureau laid out a plan of large-scale tests, alternating between mailout-only tests (not involving field follow-up for nonresponding households) and full-field operational trials in successive years. Mailout-only tests were scheduled for 2003 and 2005 and field tests in 2004 and 2006, culminating in the final dress rehearsal in 2008. Subsequently, the Bureau also created an ad hoc mailout-only test for 2006, run in addition to the 2006 field test.

In terms of directions for the residence instructions and questions on the 2010 form, major developments began with the 2005 test; questionnaires were to be sent by mail and responses by either mail or by the Internet were planned, but no follow-up interviewing was performed. Initial proposals for

Box 6-2 Residence Question and Instructions in the 2000 Census and the 2000 Alternative Questionnaire Experiment

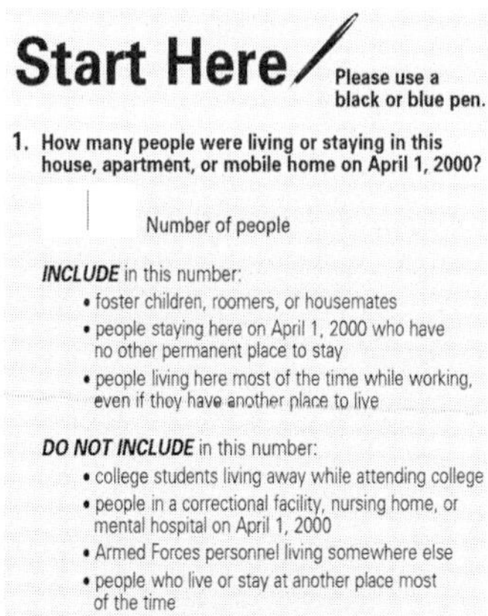

Start Here Please use a black or blue pen.

1. How many people were living or staying in this house, apartment, or mobile home on April 1, 2000?

Number of people

INCLUDE in this number:
- foster children, roomers, or housemates
- people staying here on April 1, 2000 who have no other permanent place to stay
- people living here most of the time while working, even if they have another place to live

DO NOT INCLUDE in this number:
- college students living away while attending college
- people in a correctional facility, nursing home, or mental hospital on April 1, 2000
- Armed Forces personnel living somewhere else
- people who live or stay at another place most of the time

2000 Census

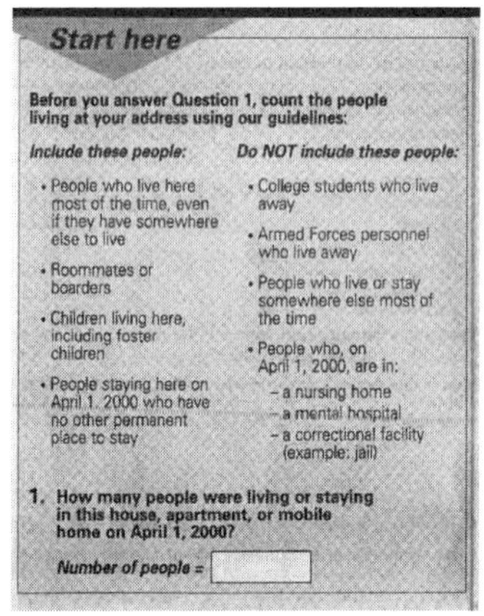

Start here

Before you answer Question 1, count the people living at your address using our guidelines:

Include these people:
- People who live here most of the time, even if they have somewhere else to live
- Roommates or boarders
- Children living here, including foster children
- People staying here on April 1, 2000 who have no other permanent place to stay

Do NOT include these people:
- College students who live away
- Armed Forces personnel who live away
- People who live or stay somewhere else most of the time
- People who, on April 1, 2000, are in:
  - a nursing home
  - a mental hospital
  - a correctional facility (example: jail)

1. How many people were living or staying in this house, apartment, or mobile home on April 1, 2000?

Number of people =

2000 AQE

Compared with the residence question on the 2000 short form, the 2000 AQE represented a package of numerous design changes, each (or any) of which could affect the outcomes (responses):

- **Order of answer space:** whether the question (and response box) appear before or after the instructions
- **Order of other components**, including structure of include/exclude lists
- **Space allocated to questions:** whether the question block is formatted to appear larger, vertically
- **Dominant graphic element:** replacement of an illustration of a pen as a major visual element with the flow-promoting orange "Start Here" triangle
- **Summary categories in instructions:** whether the instructions include a general summary category ("people who live here most of the time, even if they have somewhere else to live"), and in what order in the include/exclude lists it falls
- **Stronger instruction language:** whether an imperative statement (to count people "using our guidelines") is included
- **Typographical emphasis:** whether having the entire phrases INCLUDE and DO NOT INCLUDE capitalized (and hence accorded equal weight) differs from only capitalizing NOT (as in the AQE)
- **Formatting of lists:** whether the list of include/exclude instructions are in a single column or double-banked
- **Label placement:** whether the "number of people" label is placed before or after the response box
- **Size of answer box**
- **Location of "1", a numerical language indicator**

In addition, the two questions differ slightly in the exact wording of the bulleted instructions.

the 2005 test were presented to and discussed by the panel at its first meeting in July 2004. At that time, we raised the criticism that the 2005 test as planned did not include a control group, a benchmark against which the other alternatives could be evaluated. Feasible choices for the control group would be either the question-instruction combination used in the 2000 census itself or in the 2000 AQE.

In response, the Bureau adjusted the test plan to include six different versions of the Question 1 residence instruction/household count box, as illustrated in Figure 6-7:

(a) a version matching the 2000 AQE;[4]

(b) a modification to the AQE form, adding (and placing at the top of the include list) an item for babies and young children, reordering and revising the include/exclude lists, and reinstating (for the first time since 1960) the assurance that "excluded" persons like college students will be counted at their other place of residence;

(c) a revision of the format used in the 2004 census test, identical in content to the modified AQE except that the placement of the include and exclude lists are reversed;

(d) a "centralized" treatment that begins with a statement of objectives;

(e) a "principle-based" version that attempts a statement of the include/exclude bullet points as basic principles; and

(f) a "worksheet" approach, splitting the total "household count" into finer components, developed after the panel's July 2004 meeting consistent with ideas raised in that public discussion.

Candidate sets of coverage probe questions were developed in advance of the test, and two sets of probes—each consisting of one question intended to address possible situations of undercount and one about overcount (duplication)—were also included in the tests. These probes are shown in Figure 6-8.[5] One set pairs an undercount question—listing possible categories, but not permitting write-in space for explanations (as in probes used in previous censuses)—with a question asking whether each person "sometimes" lives or stays in another place. The second set of probes includes a query specifically about people who have recently moved (into our out of) the housing unit; it

[4]Close examination of the 2000 AQE question in Box 6-2 and the version shown in panel (a) of Figure 6-7 reveals that the two are not completely identical. The graphical "Start here" element is slightly different and combined with the directions on how to respond; the exact spacing and layout of items are slightly discrepant. The 2005 test form also has a blue tint as the questionnaire color, rather than 2000's orange.

[5]An interesting difference between these 2005 probes and the ones used in earlier censuses is the reversal of answer categories; in the 2005 versions, "No" is listed as an option before "Yes."

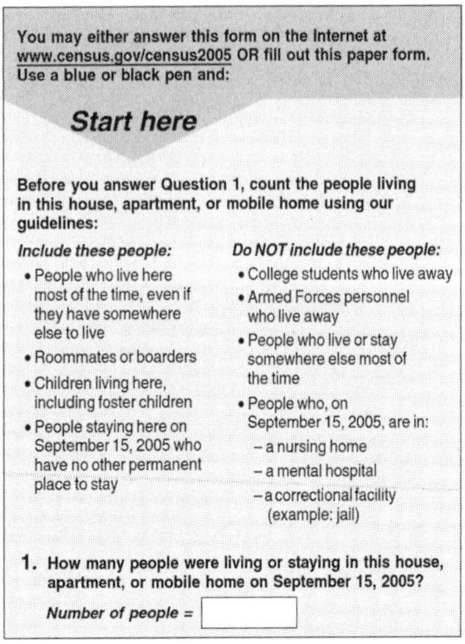
You may either answer this form on the Internet at www.census.gov/census2005 OR fill out this paper form. Use a blue or black pen and:

Start here

Before you answer Question 1, count the people living in this house, apartment, or mobile home using our guidelines:

Include these people:

- People who live here most of the time, even if they have somewhere else to live
- Roommates or boarders
- Children living here, including foster children
- People staying here on September 15, 2005 who have no other permanent place to stay

Do NOT include these people:

- College students who live away
- Armed Forces personnel who live away
- People who live or stay somewhere else most of the time
- People who, on September 15, 2005, are in:
  - a nursing home
  - a mental hospital
  - a correctional facility (example: jail)

1. How many people were living or staying in this house, apartment, or mobile home on September 15, 2005?

Number of people =

(a) 2000 AQE

You may either answer this form on the Internet at www.census.gov/census2005 OR fill out this paper form. Use a blue or black pen and:

Start here

Before you answer Question 1, count the people living in this house, apartment, or mobile home using our guidelines:

Include these people:

- Babies and children living here, including foster children
- People who stay here most of the time, even if they have somewhere else to live
- Roommates or boarders
- People staying here on September 15, 2005 who have no other permanent place to stay

Do NOT include these people *(They will be counted at the other place):*

- College students who live away from this address most of the year
- People who live or stay somewhere else most of the time
- Armed Forces personnel who live away
- People who, on September 15, 2005, were in a:
  - Nursing home, mental hospital, etc.
  - Jail, prison, detention facility, etc.

1. How many people were living or staying in this house, apartment, or mobile home on September 15, 2005?

Number of people =

(b) Modified AQE

You may either answer this form on the Internet at www.census.gov/census2005 OR fill out this paper form. Use a blue or black pen and:

Start here

Before you answer Question 1, count the people living in this house, apartment, or mobile home using our guidelines.

Do NOT INCLUDE these people (They will be counted at the other place):

- College students who live away from this address most of the year
- People who live or stay somewhere else most of the time
- Armed Forces personnel who live away
- People who, on September 15, 2005, were in a:
  - Nursing home, mental hospital, etc.
  - Jail, prison, detention facility, etc.

INCLUDE these people:

- Babies and children living here, including foster children
- People who stay here most of the time, even if they have somewhere else to live
- Roommates or boarders
- People staying here on September 15, 2005 who have no other permanent place to stay

1. How many people were living or staying in this house, apartment, or mobile home on September 15, 2005?

Number of people =

(c) Modified 2004 Test

Figure 6-7 Coverage treatment groups, 2005 National Census Test

You may either answer this form on the Internet at www.census.gov/census2005 OR fill out this paper form. Use a blue or black pen and:

Start here

Before you answer Question 1, count the people living in this house, apartment, or mobile home using our guidelines.

- We want to count people where they usually live and sleep.
- For people with more than one place to live, this is the place where they sleep most of the time.

EXCLUDE these people (They will be counted at the other place):

- College students who live away from this address most of the year
- Armed Forces personnel who live away
- People who, on September 15, 2005, were in a:
  - Nursing home, mental hospital, etc.
  - Jail, prison, detention facility, etc.

INCLUDE these people:

- Babies and children living here, including foster children
- Roommates or boarders
- People staying here on September 15, 2005 who have no other permanent place to live

1. How many people were living or staying in this house, apartment, or mobile home on September 15, 2005?

Number of people =

(d) "Centralized" Approach

You may either answer this form on the Internet at www.census.gov/census2005 OR fill out this paper form. Use a blue or black pen and:

Start here

The Census must count every person living in the United States on September 15, 2005.

Before you answer Question 1, count the people living in this house, apartment, or mobile home using our guidelines.

- Count all people, including babies, who live and sleep here most of the time.

The Census Bureau also conducts counts in institutions and other places, so:

- Do not count anyone living away either at college or in the Armed Forces.
- Do not count anyone in a nursing home, jail, prison, detention facility, etc., on September 15, 2005.
- Leave these people off your form, even if they will return to live here after they leave college, the nursing home, the military, jail, etc. Otherwise, they may be counted twice.

The Census must also include people without a permanent place to stay, so:

- If someone who has no permanent place to stay is staying here on September 15, 2005, count that person. Otherwise, he or she may be missed in the census.

1. How many people were living or staying in this house, apartment, or mobile home on September 15, 2005?

Number of people =

(e) "Principle-Based" Approach

This is the official form for all the people living or staying at this house, apartment, or mobile home on September 15, 2005. You may either answer this form on the Internet at www.census.gov/census2005 OR fill out this paper form. Use a blue or black pen and:

Start here

1. Including yourself, how many people live here and have no other place they live? *Don't forget to include babies.*

Number of people =

2. Besides those counted in Question 1, how many other people have more than one place they live, but sleep here more often than anywhere else? *Enter "0" if none.*

Number of people =

3. Besides those counted in Questions 1 or 2, how many other people were staying here temporarily on September 15, 2005 and had no other place where they live? *Enter "0" if none.*

Number of people =

(f) "Worksheet" Approach

Figure 6-7 (continued)

also asks whether and why each person "sometimes" lives or stays somewhere else, but goes further to ask the city and state of this other residence.

At this writing, the results of the 2005 National Census Test have not yet been made available. The Bureau has released a report on cognitive testing of the "worksheet" approach (Hunter and de la Puente, 2005), along with the other questionnaire design features being tried in the 2005 test (e.g., layout changes in checkbox response areas to cut down on physical space and revised instructions on the race and Hispanic origin questions). The study was based on 14 cognitive interviews, conducted in the Washington, DC area in January 2005. It concluded that the worksheet approach generally worked well; with a single exception, the numbers reported in the initial questions matched the number of people for which data were collected on the form.[6] The Bureau authors expressed concern that the worksheet form might cause some increase in duplication, particularly among commuter workers, and suggested using the phrasing "live and sleep here" in Question 1.

While the 2005 test was in progress, the Bureau scheduled a further mail-based test to be conducted in early 2006, in addition to the broad-scale field test already planned for 2006. Due to its timing, this ad hoc test did not build directly from the results of the 2005 test; the Question 1 box planned for the 2006 test instruments would use the "centralized" approach from the 2005 test. The 2006 ad hoc test was to be focused on two different cues to respondents: (1) the effect of specifying a "deadline" for return on the outside envelopes, cover letter, advance letter, and reminder postcards sent to respondents; (2) more relevant to residence data collection, variations on probes included in a section labeled "final questions for everyone" at the end of the census questionnaire. Treatment groups included a direct question on whether the number of household members specified at the beginning of the form was the same as the number for which data had been recorded throughout the form and, if not, why. Some instructions for this final question block included reminders to respondents to make sure that he or she had counted themselves on the form, as well as babies and temporary guests with no other place to live.

6–E CHANGING THE STRATEGY: GETTING THE RIGHT RESIDENCE INFORMATION

At the 1986 COPAFS residence rules conference, Mann (1987:5) commented on the fundamental difficulty of the Census Bureau's instruction-based method of collecting residence count information. "What is printed in

[6]The single exception was a respondent who included himself in the household count but failed to include himself in the listing and data collection in the rest of the form; exactly why this happened is unknown.

2. Were there any additional people staying here September 15, 2005 that you did not include in Question 1? *Mark ☒ all that apply.*

- ☐ Children, such as newborn babies or foster children
- ☐ Relatives, such as adult children, cousins, or in-laws
- ☐ Nonrelatives, such as roommates or live-in baby sitters
- ☐ People staying here temporarily
- ☐ No additional people

10. Does Person 1 sometimes live or stay somewhere else?

☐ No ☐ Yes— *Mark ☒ all reasons that apply.*

- ☐ To attend college
- ☐ To be closer to work
- ☐ While in the military
- ☐ To stay at a seasonal or second residence
- ☐ For a child custody arrangement
- ☐ While in jail or prison
- ☐ While in a nursing home
- ☐ For another reason

(a) Standard coverage probe (question 2) and query for reasons for another place of residence (question 10)

2. Were there any people staying here around September 15, 2005 that were . . .
Mark ☒ all that apply.

- ☐ Moving out?
- ☐ Just moving in?
- ☐ Staying here temporarily?
- ☐ Moving between places?
- ☐ No one moved in or out

10. Does Person 1 sometimes live or stay somewhere else?

☐ No ☐ Yes — **What type of place is it?**

- ☐ House or apartment
- ☐ College housing
- ☐ Jail or detention facility
- ☐ Nursing home
- ☐ Shelter
- ☐ Other type of place

If yes, where is it?

City ________ State ____

(b) Probe targeted at movers near Census Day (question 2) and request for city/state of other residence (question 10)

Figure 6-8 Coverage probe questions, 2005 National Census Test

Question 1 has to be clear and it has to be comprehensive. This is not an easy task for a questionnaire designer," and "the accuracy of the head count largely rests upon the" quality of Question 1 and its instructions. When instructions are read and understood by respondents (as is not universally the case), seemingly small changes in the wording and ordering of instructions can have large or unintended consequences. It is impossible to create a specific instruction for every possible living situation—like the creation of specific formal residence rules for every contingency—and difficulties may arise from mentioning some groups but not others.

At that conference, Hill (1987:3) offered an alternative approach: "listing people where they are staying on the Census date but then regrouping them into their usual residence." Though "the outlined alternative would give the more reliable overall count," Hill concluded that it would not be feasible given the technology available in 1980 and 1990. Due to the "massive matching problems" involved with searching reported usual residence information against other census records, the count might be improved but "the reliability of the final geographic allocation might be worse."

Based on our review of current survey methods, we favor an approach to the decennial census based on asking guided questions—and multiple questions, as necessary—rather than relying on instructions to convey complex definitions. Like Hill's speculative strategy from 1986, the basic goal of our approach is to shift the burden of deciding what constitutes "usual residence" from the census respondent to the Bureau. The ideal form is one that gathers information without the need for elaborate instruction—and collects sufficient information from respondents to allow the Bureau to make determinations of "usual residence" (by applying residence principles and through matching to relevant census records) during processing and editing.

This question-based approach is an immediate corollary to our recommended residence principles—in particular, the principle that determination of what constitutes "usual" residence should be made at the level of the individual. Achievement of that principle is possible only if sufficient information to make an individual-level determination is collected by the census.

6–E.1 Questions, Not Instructions

The instructions on the 2000 census form are front-loaded onto the basic household count requested in Question 1. That count is supposed to match the number of persons for whom data are recorded in the rest of the questionnaire (if not, or if the household count exceeds 12, follow-up by an enumerator is triggered). Accordingly, the structure of the opening, Question 1, section of the questionnaire is the one that would have to change the most in a revised approach.

The "worksheet" treatment fielded in the 2005 test (see Figure 6-7) is an example of the question-based structure we envision for Question 1, breaking the complex cognitive task of determining a complete household count into smaller (and ideally more tractable) components, such as those persons staying at the household temporarily on Census Day. The intended effect is to replace dense instructions with a guided series of questions. The text of the questions should be concise and so harder to cursorily scan or skip.

It should be emphasized that the specific "worksheet" form tested in 2005 is not the only possible configuration of questions, nor may it be the optimal one. We also emphasize that the same warning that applied to the development of specific residence rules applies here: just as it is confusing and undesirable to carve out a new residence rule for every possible living situation and population group, so too is it unwise to divide the task of deriving a household count across too many questions and categories.

6–E.2 The Short Form Is Too Short

In addition to a revised structure of the basic household count and listing question at the beginning of the census questionnaire, a fully question-based strategy for gathering accurate residence information requires additional queries in the body of the questionnaire. Such additional questions are necessary to obtain enough information to make an individual-level "usual residence" determination and to make sure that a respondent is being counted at the correct place.

The assertion that "the short form is too short" is a strong one and is meant to draw attention to the need for some additional data collection; to be clear, though, we do not suggest a radical lengthening of the form. Follow-up interviews like the Accuracy and Coverage Evaluation interview used in 2000, or the planned coverage follow-up operation for 2010 (see Chapter 8), can include an entire battery of residence-related questions. It is proper that these highly detailed interviews ask numerous residence questions, while the question content of the census itself is kept more limited.

Two basic conceptual checks need to be made in order to determine whether a person's enumeration location (that is, where the questionnaire finds them) is where they should be counted in the census:

- Does this person have a residence elsewhere that could be considered their usual residence?
- Is this census address correct, for this person and this building (physical structure)?

These two concerns prompt the questions we suggest adding to the census questionnaire. There is also a need for reinstating coverage probes to the census questionnaire, to jog respondents' memories and to provide further clues to census staff on the appropriate resolution of "usual residence."

"Any Residence Elsewhere" Collection

For the 1900 census—decades before self-response questionnaires, when the census was still conducted by enumerators recording entries in large, columnar ledgers—enumerators received a unique instruction. Concerning "inmates of hospitals or other institutions," the instructions directed that all such inmates should be enumerated. However, the instruction concluded, "if they have some other permanent place of residence, write it in the margin of the schedule on the left-hand side of the page." What, if anything, was done with any such information scrawled on the margins is unknown.

As described in Chapter 3, practice varied over subsequent decades as to whether such groups as hospital patients and military personnel stationed at domestic bases (but living in an off-base property) were to be counted at the institution or base or at another place. The spirit of the 1900 instruction and later experience ultimately led to the practice of "usual home elsewhere" (UHE) data collection. A major part of what limited organization exists in the formal residence rules of the 2000 census is the idea of reporting another address as a UHE, a privilege that the rules granted to some types of group quarters residents but not others. (However, as we will discuss in Chapter 7, the Individual Census Reports used for group quarters data collection asked for a UHE from *all* residents; only the information from members of permitted groups was further processed.) In recent censuses, UHE reporting has been solely limited to the group quarters population, with the exception of the provision in the 1970–1990 censuses of a checkbox for indicating that *everyone* at the address has a usual place of residence elsewhere and is only staying at the questionnaire address temporarily. Under those conditions only, the 1980–1990 versions asked for the full street address of the other place of residence.

The collection of accurate residence data is central to the census mandate, and the living situations in which people have legitimate, strong ties to more than one residence is not limited to the group quarters population. Collecting information on another place of residence, if applicable, for all persons on the census form better equips the Census Bureau to determine where each individual should be counted, as well as making the structure of the census form's residence questions more consistent with real-life settings. The Census Bureau should strive to collect alternative residence (address) information from all census respondents, not a selected subset.

> *Recommendation 6.2:* **Information on "any residence elsewhere" (ARE) should be collected from census respondents. This in-**

formation should include the specific street address of the other residence location. A follow-up question should ask whether the respondent considers this ARE location to be their usual residence, the place where they live or sleep more than any other place.

To be clear, we do not suggest by this recommendation that the decennial census questionnaire attempt to collect a complete residential history of each person in the household. Rather, what we suggest is that the census form allow the entry of a single street address of an ARE where a person may spend a significant portion of a week, month, or year. As stated in the recommendation, a follow-up question should ask whether the respondent considers this alternate address their "usual" residence. Another follow-up question could include other categories to be specified, such as whether the alternate address is a seasonal or vacation home or a family home (e.g., for college students or commuter workers).[7] The objective of this approach is not to produce a full profile of mobility in the United States, but to further the basic objective of the census: to allow census clerks to determine the appropriate location for counting, to provide specific information for the detection of duplicates or omissions in the census, and to flag cases for which personal follow-up by an enumerator would be helpful. For research purposes, a more extensive battery of residence questions would be helpful: however, that should be the goal of a revised Living Situation Survey (Recommendation 5.5) or other moderate-scale surveys (as we discuss in Chapter 8), not the census form itself.

This collection of ARE information is consistent with the approach of some foreign censuses, which collect both a respondent's *de jure* usual residence as well as the *de facto* information about where they were (and were enumerated) on Census Day. The principal justification for the addition of the ARE question is similar to that used in these foreign censuses: it is information that is used to help determine that a person's usual residence is recorded correctly and to help ensure that each person is counted once and only once.

Lowry (1987) advocated adopting this approach in the U.S. census, collecting both *de jure* and *de facto* addresses, at the 1986 COPAFS residence rules conference. Discussants at that workshop held this idea to be "definitely a good one" (Hill, 1987:4). The conference summary (CEC Associates, 1987:25) concurred with the collection of UHE information but also noted that "there may be major technical-procedural limitations on the implementation of" said recommendations:

> The Census Bureau has limited resources with which to check all those reporting a UHE back to the referenced address to ensure that they were

[7]To ease respondent burden, another feature that might be added is a checkbox to indicate that the same address specified for Person 1 should be used for another person on the form.

> indeed counted at home, and limited time before 1990 in which to acquire computer capacity to facilitate such matches. . . . The costs and complexities of introducing these procedures increase the importance of identifying those groups for which follow-up of UHE reports is most likely to be productive and cost-effective.

Computational capacity is a lesser concern now than in 1986, and the coverage evaluation research program for the 2000 census was a showcase for improved searching and matching capability. Ultimately, Bureau staff were able to learn much about the nature of potential duplicate records in the census by performing a complete match of the census returns against themselves based mainly on person name and date of birth (Fay, 2002b, 2004). This methodology is a good base for improvements in 2010 and beyond. The method was sufficiently tractable that the Bureau is considering "real-time unduplication" based on similar matches while census processing is still in progress (and while capacity exists to deploy enumerators for follow-up interviews). These developments suggest that the sheer technical capacity to collect and process additional reports of address information is at hand.

However, other logistical concerns—limited field follow-up resources (to resolve discrepancies found during matching or to validate addresses not on the Master Address File) and the time constraint of producing apportionment counts by the end of the census year—carry great weight. Although we conclude that universal provision for ARE reporting is the proper direction for the census to follow, we also recognize the need for testing and evaluation of new procedures before they are applied in the census; our recommendations in Section 6–E.4 reflect these constraints.

One of the coverage probes tested in the 2005 National Census Test (Figure 6-8, (b), Question 10) is close to an ARE question, though it asks only for the place (city/state) where the person's other residence is located. The two Question 10 probes are a starting point for part of an ARE question (or an immediate follow-up to it), asking about the nature of the other residence (e.g., vacation home or school-location address for a college student) or the reason that the person resides there some of the time (e.g., seasonal move or commuter worker). Ideally, the ARE location can be obtained along with some indication of how much time the person spends at the other location so that usual residence can be determined.

Our proposed ARE question does not go as far as some researchers of seasonal "snowbird" populations—or demographers, generally, interested in the dynamics of migration—would like. With specific eye toward collecting data on "snowbirds," Happel and Hogan (2002) suggest a question akin to: "Besides your usual place of residence, have you spent 30 days or more during the past year at another locale? If so, at which location(s) did you stay and for how long (in each location)?" The structure of this question—a full residence profile for each person—also speaks to a limitation of the request for a single

ARE per person: such an approach would still not capture the full experience of intensive travelers, migrant workers, and recreational vehicle users with ties to more than two places. However, a universal ARE option at least breaks free from the notion that respondents must specify one, single place of residence and could provide greater accuracy in the many cases—college students, hospital patients, children in joint custody arrangements, and so forth—where the tension in reporting is between two places.

Verifying Addresses

Current census methodology makes the strong—and sometimes erroneous—assumption that the census questionnaire is properly delivered to the correct housing unit (and to the people who live there). In multi-unit structures like apartment buildings, questionnaires may be placed in the wrong unit's mailbox by mistake. Particularly for large multi-unit structures with a common mail delivery point, mail carriers may also view the census questionnaires as interchangeable (for better or worse, like high-volume advertising or "junk" mail) and put them in mailboxes haphazardly. The sheer volume of the major census mailout in a decennial year can contribute to the perception of the questionnaires as interchangeable, when in fact they are intended to reflect input from a specific designated housing unit. Improper sorting or misplacement prior to giving the mail to letter carriers can lead to questionnaires being delivered to the wrong house.

The 2000 census questionnaire lacked a provision for a respondent to indicate that the address printed on the received questionnaire was erroneous, or that the questionnaire was misdelivered. This has not always been the case. The 1970 census questionnaire—the first administered primarily by mail—included a three-line address box; directly underneath, an italicized instruction read: "If the address shown above has the wrong apartment identification, please write the correct apartment number or location here." One line was provided for response. The 1980 mailing label space shared the same feature and identical copy (save that the entry space was above, not below, the printed label); three lines were provided for entry. Foreign censuses (see Appendix B) generally do not offer guidance on or examples of such a correction question, yet whether self-administered or conducted by an interviewer, they direct a respondent to write in full address information rather than relying only on a printed address label.

Providing space for respondents to advise of changes to the address printed on their census form is a relatively simple change, but it would entail some cost. Depending on its exact implementation, respondents' handwritten corrections would likely be ill suited to automatic optical character recognition; accordingly, the burden of data entry would be shifted to human clerks (working with the raw paper forms or, more likely, scanned images). However, we believe that this cost would be offset by potential gains in the quality of data reporting and assignment of reported persons and households to particular housing units. It would also provide very valuable input for census follow-up, unduplication, and coverage evaluation efforts, and it would also provide useful operational data for evaluation (e.g., quantitative evidence of the magnitude of some mail delivery problems).

> *Recommendation 6.3:* **The census questionnaire should allow respondents to correct the address printed on the form if it is wrong (e.g., address is listed incorrectly or questionnaire is delivered to wrong unit or apartment number).**

In addition to mail delivery problems, even questionnaires delivered correctly to the right address may not reach the "right" people. Prominent among these cases are seasonal residences that may be vacant for much of the year or are not consistently occupied by the same people over a year. The occupants of a home in February or March may receive the form—response to which is compelled by law—yet they have no mechanism in the current short form for indicating that their stay at the address in question is only temporary (and that they may correctly be counted elsewhere). People who move residences on or near Census Day may also face the problem of a correctly delivered questionnaire finding them at the "wrong" address. A questionnaire could be carried with the movers from the old housing unit to the new one, or it may be inadvertently forwarded.

One approach to handling this problem could be to borrow from the 1990 census example and add a follow-up question to the address correction space—to provide checkboxes to indicate that all residents of the indicated housing units are there temporarily or are in the process of moving residences. The use of such a checkbox flag could be useful in identifying cases that need follow-up from an enumerator. However, we believe a better approach would involve collecting the address information for the other residence location (as did Question 1b on the 1990 census form; see Figure 6-4); this is entirely consistent with the broader change in residence questions that we describe in the next section.

Coverage and Housing Type Probes

We are encouraged that the 2005 test suggests that the Census Bureau is considering adding separate coverage probe questions, which were absent on the streamlined 2000 census questionnaire. As we discuss in Chapter 8, the Bureau's interest in probe questions is partly operational, because responses to the probes are intended to trigger eligibility for an enhanced coverage follow-up operation; see Box 6-3. In addition to addressing these practical concerns, the use of coverage probe questions is beneficial in its own right; the questions

Box 6-3 Coverage Follow-Up Plans for the 2010 Census

Plans for the 2010 coverage follow-up operation call for an expansion of the coverage improvement follow-up operation from 2000 (Moul, 2003). The revised operation would be triggered to prompt follow-up enumeration if

- the housing unit is returned as vacant;
- the household size is large, exceeding the limit of seven family members for whom full information can be reported;
- the household count reported at the beginning of the questionnaire does not match the number of people for whom data are returned;
- interview information suggests that people who are reported as living elsewhere should be reported as belonging to this household;
- interview information suggests that people who are reported in this household may more properly be counted at another residence;
- probabilistic matching based on name and date of birth suggests that there are potential duplicate entries for members of the household.

The problems of implementing this are two-fold: first, the fraction of census returns for whom this operation would be triggered might be too large to be workable (e.g., reinterview of 30 percent rather than 10 percent), and second, it may conflict with the postenumeration survey conducted as part of coverage measurement.

provide further opportunity to elicit residence information through questions rather than relying on a preamble of instructions.

We note above that some of the probes used in the 2005 National Census Test are useful starts to follow-up questions to accompany a reported "any residence elsewhere" address. The presence of such a question, related to the basic nature of a housing unit, is particularly essential given the short-form-only orientation of the 2010 census; the standard question on housing tenure (own or rent) may be the only question focused on the housing unit itself and the people living in it. In general, we agree with the approach taken in past censuses, which was the basic goal of formulating possibilities for the 2005 test: it is essential that separate probes targeting overcount (potential duplicates) and undercount (possible omissions) be included on the census form.

6–E.3 Mode Effects

There is another compelling reason that the Bureau should adopt a question-based framework for data collection, particularly the basic count in Question 1. It is well known from research on surveys that differences in the mode of administration of a census or survey can influence responses. The experience of answering a survey questionnaire from a human interviewer standing in one's doorway is substantively different than that of an interview conducted in the less personal interface of a phone call. The rapport and trust between interviewer and respondent, the perceived level of "security" in answering private queries, and the impulse to decline to answer some questions (or cut off an interview altogther) differ between the two modes, which in turn are different from the effects of responding through the impersonal media of mailed questionnaires or the Internet. Survey modes also differ in their ability to present instructions and support information.

The decennial census has become a mixed-mode survey collection of vast proportion and will likely continue on that path. In 2000, responses were acquired through self-enumeration on paper questionnaires and through computer-assisted telephone interview if respondents called a number for assistance. Response to the census short form was also possible on the Internet though this option was not widely publicized. In 2010, mail and self-response will continue to be the bulk of the census response. However, the Bureau plans to conduct nonresponse follow-up data collection using computer-assisted personal interviewing on a handheld device; in 2000, follow-up interviews were conducted using paper forms. In the initial planning for the 2010 census, data collection by the Internet and by telephone using an automated "interactive voice response" system were expected to be major parts of the census process (National Research Council, 2004b), but this is no longer so. The interactive voice response system appeared to perform badly in the 2003 National Census Test, and the Bureau acknowledged in June 2006 that it planned to scrap Internet data collection.[8]

There is reason to believe that a question-based mode of collecting residence information may be more robust to mode differences than an instruction-based model, because response to instructions may depend crucially on medium of administration. We emphasize, though, that this is an empirical question, a supposition which requires testing of the form we call for in Chapter 8.

> *Recommendation 6.4:* **Regardless of the final structure of residence questions chosen for 2010, research must be done on response effects created by mode of administration (mail, phone, Internet, interview with handheld computers).**

6–E.4 Testing ARE in 2010

Some elements of a question-based approach to gathering residence information are already in testing, such as the coverage probe questions and the

[8]The plans to drop Internet response were discussed at a June 6, 2006 hearing of the U.S. Senate Subcommittee on Federal Financial Management, Government Information, and International Security (a subcommittee of the Homeland Security and Governmental Affairs committee).

"worksheet" prototype for Question 1 on the 2005 census test. The most complicated part of the approach—placing an ARE item on the 2010 census questionnaire—is certainly feasible for 2010. However, effective plans for how to work with ARE data in processing census returns and scheduling enumerator follow-up as necessary are vitally important to the resulting census data. In the interests of keeping the 2008 census dress rehearsal as true a rehearsal as possible (and not a major experiment), the Bureau should test a question-based approach as part of the experimentation program of the 2010 census, in order to provide a base of information for further research over the next decade and possible inclusion in 2020.

> *Recommendation 6.5:* **In the 2010 census, the Census Bureau should conduct a major experiment to test a form that asks a sufficient number of residence questions to determine the residence situation of each person, rather than requiring respondents to follow complicated residence instructions in formulating their answers. The results of this test, and associated research, should guide decision on full implementation of the approach in 2020.**

As we recommend in Chapter 7, ARE address data should be included on the questionnaires used to enumerate all elements of the group quarters and other nonhousehold population in 2010, and those data should be fully evaluated and used in unduplication screening as appropriate. As has been done in previous censuses, a sample of the household population should receive a question-based form with ARE items.

From a research standpoint, it would be useful for evaluations of the collected ARE data to compare the results of this test with those of the census coverage follow-up program, in which more detailed residence probes and rostering questions are possible than on the regular census form. If possible, field reinterviews with some test respondents would be useful, perhaps using a traditional instruction-based form to see if the same results are obtained.

It should be acknowledged that a form that implements a fully question-based approach to gathering residence information may also require a departure from recent census norms, in the physical sense. Since the start of self-administered forms in 1970, the census short form has generally been kept to a one-sheet, folded ledger-size booklet (the census long-form questionnaire, of course, included more pages). Accommodating both a modest number of new questions as well as visual features (to aid flow through the questionnaire) and adequate space for optical scanning may require a slightly longer or larger short-form booklet—possibly not an undesirable outcome, as the census questionnaire is now fairly dense and tight.

6–F A VIOLATION BY DESIGN: THE CENSUS DAY RESPONSE PROBLEM

The language of the census questionnaire and accompanying mailings, together with the basic logistics of the decennial count, combine to create a problem that appears almost trivial at first, but ultimately cuts to the core meaning of the census. The problem is that the census is intended to be a count of the resident population of the United States on Census Day, April 1, of the decennial year. However, the country does not stand still on Census Day; the count is not made solely in that span of 24 hours. Instead, questionnaires are mailed out a few weeks in advance. In order to ensure smooth processing and enable follow-up interviewing to take place in fairly short proximity to Census Day, the Census Bureau's incentive is to have the respondents answer the questionnaire as quickly as possible. Hence, the problem: what does it mean for a respondent to report the population of a household as of Census Day, if he or she is filling out and returning the questionnaire up to 2 weeks before that date?

Since 1960, the questionnaires and their accompanying instructions have offered different guidance on exactly when the form is to be completed and returned:

- Some households receiving the 1960 form in the mail were asked to return the form by mail, as part of a census experiment; the households that received those instructions were advised to "please mail the completed form within 3 days in the special envelope."
- In the most explicit instruction given in this regard, the beginning instruction on the 1970 census questionnaire read, "Please fill it out and mail it back on Census Day, Wednesday, April 1, 1970."
- The 1980 questionnaire adapted the 1970 instruction slightly; the initial instruction told the reader to "please fill out this official Census Form and mail it back on Census Day, Tuesday, April 1, 1980." Later in the instructions, this guidance was softened to mailing it back on Census Day "or as soon afterward as you can."
- The 1990 instruction directly contradicted the notion of a resident count as of Census Day by urging responses *prior* to Census Day. The first page of the form declared, "**Remember: Return the completed form by April 1, 1990**," accompanied by the softer instruction to "**please answer and return your form promptly**."[9] Despite this instruction, the census form still asked respondents to report their household members as of April 1 (as shown in Figure 6-4).

[9]The 1990 instructions also added a curious warning, implying that forms should be returned early in order to "Avoid the inconvenience of having a census taker visit your home."

None of the instructions in the 2000 census about mailing back the questionnaire were as explicit as in 1990, but the effective message was much the same. In early March, households received an advance letter: the letter, dated March 6, 2000, advised readers that "**about one week from now, you will receive a U.S. Census 2000 form in the mail**"; the letter asked the reader to "please fill it out and mail it in promptly."[10] The 2000 census questionnaire itself contained no deadline or guidance on when the form should be completed and returned; a March 13, 2000, press release on the mailout of 98 million census forms noted only that "people are asked to mail the forms back as soon as possible."[11] A week later, households received a follow-up postcard dated March 20 offering "sincere thanks" to those households that already returned the form; "if you have **not**, please fill out the form and mail it back as soon as possible."[12] Both the letters and the public relations stance suggested by press releases suggested that the 2000 census forms were to be returned as promptly as possible, even if the return was made before the Census Day reference point used in the questionnaire.

The effect of possibly failing to fully account for some births, deaths, moves, and displacements in the last week of March may be relatively small. Yet however subtle the effect may be, the practice of encouraging immediate and rapid return of census questionnaires—possibly *before* the reference date of Census Day—is a basic violation of residence principles and could undermine the credibility of the concept in general. It is an empirical question as to whether being asked to violate one rule at the outset may prompt people to disregard other instructions; that kind of experiment has not been performed.

> *Recommendation 6.6:* **To be consistent with the principle of the basic residence question on the census form—where did you live on April 1?—the Census Bureau should encourage prompt response but make clear that the form should be completed and returned on Census Day or as soon thereafter as possible.**

It would also be helpful for the major mailout of census materials to occur as close to Census Day as practicable. Too lengthy a lead time between the arrival of the questionnaire and Census Day encourages the "forecasting"

[10]The 2000 census advance letter included a second page in several foreign languages, with checkboxes that allowed the respondents to request that a special-language census form be mailed to them. The advance letter is perhaps best remembered for a processing error; every letter was misaddressed due to the accidental insertion of a prefix digit before every house number (the barcode printed on the mailing was correct, though, and the Postal Service was able to deliver the letters). See http://www.census.gov/dmd/www/pdf/letter_1.pdf [6/1/06].

[11]"Nearly 100 Million Census Questionnaires in the Mail, Census Workers Delivering the Rest." U.S. Department of Commerce News Release, March 13, 2000. Available: http://www.census.gov/Press-Release/www/2000/cb00cn24.html [6/1/06].

[12]See http://www.census.gov/dmd/www/pdf/letter_2.pdf [6/1/06]. The choice of a reminder postcard was based in part of research reported by Dillman et al. (1995).

of Census Day living situations that our recommendation tries to curb. Accounting for variation in mail delivery, the current schedule for delivery in mid-March is quite appropriate, and we do not suggest a delay in mailing out the questionnaires. Indeed, the public relations consequences of erring too far in the other direction—and having major segments of the population not receive questionnaires until after Census Day—would be severe. What we do recommend is that the Bureau not go out of its way (as with the 1990 census form instructions and the pre-Census Day follow-up letter in 2000) to encourage violation of the most basic residence precepts.

6–G RESEARCH NEEDS

Issues of questionnaire format and layout can have an effect on response. In recent decades, the Census Bureau has probed some of these issues through its census tests and through stand-alone efforts like the 1993 Living Situation Survey.

> *Recommendation 6.7:* **The Census Bureau should continue and strengthen its research on the combined effects of visual layout features and specific wording situations in the development and testing of questionnaires and their effectiveness.**

A particular, vitally necessary improvement to this research is discussed in detail in the next chapter—the need to use more small-scale experiments and field trials, rather than the current course of relying on either a very small number of cases in cognitive tests or on tens or hundreds of thousands of cases in census operational trials.

The panel's review of the evolution of the census forms being developed for the 2005 and 2006 census tests, with an eye toward the final design of the 2010 census questionnaire, revealed that the forms are being developed on the basis of inadequate experimental evidence. For instance, alternative forms being submitted for testing include multiple changes in content and layout (including content and formatting of the residence rules instruction box) so that the cause of improvements cannot be traced to a single feature. There are also discontinuities with previous versions: for example, a modified version of the question used in the 2000 AQE—and not the original—was used in the original plans for the 2005 test.

> *Recommendation 6.8:* **When designing experimental tests, the Census Bureau should always include a control form—either the questionnaire items used in the preceding census or the exact items used in immediately previous census tests—so that individual modifications can be more effectively assessed.**

To achieve our recommended goals—a small number of core residence principles and a question-based (not instruction-based) approach for gathering residence information—the census short form will have to be longer than it currently is. Determining exactly how long the form should be is a matter for further research, which should include a major experiment in conjunction with the 2010 census. In addition, we encourage the Bureau to work with outside researchers—and conduct its own experiments, as possible—on length-of-form issues, such as respondent unease with questionnaires of various lengths. The Bureau should encourage work on formats that simplify the questions; prior work (Krosnick and Berent, 1993) suggests that respondents may find it easier to answer two simple questions rather than one complicated one and that higher quality data can result.

– 7 –

Nonhousehold Enumeration

THOUGH RELATIVELY SMALL in magnitude—just below 3 percent of the total population in both the 1990 and 2000 censuses—the population living in group quarters (nonhousehold) settings includes many of the most prevalent living situations that complicate interpretation of "usual residence," among them college students, hospital and nursing home patients, and people in correctional facilities. In the difficult task of censustaking, group quarters data collection is surely one of the most challenging parts. Group quarters are hard to define, and collection of the data is highly likely to face both physical and legal impediments. Collection of data from individual residents may be flatly denied by prison wardens, college administrators, or facility administrators, citing safety or logistical concerns; the quality of data from these facilities then hinges on the quality of administrative or facility records. Legal impediments include concern over confidentiality, particularly in such settings as medical facilities or work camps where sensitivity to medical records privacy and citizenship status may loom large. We have discussed conceptual aspects of residence data collection from the nonhousehold population in Chapter 3; in this chapter, we describe operational improvements to nonhousehold enumeration and the research needed to attain them.

It is important to get group quarters enumeration right in the decennial census, not only because of the consequent problems of omission and duplication, but also because the decennial census is currently the only comprehensive data source on this segment of the population. As of this writing, most group quarters types have been included in the long-form replacement American Community Survey (ACS); however, it is unclear how accurate ACS-based data on group quarters will be. As we will discuss below, obtaining complete long-form data for group quarters residents in the census has been difficult in the past, and it remains to be seen how successful a long-form-only collection will be for group quarters facilities.

7–A IMPLEMENTATION PROBLEMS IN THE 2000 CENSUS

Our predecessor panels on the 2000 census and the 2010 census plan (National Research Council, 2004c:152–155; 2004b:150–152) reviewed the many implementation problems that plagued group quarters data collection in the 2000 census. In this section, we draw from and expand on their analyses in our own account of the group quarters process in 2000 (see also Jonas, 2003; Abramson, 2003).

Failure to Reconcile Group Quarters Roster with MAF The Census Bureau's inventory of group quarters was developed independently from the Master Address File (MAF). The two were developed by separate internal divisions of the Bureau, and were not cross-checked with each other prior to November 1999. An adjunct of the Local Update of Census Addresses Program was instituted to allow local and tribal governments to review group quarters listings, but the effectiveness of that program was severely compromised by an 18-month-late start. Since the program began just 4 months prior to Census Day, participation was low given competing demands on local resources for census outreach and other activities. The failure to adequately reconcile the group quarters roster with the MAF and with local authorities hurt the Bureau's ability to use an accurate, nonoverlapping, and comprehensive address list as the base of the 2000 census.

Poor Handling of Geographic Location The group quarters roster was also not well reconciled with the Census Bureau's Topologically Integrated Geographic Encoding and Referencing geographic database. For unknown reasons, a number of group quarters—including very long-established prisons and college dormitories—were given inaccurate geographic location codes. These inexplicable errors led to sizable local shifts in population, as group quarters facilities were inadvertently assigned to neighboring cities and counties. These group quarters discrepancies account for a large fraction of the population counts that local jurisdictions challenged in the Census Bureau's Count Question Resolution Program.[1]

Of these discrepancies, the highest-profile case arose when the Census Bureau acknowledged that 2,696 students at the University of North Car-

[1]Though the Count Question Resolution Program could result in the Census Bureau issuing an errata statement and a certificate of revised population, a condition of the program was that the revised counts could not be used for apportionment or redistricting.

olina at Chapel Hill—representing 1,583 dormitory rooms in 26 separate buildings—had been double-counted. These dormitory rooms were included in the U.S. Postal Service's Delivery Sequence File (a primary source of updates to the MAF) and subsequent operations provided conflicting information on whether the rooms were valid housing units, but not enough negative information for them to be deleted.[2] The admission was sensitive because North Carolina edged Utah (by 857 residents) for the 435th and final seat in the U.S. House of Representatives; Utah had already conducted two lawsuits challenging the census count.[3]

Though the North Carolina case was the highest profile, group quarters also accounted for other sizable discrepancies (U.S. Government Accountability Office, 2005), including:

- Dormitories at Morehead State University were incorrectly placed outside the city limits of Morehead, Kentucky, in the original 2000 census counts; the city's population was revised from 5,914 to 7,593. Similarly, misplacement of dormitories at the University of Wisconsin-Whitewater led to a population shift of 1,699 between Jefferson and Walworth Counties.
- The Crossroads Correctional Center in Cameron, Missouri—a 1,500-capacity prison completed in March 1997—was omitted from the city's population (as well as that of DeKalb County); correcting the error raised the city's total from 8,312 to 9,788.
- The Oxford Federal Correctional Institution is physically located in Grand Marsh, Wisconsin, in New Chester town of Adams County. However, the count at this prison was assigned to the town of Packwaukee in Marquette County; Packwaukee and New Chester towns do not have a common boundary.
- In a rare case of a complaint by local officials of an *overcount* in their area, officials in the town of Fredonia, Wisconsin, argued that the census count of 886 in one census block in the northeastern section of the town was not credible. The block in question was home to the 150-year-old dairy farm owned by the town chairman, a 6–7 house subdivision, a restaurant, and a campground—perhaps close to 52 people, as was the case in the 1990 census, but nowhere near 886. Ultimately, the Count Question Resolution Program certified the block's true population as

[2]The U.S. Government Accountability Office (2005:34) notes that—as part of the Count Question Resolution process in the North Carolina case—the Bureau concluded that "a similar issue was not problematic elsewhere in the country" on the strength of a search for the text string "dorm" in the address field of the decennial census MAF.

[3]Using the final population figures from the Count Question Resolution Program, North Carolina would retain the 435th seat, but by a margin of only 87 residents (U.S. Government Accountability Office, 2005).

43; the other 843 people were dormitory residents at Concordia University in the city of Mequon—in the same county, but tallied some 18 miles away from the proper location (Cole, 2001a,b).

Poor Levels of Full, Self-Report Data A higher-than-expected proportion of group quarters enumerations were obtained from administrative or facility records; the means by which group quarters questionnaires were completed in 2000 are described in Table 7-1. Of the 83 percent of enumerations for which enumerators indicated the source of data, 59 percent were filled out from administrative data, 30 percent were filled out by the resident, and 12 percent were filled out by an enumerator interviewing the resident. Types of group quarters with high percentages of enumerations obtained from administrative data included nursing homes, hospitals, group homes, and prisons. These types of group quarters had especially high rates of missing data for long-form items.

Lack of Coverage Measurement and Use of Reported "Usual Home Elsewhere" Addresses The Census Bureau opted not to include group quarters residents in the 2000 Accuracy and Coverage Evaluation program, principally because of low match rates in the 1990 Post-Enumeration Survey for group quarters residents in comparison with the rest of the household population. That is, the Bureau found that it was difficult to match group quarters residents interviewed in a follow-up survey (conducted a few months after the census) to their census records. This low match rate was attributed to high short-term mobility in the group quarters population (e.g., college students returning from their studies and shelter residents moving from one locale to another). Low match rates complicate the creation of population-adjusted census estimates using dual-systems estimation (the core methodology of the Accuracy and Coverage Evaluation Program), and so group quarters were dropped from consideration for the coverage evaluation program in 2000.

Because of the difficulty in matching follow-up survey records with samples from the census, dual-systems estimation is most likely not the best means of assessing group quarters coverage. However, in the absence of estimates of omissions and erroneous enumerations in group quarters settings, the 2000 census plan did not include any systematic or comprehensive review of the coverage in group quarters, whether through rigorous comparison with facility or administrative records or through structured observational studies.

Design choices in the Bureau's Non-ID Process—the procedure by which the Bureau processed all census forms without a MAF identification number, including "Be Counted" forms and all group quarters forms indicating a "usual home elsewhere"—also led to a major lost opportunity for understanding residence reporting problems. As shown in Box 3-1, the Individual

Table 7-1 Mode of Completion, Group Quarters Individual Census Reports, 2000 Census

Group Quarters Type	Percent				Number
	Self-Response	Administrative Records	Enumerator Interview	Unknown[a]	
Colleges and universities	57.5	30.2	5.5	6.7	2,028,150
Correctional institutions	15.3	56.3	4.4	24.0	1,930,233
Nursing homes	5.0	72.8	15.1	7.1	1,707,039
Service-based facilities and other group quarters[b]	25.3	41.3	23.3	10.1	669,702
Group homes	9.4	59.5	16.0	15.1	415,205
Hospitals	8.8	65.8	9.8	15.6	216,403
Juvenile institutions	23.8	48.8	9.9	17.5	122,291
Military facilities	36.9	37.6	5.7	19.7	279
Total	25.8	51.7	10.0	12.5	7,089,302

[a] Enumerators were supposed to enter a 1, 2, or 3 in a box on the back of group quarters forms to indicate that they were self-responses, enumerator interviews, or administrative record entries, respectively. This "unknown" category represents those forms on which the box was left blank or an invalid entry was made.

[b] Service-based facilities include shelters and soup kitchens; "other group quarters" include work dormitories, religious group quarters, and crews of maritime vessels.

SOURCE: Jonas (2003:Table 6.1b).

Census Reports (ICRs) to be filled in by group quarters respondents asked for address information on the place where they live most of the time, regardless of whether the Bureau considered their group quarters type eligible for "usual home elsewhere" consideration.

In principle, these group quarters questionnaires were supposed to be twice-filtered before Non-ID Process clerks geocoded the records and attempted to find matches on the MAF—first to determine if the Bureau considered the particular group quarters type in question as eligible for a "usual home elsewhere" and second on the basis of the response to a screening question on the form ("Do you have a place where you live or stay MOST OF THE TIME?"). In practice, though, the filtering by the screening question was not performed at all. More significantly, the filtering by group quarters type was only done *after* the reported address data had been read and matched to the MAF by census clerks (though before field verification would have been used to resolve the true location). Hence, as shown in Table 7-2, only 659,000 out of 2.9 million group quarters forms *should* have gone through the Non-ID Process; however, all 2.9 million went through the first clerical geocoding stage. Only then, when group quarters type was considered, were almost 2 million of them dropped from the process (and tabulated at the group quarters locations). Another 389,000 should have been excluded by the screening questions, but were not.

From the standpoint of conducting the operation as planned, the failure to filter the group quarters questionnaires before Non-ID geocoding and matching was a significant mistake and a burden; the workload of the operation was multiplied. But from the perspective of research and evaluation, a geocoded set of claimed "usual home elsewhere" responses—which could be compared with the actual group quarters locations—would have been a trove of information on the nature of potential census duplicates and could have identified particular problem areas. That this type of file was (albeit inadvertently) constructed, then discarded when the filter was applied late, was a highly regrettable lost research opportunity.

Ineffective Processing of Group Quarters Questionnaires Instead of a bar code tracking system for residents of particular group quarters sites, the Bureau relied on a total count of questionnaires from the group quarters site logged onto a control sheet. These control sheets were easily separated from their questionnaires: a special review of approximately 700,000 questionnaires had to be initiated in May 2000 when they became separated from their control sheets and hence could not be identified with their proper location. Worse, a small but undetermined number of group quarters questionnaires were never returned to the appropriate local census office and never included in the census.

Table 7-2 Group Quarters Questionnaire Records in the Non-ID Process by Form Type, 2000 Census

Record Type	ICQs	ICRs	MCRs	SCRs	Total
GQ records entering the Non-ID Process	8,551	2,232,674	630,252	69,801	2,941,278
Records filtered out due to invalid GQ type	222	1,862,295	30,225	0	1,892,742
Records remaining in Non-ID Process	8,329	370,379	600,027	69,801	1,048,536
Records that would have been filtered based on screening questions	1,432	345,524	39,251	2,763	388,970
Records that actually belonged in Non-ID Process, as planned	6,897	24,855	560,776	67,038	659,566

NOTES: Individual Census Questionnaires (ICQs) were used only for enumeration at soup kitchens and mobile food vans, and were intended to be filled out by enumerators (either through an interview or by observation). Individual Census Report (ICRs) are the basic group quarters data collection form, along with Military Census Reports (MCRs) and Shipboard Census Reports (SCRs); see Box 3-1.

SOURCE: Census Bureau tabulation of Non-ID File; Jonas (2003:Table 8.2a).

High Levels of Imputation In July 2000 two special telephone operations were implemented to follow up group quarters with no recorded population (presumably refusals) and group quarters for which the reported resident count on census questionnaires fell far below the approximate population count (obtained in advance visits to special places conducted in February–March 2000). Population counts were obtained for these group quarters, and the results used to impute group quarters residents. More than 200,000 group quarters records (almost 3 percent of the group quarters population) were wholly imputed as a consequence of the telephone follow-up and the reconciliation of multiple population counts on the group quarters control sheets.

Failure to Unduplicate Within the Group Quarters Population Some group quarters residents were mailed a housing unit questionnaire. If they returned it and the address was matched to a group quarters address, they were added to the appropriate group quarters count, but there was no provision to unduplicate such enumerations with enumerations obtained through the group quarters enumeration procedure. From a clerical review of a sample of cases in selected types of group quarters (excluding prisons, military bases, and service-based facilities such as soup kitchens), an estimated 56,000 group quarters enumerations were duplicates of other enumerations within the same group quarters.

Our predecessor panel (National Research Council, 2004c:155) concluded that the 2000 census "procedures for enumerating group quarters residents and processing the information collected from them were not well controlled or carefully executed" and resulted in poor data quality. The National Research Council (2004b) concurred that group quarters enumeration needs complete revision for the 2010 census, as do we.

> **Finding 7.1:** As implemented in the 2000 and recent censuses, group quarters enumeration is unacceptably bad. Failure to reconcile the group quarters roster with the MAF contributed to a host of census errors. Group quarters frames were constructed without sufficient standardization and awareness of diversity in housing unit and group quarters stock, and data from the 2000 census long-form sample were particularly marred by extremely high levels of item nonresponse.

It should be reiterated that group quarters represent only a very small portion of the overall population, but that the decennial census remains the sole source of comprehensive data on the size of all segments of this diverse group (which includes some especially policy-relevant subgroups). The problematic manner in which group quarters data were collected in 2000 raises the question of whether the effort is given the right allocation of resources in the Bureau's census plans.

7–B RETHINKING THE CONCEPT

The Census Bureau has taken some positive steps with regard to group quarters enumeration, the most significant being an internal effort to revise and clarify its definitions of group quarters. These revised definitions were crafted for use in the 2006 census test and the 2008 dress rehearsal, and are described in Box 7-1.

The redefinition effort reflects serious work to provide more meaningful categories and to be more consistent with terms used by practitioners. When the panel reviewed the definitions, they were still incomplete, with some major categories—among them military and seaborne quarters—yet to be fully developed. The new definitions were not ready for full testing in the 2004 field test, but were expected to be given a full airing in the 2006 census test and the 2008 dress rehearsal. We encourage the redefinition efforts and look forward to continued refinement. We concur with Drabek (2005:2) that it is particularly useful that the Bureau continue to work with experienced staff of all types of group quarters facilities in order to have a working vocabulary of terms that are relevant to communities and to data users. Particularly in elderly and health care, the Census Bureau needs to make sure that its definitions are based on the services that the facilities are licensed to provide, rather than rely on the name of an institution to provide a categorization.

Though we support the group quarters redefinition effort in the short term, we strongly encourage the Census Bureau to broaden its focus. The lines between what the Bureau has traditionally walled off as "group quarters" as distinct from the "household" population are becoming increasingly blurred, and it is useful for the Bureau to reexamine the very concept of what it means to be a "group quarters."

> **Finding 7.2:** There is sufficient diversity in what the Census Bureau has treated as the "group quarters" population that the term "group quarters" no longer makes conceptual sense. Its compartmentalization as a separate list and a separate operation—trying to force this entire segment of the population to respond to the census using a single form—is fundamentally flawed.

To a very limited degree, two recent censuses attempted (or at least signaled an intent) to provide a more substantive split in the usual "group quarters" population. In 1980, enumerators assigned to group quarters were instructed to transcribe information from ICRs onto census short or long forms as appropriate, using "CENSUS USE ONLY" boxes on the form to indicate

Box 7-1 2006 Census Test Group Quarters Definitions

Draft definitions provided by the Census Bureau as of March 2005. The list of definitions is incomplete; military quarters, domestic violence shelters, and crews of maritime vessels are among the group quarters types left to be specified before the 2008 dress rehearsal.

Group Quarters

A group quarters is a place where people live or stay that is normally owned or managed by an entity or organization providing housing and/or services for the residents. These services may include custodial or medical care as well as other types of assistance, and residency is commonly restricted to those receiving these services. People living in group quarters are usually not related to each other.

Group quarters include such places as college residence halls, residential treatment centers, skilled nursing facilities, group homes, military barracks, correctional facilities, workers' dormitories, and facilities for people experiencing homelessness.

Correctional Facilities for Adults

1. *Correctional Residential Facilities:* These are community-based facilities operated for correctional purposes. The facility residents may be allowed extensive contact with the community, such as for employment or attending school, but are obligated to occupy the premises at night. Examples are halfway houses, restitution centers, and prerelease, work release, and study centers.
2. *Federal Detention Centers:* Standalone, generally multilevel, federally operated correctional facilities that provide "short-term" confinement or custody of adults pending adjudication or sentencing. These facilities may hold pretrial detainees, holdovers, sentenced offenders, and Immigration and Customs Enforcement (ICE) inmates, formerly called Immigration and Naturalization Service inmates. These facilities include: Metropolitan Correctional Centers, Metropolitan Detention Centers, Federal Detention Centers, Bureau of Indian Affairs Detention Centers, ICE Service Processing Centers, and ICE contract detention facilities.
3. *Federal and State Prisons:* Adult correctional facilities where people convicted of crimes serve their sentences. Common names include: prison, penitentiary, correctional institution, federal or state correctional facility, and conservation camp. The prisons are classified by two types of control: (1) "federal" (operated by or for the Bureau of Prisons of the Department of Justice) and (2) "state." Residents who are forensic patients or criminally insane are classified on the basis of where they resided at the time of enumeration. Patients in hospitals (units, wings, or floors) operated by or for federal or state correctional authorities are counted in the prison population. Other forensic patients will be enumerated in psychiatric hospital units and floors for long-term nonacute patients. This category may include privately operated correctional facilities.
4. *Local Jails and Other Municipal Confinement Facilities:* Correctional facilities operated by or for counties, cities, and American Indian and Alaska Native tribal governments. These facilities hold adults detained pending adjudication and/or people committed after adjudication. This category also includes work farms and camps used to hold people awaiting trial or serving time on relatively short sentences. Residents who are forensic patients or criminally insane are classified on the basis of where they resided at the time of enumeration. Patients in hospitals (units, wings, or floors) operated by or for local correctional authorities are counted in the jail population. Other forensic patients will be enumerated in

Box 7-1 (continued)

psychiatric hospital units and floors for long-term nonacute patients. This category may include privately operated correctional facilities.

Group Homes and Residential Treatment Centers for Adults

5. *Group Homes Intended for Adults:* Group homes are community-based group living arrangements in residential settings usually consisting of three or more clients of a service provider. The group home provides room and board and supportive services, such as assistance with daily living skills, and social, psychological, or behavioral programs. Clients are generally not related to the caregiver or to each other. Group homes do not include residential treatment centers or facilities operated by or for correctional authorities.
6. *Residential Treatment Centers for Adults:* Residential facilities that provide treatment on-site in a highly structured live-in environment for the treatment of drug/alcohol abuse, mental illness, and emotional/behavioral disorders. They are staffed 24 hours a day. The focus of a residential treatment center is on the treatment program. Residential treatment centers do not include facilities operated by or for correctional authorities.

Juvenile Facilities

7. *Correctional Facilities Intended for Juveniles:* Includes specialized facilities that provide strict confinement for its residents and detain juveniles awaiting adjudication, commitment or placement, and/or those being held for diagnosis or classification. Also included are correctional facilities where residents are permitted contact with the community, for purposes such as attending school or holding a job. Examples are residential training schools and farms, reception and diagnostic centers, group homes operated by or for correctional authorities, detention centers, and boot camps for juvenile delinquents.
8. *Group Homes for Juveniles:* Includes community-based group living arrangements for youth in residential settings usually consisting of three or more clients of a service provider. The group home provides room and board and supportive services, such as assistance with daily living skills, and social, psychological, or behavioral programs. Clients are generally not related to the caregiver or to each other. Examples are maternity homes for unwed mothers, orphanages, and homes for abused and neglected children in need of services. Group homes for juveniles do not include residential treatment centers for juveniles or group homes operated by or for correctional authorities.
9. *Residential Treatment Centers for Juveniles:* Includes facilities that primarily serve youth that provide services on-site in a highly structured live-in environment for the treatment of drug/alcohol abuse, mental illness, and emotional/behavioral disorders. These facilities are staffed 24 hours a day. The focus of a residential treatment center is on the treatment program.

Residential School-Related Facilities

10. *College/University Housing:* College/university housing includes residence halls and dormitories owned, leased, or managed by a college, university, or seminary. Fraternity and sorority housing identified by the college or university are included as college housing. Students attending the U.S. Naval Academy, the U.S. Military Academy (West Point), the U.S. Coast Guard Academy, and the U.S. Air Force Academy are counted in military group quarters.

Box 7-1 (continued)

11. *Residential Schools for People with Disabilities:* Includes schools that provide education programs and care for students with disabilities on-site in a live-in environment. The focus of a residential school for people with disabilities is on the education programs. Examples are residential schools for the blind, for the deaf, and for the developmentally disabled.

Health Care Facilities

12. *Hospitals with Patients Who Have No Usual Home Elsewhere:* Includes hospitals if they have any patients who have no exit or disposition plan, or who are known as "boarder patients" or "boarder babies." All hospitals are eligible for inclusion in this category except psychiatric hospitals, units, wings, or floors operated by federal, state, or local correctional authorities. Patients in hospitals operated by these correctional authorities will be counted in the prison or jail population. Psychiatric units and hospice units in hospitals are also excluded. Only patients with no usual home elsewhere are enumerated in this category.
13. *In-Patient Hospice Facilities:* Includes in-patient hospice facilities (both free-standing and units in hospitals) that provide palliative, comfort, and supportive care for the terminally ill patient and their families. All patients are enumerated in this category.
14. *Mental (Psychiatric) Hospitals and Psychiatric Units in Other Hospitals:* Includes psychiatric hospitals, units and floors for long-term, nonacute care patients. The primary function of the hospital, unit, or floor is to provide diagnostic and treatment services for long-term nonacute patients who have psychiatric-related illness. All patients are enumerated in this category.
15. *Nursing Facilities/Skilled-Nursing Facilities:* Includes facilities licensed to provide medical care with 7-day, 24-hour coverage for people requiring long-term nonacute care. People in these facilities require nursing care, regardless of age. Either of these types of facilities may be referred to as nursing homes.

Service-Based Enumeration Facilities

16. *Emergency and Transitional Shelters (with Sleeping Facilities) for People Experiencing Homelessness:* Facilities where people experiencing homelessness stay overnight. These include: (1) shelters that operate on a first-come, first-serve basis where people must leave in the morning and have no guaranteed bed for the next night; (2) shelters where people know that they have a bed for a specified period of time (even if they leave the building every day); and (3) shelters that provide temporary shelter during extremely cold weather (such as churches). This category does not include shelters that operate only in the event of a natural disaster. Examples are emergency and transitional shelters; missions; hotels and motels used to shelter people experiencing homelessness; shelters for children who are runaways, neglected, or experiencing homelessness; and similar places known to have people experiencing homelessness.
17. *Soup Kitchens, Regularly Scheduled Mobile Food Vans, and Targeted Nonsheltered Outdoor Locations:* Includes soup kitchens that offer meals organized as food service lines or bag or box lunches; street locations where mobile food vans regularly stop to provide food to people experiencing homelessness; and targeted nonsheltered outdoor locations where people experiencing homelessness live without paying to stay. Targeted nonsheltered outdoor locations must have a specific location description; for example, "the Brooklyn Bridge at the corner of Bristol Drive" or "the 700 block of Taylor Street behind the old warehouse."

Box 7-1 (continued)

Other Facilities

18. *Living Quarters for Victims of Natural Disasters:* These are temporary group living arrangements established as a result of natural disasters.
19. *Religious Group Quarters:* These are living quarters owned or operated by religious organizations that are intended to house their members in a group living situation. This category includes such places as convents, monasteries, and abbeys. Living quarters for students living or staying in seminaries are classified as college student housing not religious group quarters.
20. *Workers' Group Living Quarters and Job Corps Centers:* Includes facilities such as dormitories, bunkhouses, and similar types of group living arrangements for agricultural and nonagricultural workers. This category also includes facilities that provide a full-time, year-round residential program offering a vocational training and employment program that helps young people 16 to 24 years old learn a trade, earn a high school diploma or GED, and get help finding a job. Examples are energy enclaves in Alaska, migratory farmworker camps on farms, construction workers' camps, and Job Corps centers and vocational training facilities.

that it was a group quarters and whether or not the person was an inmate of an institution. "Inmates" were defined as "persons under care or custody in institutions at the time of the enumeration, regardless of their length of stay or the number of persons in the institution," and institutions were taken to include homes for the handicapped, nursing homes, orphanages, and correctional institutions. The accuracy of returns from this process is perhaps questionable given that it required another task for enumerators to complete (repeatedly). However, the general concept of a split between the institutionalized and noninstitutionalized populations—as is used in some foreign censuses as well as in the U.S. Current Population Survey—is worthy of consideration, and may be more meaningful than the current group quarters distinction.

Likewise, the initial residence rules for the 1990 census indicate that the Bureau envisioned a different split (Jones, 1987:18):

> [The Bureau planned] to differentiate between long-term and short-term facilities. We will count persons residing in long-term facilities (for example, penitentiaries [and] mental hospitals) as residents of the institution. The Census Bureau will count persons residing in short-term facilities (for example, county jails and general hospitals) at their usual place of residence if they have one.

This ultimately did not happen, and the people were counted at the facility location; moreover, as expressed several times in Chapter 3, this type of blunt categorization of whole *facilities* by term length is problematic given the variation of length of stay in places like local jails. However, the concept of treating people in nonhousehold settings on the basis of their length of stay is consistent with our principle of making residence determinations at the level of the individual, and an enumeration based on a length-of-stay distinction may be more relevant for policy needs.

7–C ALLOW "ANY RESIDENCE ELSEWHERE"

We strongly support a system in which all group quarters and nonhousehold residents are approached and enumerated in the same manner as the general household population, to the greatest extent practicable. Accordingly, Recommendation 6.2 applies to all census respondents: the panel recommends that "any residence elsewhere" (ARE) address information be collected for all group quarters and nonhousehold residents, just as we advocate its collection in the main household census form.

As with the main household population, the physical collection of ARE data from nonhousehold respondents in 2010 is certainly feasible; indeed, "usual home elsewhere" (UHE) was asked on all group quarters ICRs in 2000, but only deemed valid for specific group quarters types. The Census Bureau's failure to analyze the UHE data it collected on group quarters forms in 2000, particularly having progressed through the geocoding of reported addresses, looms large as a lost opportunity. That failure is a principal reason that we encourage the analysis of ARE data as a large-scale experiment in the 2010 census (and a future direction for 2020), rather than as part of the 2010 count. Unlike the 2000 experience, ARE information gathered in 2010 should be collected from all group quarters residents, and those data should be analyzed and evaluated extensively, to establish—for example—the degree to which college students' reported address information matches (or fails to match) reports from their parental households.

7–D CONDUCTING THE COUNT

Consistent with our support for treating household and nonhousehold populations alike, the direct enumeration of people in nonhousehold settings—questionnaires distributed to and filled out by respondents or administered by enumerator interview—is preferable to other means of data collection. But two important caveats are in order: they concern the use of facility and administrative records and the need for different forms for different purposes.

7–D.1 Facility and Administrative Records

The first of these arises from the sobering, underlying message of Table 7-1. The 2000 census mounted a vigorous and highly visible partnership with community organizations; even with these strong efforts to boost awareness of and cooperation with the census, only about one-half of census records for the nonhousehold population were obtained from direct enumeration. Instead, the dominant source of reported data (among those questionnaires where enumerators coded the source) was reference to facility and administrative records.

Though direct enumeration is absolutely preferable to other means of other data collection, it is likewise absolutely unrealistic to assume that questionnaire delivery or enumerator access to all parts of the nonhousehold universe will be granted. Instead, participation will vary from place to place. It is admittedly anecdotal evidence, but an observation report filed by a Census Bureau employee (Jones, 2000) during the conduct of the 2000 census usefully highlights these varying levels of participation, commenting on the group quarters enumeration at three Seattle-area universities. One university participated fully, providing comprehensive lists of university housing (including fraternity and sorority houses and student apartments) and directing its own staff to complete the basic roster of all persons living in those units.[4] That university also permitted enumerators (with student escorts) to knock on dormitory room doors to collect questionnaires and to set up tables in common areas. At the other extreme, another university cited privacy concerns: the university blocked access to its housing records and denied enumerators permission to enter dormitories, though it did distribute ICRs to current students. The third university fell in between these positions: it permitted access to housing records but denied access to the dorms, instead only permitting enumerators to sit outside campus cafeterias at lunchtime.

The Census Bureau needs to try to achieve direct enumeration whenever and wherever possible—in prisons, colleges, and hospitals alike. To do so, it should continue to aggressively pursue partnerships with relevant authorities and to streamline processes by which institution staff may be sworn in as census agents for purposes of administering questionnaires. However, the Bureau also needs to confront the reality that it will have to rely on facility or administrative data in many (if not most) nonhousehold settings.

Albeit on a much smaller scale than the group quarters portion of the full decennial census, both of these models are not foreign to the Census Bureau's current repertoire of surveys. For example, the biennial Census of Juveniles in Residential Placement (CJRP), conducted on behalf of the Office of Juvenile Justice and Delinquency Prevention (discussed in Section 3–D.3), is an example of a data collection effort aimed at collecting administrative data from facilities. The questionnaire is oriented around a roster and provides a mechanism by which facilities can provide data to the Census Bureau in electronic

[4]The enumeration of the student apartments hit a snag, though, in that the university-completed roster included only current students. The apartment complexes also housed spouses and children of current students as well as nonstudent roommates.

(spreadsheet) form if that is more convenient for them.[5] The specific implementation of the CJRP instrument is not necessarily something we endorse—indeed, the list of "include" and "exclude" instructions on the instrument is particularly elaborate and potentially confusing, but the general approach of tailoring the collection method to the anticipated means by which facilities can respond is a useful one.

Because administrative and facility records have been, and will almost certainly continue to be, a major source of data on the nonhousehold population, it is imperative that the Census Bureau undertake a continuing research effort to assess the accessibility of facility records at group quarters facilities and to determine whether the existing data systems meet census data collection needs.

In a sense, what we envision is analogous to the development of the Census Bureau's MAF. Following the 1990 census, the Bureau elected to take the 1990 address list as a base and to continue to update and edit it over subsequent years, rather than follow past practice and rebuild the list from scratch 10 years later. The 1990 and 2000 censuses relied on pre-census facility visit operations to establish contact with group quarters sites and generate preliminary population estimates for workload planning purposes. The research effort we envision would maintain facility listings as a continuous resource, much like the MAF. Through queries to the facilities, it should be determined whether the facility's records can provide the data of interest—the short-form census items like name and race, plus whether alternate address information (ARE) is known for each person in the facility and whether stays can be characterized as short or long term.

7–D.2 Different Forms for Different Settings

The second caveat to our general preference for direct enumeration whenever possible is that too much homogeneity in the style and substance of the questionnaire used for the nonhousehold population may be harmful. For instance, soldiers and sailors have a different vocabulary related to the nature of their residence and their length of stay at a location than do college students, who in turn probably respond differently to questions and probes than people incarcerated in a county jail. In 2000, the Military and Shipboard Census Reports used to collect information from soldiers and sailors differed slightly (with terminological differences) from the ICRs used for the rest of the nonhousehold population. Though it is inadvisable to push this guidance to extremes and try to develop a form for every population type—just as it is self-defeating to attempt to delineate a specific residence rule for ev-

[5]The questionnaire is available at http://ojjdp.ncjrs.org/ojstatbb/cjrp/asp/methods.asp [6/1/06].

ery group—the Census Bureau should consider further opportunities to tailor forms to best reach large segments of the nonhousehold population.

> *Recommendation 7.1:* **The Census Bureau should produce a small number of alternative census forms that collect a common core of information for different types of residence settings, such as those that are known to have long lengths of stay rather than short term stays. The Census Bureau should also develop a spreadsheet-type ledger form that reflects the reality that some "responses" will have to be obtained from facility administrative records or a central "gatekeeper."**

These custom nonhousehold forms should include ARE queries and other relevant data items, such as time spent at the location and expected length of stay.

The precise tailoring of census forms for the nonhousehold population is arguably more of an issue for the ACS than the short-form-only 2010 census. Subjecting group quarters persons to the full battery of long-form items leads to such incongruities as asking prisoners about their employment status last year or long-term-care residents about their commute to work. We discuss other aspects of the ACS in Section 8–C.

7–E COUNTING PRISONERS IN THE CENSUS

As we describe in Section 3–D, a particular issue involving the nonhousehold population that has drawn considerable attention in the buildup to the 2010 census is whether prisoners should be counted at the prison location or at some other place. A provision in the Census Bureau's 2006 appropriations "direct[ed] the Bureau to undertake a study on using prisoner's permanent homes of record, as opposed to their incarceration sites, when determining their residences." The Bureau was obligated to report back to the congressional appropriations committees within 90 days, and it did so in a report on February 21, 2006 (U.S. Census Bureau, 2006b). Prior to the release of the Bureau's report, two reports from proponents of a change in the Bureau's stance on tabulating prisoners were transmitted to the Bureau (Wagner et al., 2006; Levingston and Muller, 2006).

The debate over counting prisoners has drawn the attention of editorial pages, most prominently the *New York Times*. In September 2005, an editorial described the Census Bureau's practice of "[counting] incarcerated people as 'residents' of the prisons where most are held for only a short time, instead of counting them in the towns and cities where they actually live" as "a longstanding quirk" and "a troubling flaw." "Instead of waiting until the next census in 2010," the *Times* concluded, "the Census Bureau should simply change its procedures now" (*New York Times*, 2005). In April 2006, the

Times returned to the topic, contending that the Bureau's report to the appropriations committee was "stonewalling," "an obtuse and evasive report that supports the bad old status quo." "The report sets up a straw man by suggesting that the desired change might require the costly and invasive procedure of interviewing every inmate." However, the *Times* contends, "all that is really necessary" to implement "the common-sense idea of counting inmates at their home rather than at prison" is "to treat inmates like everyone else. That means giving them questionnaires that ask, among other things, for their home addresses and interviewing them only when a form is not returned or when some other problem occurs" (*New York Times*, 2006).

Others perceive the matter differently. The town of Jackson, Michigan, and the surrounding Blackman Township is home to the Southern Michigan Correctional Center as well as several other state correctional prisons (most on the site of the former State Prison of Southern Michigan). Thus, about 32 percent of the township's total population of 22,500 are listed as being in institutionalized group quarters. "As we peruse Blackman [Township]'s census data," editorialized the *Jackson Citizen Patriot* (2006), "we can't be sure any of those results have not been skewed in one way or another by the inmate populace." One possible approach for "the Census Bureau is to count inmates, not where they are incarcerated, but where they lived before." However, the paper urged, "we do not favor that approach because the inmates are, well, not living there and not using the services offered by their hometowns"—not using the police, fire, and other municipal services that might be required of a prison-hosting community.

That the problem of counting prisoners is complex is undeniable. Including thousands of people in legislative districts even though they are legally prohibited from voting can create distortions in representation, and both the urban areas from which many prisoners originate and the rural areas which house prisoners can make strident (and valid) arguments for state and federal funds to compensate for current and future services.

> **Finding 7.3:** Major growth in the prison population, accompanied by expansion in the number of correctional facilities maintained by the federal government and the states, has prompted challenges to the Census Bureau's "usual residence" standard regarding the counting of the incarcerated population.

Our guidance on how the prisoner population should be counted in the census is consistent with both our recommendation of a core set of residence principles and our proposal to broadly change group quarters enumeration. Under the panel's recommended principles for determining residence (see Chapter 6), federal and state prisoners would be counted at the prison location because that location is the place where the prisoner lives and sleeps more than any other place. This is consistent with current Census Bureau practice. We do not rule out a principle being added to our suggested listing to count prisoners at a location other than a prison; however, the information necessary for such a decision does not now exist.

We differ from proponents of an immediate change in Census Bureau policy on counting prisoners because our focus is broader: our desired objective is the collection of secondary ("any residence elsewhere") information for *all* census respondents. This broader scope leads us to recommend a major experiment as part of the 2010 census as the research base, rather than trying to force massive changes in the design and operations of the 2008 dress rehearsal. Accordingly, and consistent with our other recommendations on changes to nonhousehold enumeration, we recommend:

> *Recommendation 7.2:* **A research and testing program, including experimentation as part of the 2010 census, should be initiated by the Census Bureau to evaluate the feasibility and cost of assigning incarcerated and institutionalized individuals, who have another address, to the other location.**

Some elaborations on our approach, and on the information needed to inform a change in the principle applying to the counting of prisoners, are in order.

Our principles hold that determination of usual residence should be made at the level of the individual; this would mean that persons in prison need not have their residency fixed solely by virtue of their location in a structure identified as a prison. Using the panel's recommended question-based approach and revised nonhousehold enumeration operations, the census could obtain individual-level information on time spent in prison and expected date of release. The Bureau would then have the information needed to make individual-level judgments on the most appropriate counting location.

Wagner and Lotke (2004), Wagner et al. (2006), and Levingston and Muller (2006) all favor direct enumeration—individual questionnaire distribution and interviewing—as a preferred means for gathering information from prisoners. We agree, as we stated above. Wagner et al. (2006) cite press accounts of two notable success stories—direct enumeration, even in "supermax" security areas, of Folsom State Prison (California) and the Utah State Prison.[6] However, particularly in the case of prisons, complete direct enumeration of inmates is an ideal but almost certainly unattainable goal. In 2000, less than 20 percent of

[6] A "supermax" facility or unit "segregates inmates, restricts their movement, and limits their direct access to other inmates or prison staff." As of 1997, 57 pure-"supermax" prisons were open in the United States, 16 of them in Texas alone (Cullen and Sundt, 2000:498). In the enumeration of a supermax unit in Utah, a prison official commented that "rather than pass forms back and forth . . . we had [census takers] ask them the questions through the cell" (Wagner et al., 2006:25).

the population in correctional facilities was enumerated through self-response to a questionnaire or through enumerator interview (see Table 7-1). Even with more prominent efforts to facilitate direct interviews in prisons, it is realistic to assume that a major share of the prison population in 2010 will have to be counted through reference to administrative and prison records.[7] Hence, any prospect for counting prisoners at locations other than the presence depends vitally on the completeness, consistency, and accessibility of records maintained by individual prisons or by state and federal departments of corrections. The quality of these data resources is not well known, and this creates the critical research need—determining how well corrections department data sources match the information that can be gathered on a Census Bureau questionnaire.

The necessity for reliance on administrative and facility records accentuates a critical flaw in the argument for an immediate change in prisoner counting policy: the alternative to counting prisoners at a place other than prison is not well defined. Both *New York Times* editorials noted above give the desired counting location as (an unqualified) "home" (*New York Times*, 2005, 2006), a word that has numerous interpretations as we have described throughout this report. Levingston and Muller (2006) refer to state corrections departments' coding of "home of record," while the 2006 appropriations language mandating a Census Bureau study directed them to look at "permanent homes of record." Other words abound: an expert cited by Levingston and Muller (2006) considers how prisoners can be mapped to a "return address" (a street address to which they will return after release); the September 2005 *New York Times* editorial briefly mentions counting prisoners at their "preprison residences" (*New York Times*, 2005); and Rep. Jose Serrano (D-New York) fused several of these concepts in arguing that prisoners should be counted at "their last known permanent residenc[e] where they are most likely to return to upon release" (*East Oregonian* and Associated Press, 2006). Finally, Wagner et al. (2006) characterize the alternative as counting prisoners "where they live, not where they are temporarily confined." These are more than mere semantic differences, in that they blend elements of past residential history ("permanent," or the last-known "preprison residence"), prospective intent that is perhaps years in the future ("return"), and enduring ties to family or community (generally, "home" and "where they live"). A clear idea of what residence information is or is not coded in administrative records is essential to defining questions to be used in direct interviews when those are possible.

In addition to the basic meaning of address record information that is accessible through facility records, it is also necessary to consider what level of geographic resolution is available. The interviews conducted by Levingston

[7] An earlier Brennan Center for Justice (Allard et al., 2004:7) recommended adopting a "uniform prison enumeration process" rather than mixing self-enumeration and administrative fill-in.

and Muller (2006:9, 11) suggest different geographic resolutions in the prison files: some may yield street addresses (and individual housing units), but other records might be limited to the "sentencing district" or the "county of conviction." An analysis described in Levingston and Muller (2006:9) compared on-file residence locations with the addresses to which prisoners moved on parole; this work suggested agreement at the level of the "neighborhood."

Though an "enduring ties" argument is frequently invoked to argue for changes to the Bureau's prisoner counting policy, the strength of those ties merits empirical assessment. For example: Has the property (to which a prisoner is connected) changed ownership? Do respondents at the address have any contact with or relation to the prisoner? With what frequency would assigning prisoners to specific addresses add to the "household" count people whose very crimes have severed any "enduring ties" that may have existed (e.g., domestic or child abuse)? Levingston and Muller (2006) conjecture that about 20 percent of the prison population would not have addresses that are "meaningful, accurate, available, and verifiable," and that procedures for counting the other 80 percent should not be unduly impeded by these "exceptional" populations.[8] Assessment of these ideas requires gathering actual data such as through an experiment in 2010, as we recommend.

The base residence information contained in state corrections department databases—and even the format of the data—is not well known. The last comprehensive survey of the information systems maintained by state corrections departments was conducted by the Bureau of Justice Statistics in 1998 (Bureau of Justice Statistics, 1998). More recently, Wagner and Lotke (2004) estimate that about 30 states maintain some form of electronic record of home address information. Collection of these data is likely done "by asking prisoners to self-report a last address during intake." However, it is not clear exactly how that address is defined (e.g., last known address, specified family address, etc.) and queried in individual states' questionnaires or how accurate the address data may be. Wagner et al. (2006:16) add that "often the addresses [on file] are not updated until the time of release." Among those states that do maintain address data, the accuracy and geographic resolution is unclear:

- Clement and Keough (2004:10) analyzed Rhode Island Department of Corrections data to study patterns created by the state's laws that deny convicted felons the right to vote (even when they are on parole, rather than in prison). The Rhode Island data include "last known self-reported address." Excluding cases that were not Rhode Island residents, 4.4 percent of these self-reported addresses (688 records) were

[8]The four "exceptional" populations specifically referenced by Levingston and Muller (2006:14) are someone whose family has moved to another address following his or her incarceration; someone who will never leave prison (e.g., serving a life sentence); someone who had no address at the time of incarceration; and a noncitizen whose official home is in another country.

unable to be geocoded to the *municipality* level, and an additional 7.4 percent were unable to be associated with specific tracts or neighborhoods in urban Providence.

- Wagner et al. (2006:16) observe that the North Carolina Department of Corrections "appears to have a virtually complete dataset of home residences for its population." However—at least in the publicly accessible data—the information seems to be limited to county of conviction.

For the remaining 20 states, it is not clear what information would be available from administrative records save perhaps for court records from time of sentencing, which would suffer as a measure of the "current" prison population.

The evidence of political inequities in redistricting that can arise due to the counting of prisoners at the prison location is compelling. Short of counting prisoners at some location other than the prison—for which there is currently insufficient information as well as the lack of any principled way to do so—a partial remedy might be to provide tract- or block-level counts of prisoner populations as part of the Bureau's data products for redistricting, a compromise position noted by both Wagner et al. (2006) and Levingston and Muller (2006).

In spirit, this approach follows that already taken by the state of Kansas, which disagrees with the Census Bureau's default placement of college students and military personnel (see Box 7-2). Similar to the Kansas model, at least two recent state legislative initiatives mandating "adjustment" similar to the Kansas model—but including prisoners as a target group—have recently been mounted. The "Prisoner Census Adjustment Act" has been introduced in both the 2004 and 2005 sessions of the Illinois state legislature. The act would require that "each State and local governmental entity in this State that operates a facility for the incarceration of persons convicted of a criminal offense, including a mental health institution for those persons, or that places any person convicted of a criminal offense in a private facility to be incarcerated on behalf of the governmental entity," transmit data on prisoners already counted in the federal decennial census to the Illinois secretary of state. These data include prisoner name, age, gender, race, and "the last address at which the person resided before the person's current incarceration." Federal prisons in Illinois would be requested, though not obligated, to provide the same data. The secretary of state would then be responsible for producing tract-level data "adjust[ing] all relevant population counts reported in the census, including populations by age, gender, and race, as if the person resided at [the preincarceration] address on the day for which the census reports population" and eliminating them from the facility count. These adjusted data would then be the official series to be used for legislative redistricting. Identical legislative initiatives have been proposed in the New York state assembly, and they were proposed (but not enacted) in Texas in 2001.

Box 7-2 Kansas Census Adjustment

Article 10, Section 1 of the Kansas state constitution dictates that state senatorial and representative districts be apportioned using population totals that count "military personnel stationed within the state" and "students attending college and universities within the state" at their permanent residence. Military personnel and college students currently stationed in the state who are not Kansas residents are to be excluded from the counts. [Enabling legislation would later define "permanent residence" as "a fixed place of abode or fixed domicile which a person intends to be such person's residence and to which such person presently intends to return" and "resident" as "a person who declares that he or she is a resident of the state of Kansas and has a present intent to remain in the state."]

Added to the constitution through a popular referendum in 1988, the constitutional provision changed the apportionment date from the seventh year to the second year of each decade (e.g., 1992, 2002, and so forth), and indicated that the apportionment is to be done by modifying figures from the U.S. decennial census. Kansas' secretary of state reported the adjusted 2000 census totals to the state legislature in July 2001 (Thornburgh, 2001).

Respondents were advised that "your answers should reflect your residence as of census day: **April 1, 2000**." Respondents were asked for address information of both their "current college/military address' and their "permanent residence." Notes in Thornburgh (2001) indicate that "many adjustment questionnaires were returned with duplicate responses to the current and permanent address questions," suggesting some confusion with the questionnaire instructions. The questionnaire also included a question on whether the respondent would be 18 years or older by April 1, 2000, and a question on race and Hispanic origin (the latter questions asked using the same categories and one-or-more options as the 2000 census). Questionnaires were distributed by staff at colleges and military installations; some schools preferred to provide administrative data electronically rather than distribute questionnaires, and this was permitted if the provided information was deemed to satisfy the needed criteria. Questionnaires and data were to be returned by June 1, 2000. All Kansas colleges and military installations participated, although Fort Leavenworth initially refused; subsequently, Fort Leavenworth refused to actively participate in distributing and collecting questionnaires, but did permit state officers to do that work.

Addresses (both "permanent" and "current") reported on the questionnaires were geocoded and assigned to census blocks; counts aggregated by census block, age (17 and younger, 18 and older), and race were then added to or subtracted to block-level counts from the 2000 census P.L. 94-171 redistricting data file. Thornburgh (2001) notes particular difficulty in the 2000 processing, since the Kansas questionnaire results were tabulated by 1990 census block and the P.L. 94-171 data were in the new 2000 census tabulation geography; the new boundaries were not available early enough for state processing. The state also encountered 259 census blocks where—post-adding or subtracting students and prisoners—the population count was negative; 818 blocks had negative counts when the block count was further subdivided by race. The state was able to resolve some of the largest such deviations and issued a letter to the Census Bureau regarding the others.

The Kansas adjustment in 1990 yielded a total net decrease of 32,194 (1.30%) for the state apportionment population; the 2000 adjustment yielded a net decrease of 16,161 (0.60%).

With prisoner counts at the time of redistricting, state redistricting bodies would then have the capacity to decide whether to include or exclude prisoners from proposed districts. The states' interest in having such a separate prisoner count should be assessed by the Bureau as part of its work with state officials to determine the layout of the standard P.L. 94-171 redistricting data file in 2010.

– 8 –

Operations, Research, and Testing

DEVELOPMENT OF RESIDENCE PRINCIPLES as a conceptual base for the census is essential, as is adjusting strategies for eliciting residence and roster information from respondents. However, it is also important to carry those principles over to the design of the specific techniques and methodologies used in fielding the census. It is also true that improvements to census operations and methodologies depend critically on an effective and vigorous program of research and testing. The types of research that we believe are essential to the determination of accurate resident counts—and to census quality, generally—can be roughly partitioned into groups:

- quantitative, analytical work on diverse living situations to improve understanding of the context of residence in the census;
- a program of experiments to accompany the 2010 census, to evaluate the efficacy of major changes to residence data collection in 2020 and beyond;
- improved research on the conduct of relevant census operations, such as group quarters enumeration, address list development, and unduplication methods; and
- testing and experimentation at an intermediate level—between small-sample cognitive testing and full-scale test censuses—that improves understanding of operational concerns.

We discussed the first of these in Chapter 5 and suggested a major test of a question-based approach collecting residence information (including universal provision for "any residence elsewhere" [ARE] reporting) in Chapter 6.

This chapter focuses on the latter two threads of research, including direct recommendations for improvement of some census operations, and offers comments on the Census Bureau's research program generally.

8–A MASTER ADDRESS FILE

The concept of the Master Address File (MAF)—the Census Bureau's complete inventory of known living quarters and business addresses in the United States—is a surprisingly new one. It was only after the 1990 census that the Census Bureau elected to maintain a continuous address list, rather than scrapping the address list after one census and building it anew prior to the next. As currently implemented, the MAF contains a mailing address for each of the living quarters on the list, if one exists; it also contains an intricate set of logical flags and indicators that denote the operations that added or edited each address.

> **Finding 8.1:** An accurate MAF is crucial to the quality of the decennial census, as well as the Census Bureau's other major survey programs. Together with the Topologically Integrated Geographic Encoding and Referencing (TIGER) system database, the MAF provides the key linkage between personal census form responses and specific geographic units. Inaccuracy in MAF and TIGER detracts from the quality of the decennial census, producing errors of inclusion and omission.

A full analysis of MAF and its construction is beyond the scope of this panel, but it is germane because residence rules concerns should be reflected in several ways:

- *Scope*: Work on MAF should make use of local area expertise in building a full list of residence locations, especially for irregular housing stock such as converted apartments, multi-use buildings, and small multi-unit structures. The work should also address the "seasonality" of housing stock: that is, it would be useful to have some flag or assessment of whether an address is purely a seasonal home (e.g., a time-shared property), a part-time seasonal home (e.g., rented out for part of the year), or a full-time residence.
- *Design:* What are the useful flags to include? What new update operations should be done to ensure completeness (e.g., a mechanism for including hotel living quarters)?
- *Evaluation:* In what geographic locations, and for what types of housing stock, is error in the MAF most prevalent? A major problem and frustration in 2000 was the lack of an audit trail; because the logical flags on

the MAF were not time stamped in any way, it was difficult if not impossible to tell how and when various updating operations touched address records. Such an audit trail is essential for effective evaluation. Consideration should also be given to detailed case history (ethnography) of a sample of addresses.

We endorse the recommendations of previous Committee on National Statistics panels (National Research Council, 2004c,b), and especially the recommendation that a detailed plan for continuous updating of MAF records by state, local, and tribal governments be developed and implemented (National Research Council, 2004b:Rec. 3.1). In particular, the Bureau should continue to find ways to obtain help from local and tribal authorities in obtaining information on unusual housing stock, such as multiple housing units inside family homes and leased hotel or motel quarters.

In Recommendation 6.3 we urge the Bureau to permit respondents to indicate cases where they believe that the census questionnaire reached them in the wrong place. That recommendation serves several important purposes, among them the production of valuable operational data during the conduct of the census (allowing diagnosis of significant addressing problems) and alleviating the frustration of respondents who receive a misaddressed form. However, the recommendation also interacts with the MAF development process in important ways. It is the constant maintenance and updating of MAF that helps to make the recommendation tractable; in principle, fewer gross addressing errors should result from a continually updated MAF, and so processing respondent-corrected information should not overwhelm follow-up resources. But it is also important that the stream of respondent-corrected data not be simply thought of as an operational diagnostic: the real value of respondent-suggested corrections may come from the use of those suggestions to detect and correct errors in the MAF itself. It is reasonable to expect that many of the respondent-suggested changes will be relatively routine, such as questionnaires for individual apartments in large residential buildings being placed in the wrong mailbox. They may also provide clues to larger problems: among these are the renaming of streets, multiple names attached to the same street segment, or the repartitioning and relabeling of individual units in multi-unit structures. We urge that the Census Bureau take advantage of this information, not putting absolute weight on the precoded MAF entries or on the respondent-reported information.

Recommendation 8.1: **Pursuant to Recommendation 6.3, respondent-corrected address information should be one source of information to update the MAF.**

Relevant to the goal of the unified address list and group quarters roster, participants in the 2010 census local geographic partnership program should be allowed to review address listings for group quarters in their jurisdictions, not just the household population listings. In addition, the Census Bureau should consider an institutional Local Update of Census Addresses program under which colleges and universities, medical facilities, and other group quarters locations can review the Bureau's address listings for their facilities.

8–B UNDUPLICATION METHODOLOGY

The panel's charge is on how residence rules and related concepts affect both undercount and overcount in the census, as suggested by the titular goal of counting people once (no omissions), only once (no duplicates), and in the right place. A major focus of the coverage evaluation effort following the 2000 census was on duplication in the census, given that Bureau analysts estimated that the census represented a net overcount. Unduplication operations in the 2000 census took several forms and occurred at different stages of the census process, as outlined in Box 8-1.

Emboldened by the 2000 census work on unduplication—capped by the insights gleaned from a complete match of the full set of census records to itself (using a probability model based on name and date of birth) to detect duplicates—current plans for 2010 call for new refinements. In particular, the Bureau is considering something close to "real-time" unduplication in the processing of returns.

Another National Research Council panel is currently studying the emerging coverage evaluation program of the 2010 census, and we have viewed the exact mechanics of unduplication as the province of that panel. However, we briefly offer some suggestions and comments on emerging unduplication methods as relevant to our charge. Both the 1990 Post-Enumeration Survey and the 2000 Accuracy and Coverage Evaluation Program were designed to focus on estimates of net coverage error—aggregate measures of undercount and overcount for particular populations—and not on the individual sources that contribute to that under- or overcoverage. We concur with the Panel to Review the 2000 Census (National Research Council, 2004c:Finding 1.7) that development of methodology for examining the components of gross census error (both omissions and undercounts) is vital, rather than a pure focus on net coverage.

> *Recommendation 8.2:* **A comprehensive assessment of the components of gross coverage error (both undercount and overcount) should be added as a regular part of the census evaluation program.**

Box 8-1 Unduplication in the 2000 Census

Aside from precensus editing of the MAF, the only unduplication program explicitly planned to take place as part the conduct of the 2000 census was application of the Census Bureau's primary selection algorithm (PSA). The census provided multiple chances for inclusion (among them, the regular census mailout, Internet response, nonresponse follow-up, unaddressed "Be Counted" forms, and foreign language forms), and the PSA's function was to determine which persons and information to retain from the set of records bearing the same MAF ID number.

In all, 8,960,245 MAF IDs had more than one eligible return (representing just less than 8 percent of the IDs on the Decennial Response File, the rawest compilation of collected census data); more than 95 percent of these IDs had exactly two returns associated with them, and 55 percent of those had two enumerator returns associated with them. The exact mechanics of the PSA are confidential, so that only a brief executive summary of the Bureau's evaluation of the PSA's performance is publicly available (Baumgardner, 2002), with additional results presented by Alberti (2003). What is known is that the algorithm involves grouping the set of people on a set of records into interim PSA households, with some checking of duplicates using person matching; it is also known that the census residence rules are *not* used in analyzing the person records possibly associated with a household, since Baumgardner (2003:iii) comments that the "[PSA] itself cannot take those rules into account when making decisions."

The Bureau carried out an ad hoc operation to identify duplicate housing units in June 2000. Internal evaluations from the first few months of the year compared the count of housing units on the MAF to estimates generated by using building permits and other sources; those analyses suggested sizable duplication in the MAF records. The operation flagged 2.4 million housing units (containing 6 million person records) as potential duplicates; these were temporarily removed from the census file. After further review, 1 million housing units (2.4 million people) were reinstated to the census file, and the rest were permanently deleted.

Estimation of erroneous enumerations, including duplicate records, were a major focus of the Bureau's work in the postcensus Accuracy and Coverage Evaluation (A.C.E.) Program. Bureau staff performed person-record matches based on name and birthdate in two waves. The Person Duplication Studies (summer 2001) matched the A.C.E. samples (two samples of approximately 700,000 records each, one of which is a direct extract from the census for selected blocks) to the nationwide census results. The Further Study of Person Duplication (summer 2002) did the same level of matching, but with revised methodology.

Subsequently, Bureau researchers have extended the scope and methodology of the work, matching the entire census person-level file to itself to identify potential duplicates. This work has raised the possibility of incorporating real-time unduplication into the census process in 2010, performing the same type of nationwide matching for batches of records to identify candidates for field follow-up. The 2006 and 2008 operational tests are intended, in part, to help resolve some remaining questions about the operation, such as the ideal timing of the operation and the sequencing of a coverage follow-up interview process (meant to consolidate multiple operations from 2000, as well as provide input to unduplication) with the coverage evaluation interviews.

Several other aspects of coverage merit comment:

- The Census Bureau's techniques for person unduplication in the 2010 census, including real-time processing of returns, must be fully tested in the 2006 census test and the 2008 dress rehearsal. The final shape of the unduplication program must be based on the empirical evidence gathered by these tests.
- Perfect information is an unobtainable standard in census data collection. While electronic checks for duplication (making use of probability models for the likelihood that records are duplicates) can introduce error into census processes, so too can field verification of previously collected census data be a source of error. Accordingly, the Census Bureau should take a balanced approach to census unduplication methodology; electronic checks that can facilitate real-time corrections should be considered, and field verification should not be given undue weight as a "gold standard" for data precision.
- To provide additional information for evaluation, the Census Bureau should consider performing complete follow-up of households flagged by unduplication algorithms for a sample of local census office workload or for sample census blocks.
- The Census Bureau should test modifications of its computer algorithms for person unduplication to include matching on address (and ARE address), as well as name and date of birth.

8–C CLASHING RESIDENCE STANDARDS: THE CENSUS AND THE AMERICAN COMMUNITY SURVEY

Estimates from the American Community Survey (ACS), now in the full stages of data collection, will replace the previous census long-form sample. Collection of ACS data began in selected test sites in 1996, ultimately including about 30 test sites (counties or groups of counties) prior to the 2000 census. As a formal experiment in 2000, the Census Bureau fielded a large survey based on the sampling and residence rule anticipated for the ACS. For this survey, called the Census 2000 Supplementary Survey (C2SS), the Bureau added 1,200 counties to the initial 30 test sites and sampled 891,000 housing units. Satisfied with the operational feasibility of conducting the ACS (Griffin and Obenski, 2001), the decision to replace the census long form with the ACS (and make the 2010 census a short-form-only collection) was finalized; within the Census Bureau organizational hierarchy, responsibility for the ACS was shifted from the demographic surveys to the decennial census directorate. Data collection continued at the C2SS levels through 2004, expanding to full-scale coverage of the household population (all counties) in January 2005 and incorporating group quarters in January 2006. In its full-scale operations, the ACS is designed for a sample size of 3 million addresses per year.

The advent of the ACS raises a variety of challenges related to the construction and interpretation of statistical estimates based on ACS data. As is the case with coverage evaluation, the estimation challenges associated with the ACS are the focus of a separate National Research Council panel. While a complete examination of the ACS is the purview of that panel, and not ours, it is necessary to have some overview of the basic structure of ACS data collection and estimates in order to understand concerns about the residence concept of the ACS. We describe two basic concepts of the ACS design in brief; for additional detail, see National Research Council (2004b), U.S. Census Bureau (2006a), and U.S. Government Accountability Office (2004b).

The first basic concept about the ACS that must be understood in the context of residence is the nature of estimates produced by the survey. Though the ACS is a large survey, its sample size of 3 million addresses per year does not match the (roughly) one-sixth sample of the nation's households that received the census long form in previous censuses. In order to reduce their inherent variability, ACS estimates are constructed by aggregating monthly survey data over 1 or more years. Under the current system, small geographic areas with populations of less than 20,000 have estimates that are produced by combining data within a 60-month interval. Those areas with populations between 20,000 and 65,000 receive estimates calculated from data in a 36-month window (separately, a 60-month estimate is calculated for those areas as well). Finally, large areas with 65,000 population or greater yield estimates based on 12 months of data (36-month and 60-month estimates are also available). This system of overlapping (and, in some sense, competing) estimates creates challenges for comparison between areas and for assessing change in trends over time. The extent of these challenges is just now beginning to be understood, and their practical effects will begin to become known in the summer of 2008, when the first multiyear estimates (from 3 years of full-scale ACS collection) are released.

The second fundamental concept of the ACS is that it is a continuous measurement and data collection operation, using three consecutive modes of response. Every month, questionnaires are mailed to a new sample of 250,000 addresses.[1] All housing units for which a form is not returned by mail within the first month are contacted by phone during the second month (provided that phone numbers are available in MAF records); data are collected using a computer-assisted telephone interviewing (CATI) instrument. At the end of the second month, the remaining nonresponding housing units (plus any housing units that were unable to be reached by mail) are eligible

[1]The sampling process is constrained so that individual households are not included in the ACS more than once every 5 years.

for personal visit follow-up. Only a sample of these units are approached for data collection by computer-assisted personal interviewing (CAPI). The important point about this structure is the continuity of the operation. A new 250,000-address sample begins this three-stage process each month, but the actual data collected and processed during a particular month are a blend of first-month mailed questionnaires, second-month CATI interviews, and third-month CAPI interviews. The reference point for each of these interviews is the time of the actual interview (or determination of vacant housing unit status). That is, a CAPI interview conducted in July asks the respondent to refer to July and not to May, when that household entered the ACS process.

Bearing these design aspects of the ACS in mind, we now consider a vitally important distinction between the ACS and the decennial census: their underlying residence standards. Both the decennial census and the ACS share the basic premise that each person has only one residence at any specific point in time (Griffin, 2005). On the continuum from *de jure* to *de facto*, the decennial census uses a hybrid "usual residence" concept that is close to the *de jure* end of the scale. In comparison, the residence concept used in the ACS is close to—but not purely—a *de facto* rule. Box 8-2 excerpts the residence rules for the ACS, as they are described in technical documentation on the design and methodology of the survey (U.S. Census Bureau, 2006a).

The ACS residence standard is described as a "current residence" or a "2-month rule," but it is considerably more complex than the succinct terms suggest. The "2-month rule" diverges from a true *de facto* standard because it does not specify a complete count of all people at the contacted housing unit at the time of the interview: it is meant to exclude short-term visitors, "people only staying [at the sample housing unit] for a short period of time." As it is articulated, the 2-month rule is meant to be prospective and retrospective—a person's "expected length of stay," actual or intended, is the qualification for current residence. The rule also differs from a *de facto* standard because it permits people to be counted at the household even if they are temporarily away. As long as people "are away from the housing unit for two months or less," they are considered current residents; however, "people who have been or will be away for more than two months" are not deemed current residents. A more subtle point about the phrasing of the ACS rule is that it makes reference to a "short period of time" being synonymous with "two consecutive months," but otherwise does not specify a consecutive timespan.

As Box 8-2 also indicates, the ACS residence standard includes specific exceptions to the 2-month reference period for some living situations. The ACS' handling of boarding school students (counting them at their parental homes) is consistent with practice in the 2000 census; however, the ACS' handling of commuter workers—counting them as current residents of their "family residence," no matter where they might be or where they spend most of their time—differs from the census. In two other situations, a pure *de facto*

Box 8-2 Residence Rules for the American Community Survey

Housing Units. The ACS defined the concept of current residence to determine who should be considered residents of sample (housing units [HUs]). This concept is a modified version of a *de facto* rule where a time interval is used to determine residency. The basic idea behind the ACS current residence concept is that everyone who is currently living or staying at a sample address is considered a resident of that address, except people only staying there for a short period of time. People who have established residence at the sample unit and are away from this unit for only a short period of time are also considered to be current residents. For the purposes of the ACS, the Census Bureau defines this "short period of time" as two consecutive months, and the ACS current residence rule is often described as the "2-month rule." Under this rule, anyone who is living for more than 2 months in the sample unit when the unit is interviewed (either by mail, telephone, or personal visit) is considered a current resident of that sample unit. This means that their expected length of stay is more than 2 months, not that they have been staying in the sample unit for more than 2 months. For the ACS, the Census Bureau classifies an HU in which no one is determined to be a current resident, as vacant. In general, people who are away from the sample unit for 2 months or less are considered to be current residents, even though they are not staying there when the interview is conducted, while people who have been or will be away for more than 2 months are not considered to be current residents.

Residency is determined as of the date of the interview. A person who is living or staying in a sample HU on interview day and whose actual or intended length of stay is more than 2 months is considered a current resident of the unit. That person will be included as a current resident of the unit unless he or she, at the time of interview, has been or intends to be away from the unit for a period of more than 2 months. There are three exceptions to this rule.

- Children (below college age) who are away at boarding school or summer camp for more than 2 months are always considered current residents of their parents' home.
- Children who live under joint custody agreements and move between residences are always considered current residents of the sample unit where they are staying at the time of the interview.
- People who stay at a residence close to work and return regularly to another residence to be with their family are always considered current residents of the family residence.

A person who is staying at a sample HU when the interview is conducted but has no place where he or she stays for periods of more than 2 months is also considered to be a current resident of the sample HU. A person whose length of stay in the sample HU is only for 2 months or less and has another place where he or she stays for periods of more than 2 months is not a current resident of the unit.

Group Quarters. Residency in group quarters (GQ) facilities is determined by a purely *de facto* rule. All people staying in the GQ facility when the roster of residents is made and sampled are eligible to be selected to be interviewed in the ACS. The GQ sample universe will include all people residing in the selected GQ facility at the time of interview. Data are collected for all people sampled regardless of their length of stay in the GQ facility. Children (below college age) staying at a GQ facility functioning as a summer camp are not considered to be GQ residents.

Box 8-2 (continued)

Reference Period. As noted earlier, the survey's reference periods are defined relative to the date of the interview. The survey questions define the reference periods and always include the date of the interview. When the question does not specify a time frame, respondents are told to refer to the situation on the interview day. When the question mentions a time frame, it refers to an interval that includes the interview day and covers a period before the interview. For example, a question that asks for information about the "past 12 months" would be referring to the previous 12 months relative to the date of the interview.

SOURCE: Excerpted from U.S. Census Bureau (2006a:6-2–6-3).

standard is used for the ACS: children in joint custody living arrangements are intended to be counted where they are at the time of the interview, and all persons included in the group quarters component of the ACS are counted where they are found (no group quarters type is allowed to report a "usual home elsewhere," as in the 2000 census).

We have described the ACS residence concept in theory, but it is critical to consider how this concept is conveyed on the ACS questionnaire itself. The questionnaire booklet begins with collection of initial information from the respondent on page 1, in a box labeled "Start Here"; see Figure 8-1. Full data collection begins with constructing a "List of Residents," laid out on a double-page spread on pages 2–3; a portion of that list, with instructions, is reproduced in Figure 8-2. What is immediately conspicuous about the presentation is that the entire first page of the questionnaire, including the household count question in the "Start Here" block, contains no reference to 2 months or any other time period. The opening set of bulleted points indicates only that the ACS questionnaire is asking for "basic information about the people who are living or staying at the address" in question, with no further explanation. The basic household count question—"How many people are living or staying at this address?"—is the counterpart to Question 1 on the census. But, unlike the census form, the ACS version of the question is not accompanied by any guidance on how this count is to be computed (e.g., people to be included or excluded). In addition, the ACS question contains no reference date (such as "April 1, 2000" on the census form), although the form does collect "today's date" on the preceding line. Census Bureau staff performed limited cognitive testing of a version of the ACS questionnaire (31 interviews); this testing suggested considerable confusion on both the scope of "living and staying at this address" and the reference date, but the Bureau analysts recommended no change to the household count question. They suggested only that the date question be amended to read "Today's Date" (DeMaio and Hughes, 2003).

In designing the ACS questionnaire, the Bureau chose a ledger-type approach for collecting the "List of Residents"; data on each of five household

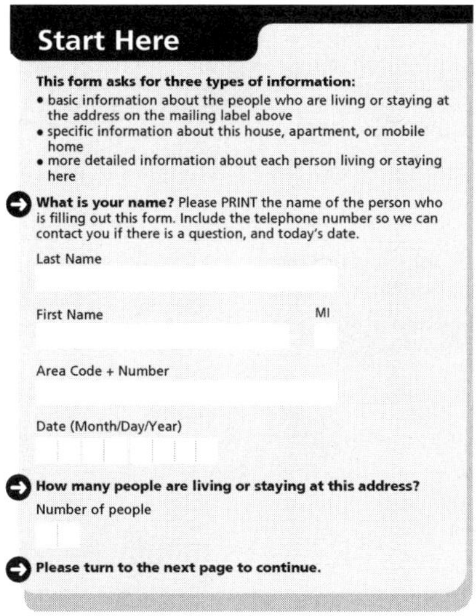

Start Here

This form asks for three types of information:
- basic information about the people who are living or staying at the address on the mailing label above
- specific information about this house, apartment, or mobile home
- more detailed information about each person living or staying here

What is your name? Please PRINT the name of the person who is filling out this form. Include the telephone number so we can contact you if there is a question, and today's date.

Last Name

First Name MI

Area Code + Number

Date (Month/Day/Year)

How many people are living or staying at this address?

Number of people

Please turn to the next page to continue.

Figure 8-1 Introductory household count question, 2005 American Community Survey

members are gathered along the rows of a two-page spread in the questionnaire booklet. A consequence of the design is that the instructions provided in order to complete the "List of Residents" (Figure 8-2) are extremely brief, limited to one thin column along the left-hand edge of the listing. The instructions take the form of two include statements ("LIST") and one exclude statement ("DO NOT LIST").

As articulated in Box 8-2, the ACS' current residence standard is arguably as intricate as the usual residence standard of the decennial census, and it includes explicit handling of particular living situations. Yet what is striking about the ACS instructions is that they convey very little information about the underlying residence concept and are actually confusing. The first bulleted instruction does try to concisely capture the ideas of a current residence ("who *is* living or staying here") and a prospective 2-month window. However, the second bulleted point is puzzling because it abruptly (and literally) switches to a "usual" residence perspective, directing that "anyone else who is staying here who does not have another usual place to stay" should be listed in the household. The nuance and complexity surrounding the interpretation of the word "usual" has been the motivation for most of this report, and all of that complexity certainly applies to its use in the ACS questionnaire. It is particularly jarring in this context because "usual" is especially ill defined and

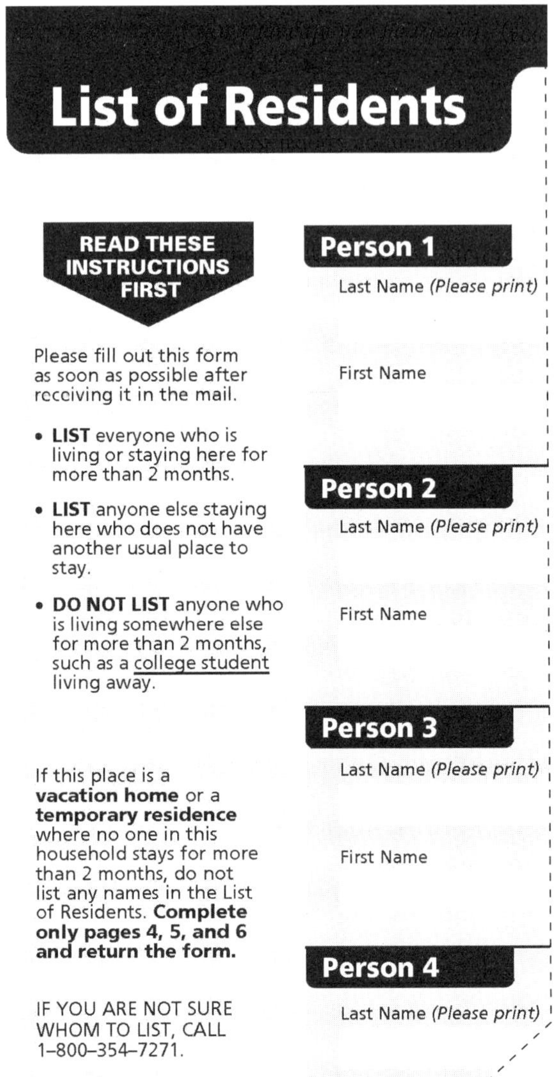
List of Residents

READ THESE INSTRUCTIONS FIRST

Please fill out this form as soon as possible after receiving it in the mail.

- **LIST** everyone who is living or staying here for more than 2 months.
- **LIST** anyone else staying here who does not have another usual place to stay.
- **DO NOT LIST** anyone who is living somewhere else for more than 2 months, such as a college student living away.

If this place is a **vacation home** or a **temporary residence** where no one in this household stays for more than 2 months, do not list any names in the List of Residents. **Complete only pages 4, 5, and 6 and return the form.**

IF YOU ARE NOT SURE WHOM TO LIST, CALL 1–800–354–7271.

Person 1
Last Name *(Please print)*
First Name

Person 2
Last Name *(Please print)*
First Name

Person 3
Last Name *(Please print)*
First Name

Person 4
Last Name *(Please print)*

Figure 8-2 Excerpt of household roster question and instructions, 2005 American Community Survey

NOTE: Space is provided for full data collection for five household members; additional space to write names of an additional seven household members is provided under the main List of Residents.

confusing when it is inserted into one of the very few instructions that are supposed to explain a "current" residence concept.

According to the rule, people who are not present at the time of the interview but are away for 2 months or less are supposed to be considered current residents and included in the ACS household. The third bulleted instruction—to exclude "anyone who is living somewhere else for more than 2 months"—addresses the converse situation: it emphasizes that long-term absentees should be omitted but does not speak directly to short-term residents who should be included. The semantics of this instruction are also interesting because it uses the strong condition "living somewhere else" rather than "staying" (as in the second bullet) or "living or staying" (first bullet). This wording raises, for instance, the problem of family members who are away from home for physical rehabilitation or other such programs, possibly for 2 or more months: a respondent may consider these family members as staying somewhere else for a time, but not necessarily living somewhere else.

The ACS current residence standard lays out several exceptions to the general 2-month rule, none of which is referenced in the instructions. Indeed, the only specific living situation included in the instructions is the prominently underlined reference to college students in the third instruction. This lone example is interesting and potentially confusing, depending on when the survey is administered. Assume a calendar where college classes end in mid-May and resume in late August or early September. By the letter of the 2-month rule, college students who have returned to their parents' homes at the end of classes ought to be reported as current residents if the ACS is administered in late May or June (the students are expected to be at the home location for just over 2 months) or in July or August (the students have been at the home location for 2 months or more). In concept, college students could also be reported as current residents in an interview at their parents' homes in March or April—the students are away right now, but will return within 2 months. Yet what stands out from a cursory look at the instructions is a connection between "do not list" and "college student[s]." The counting of college students seems to be an instance where the ACS attempts to retain the "usual residence" character of the decennial census, though that may contradict the survey's own "current residence" orientation.

Regarding the presentation of basic residence concepts on the questionnaire itself, two additional points should be made. First, the Bureau provides a companion booklet—"Your Guide for the American Community Survey"—that is intended to walk respondents through the questions. However, that booklet is keyed only to the numbered questions, the first of which is "What is this person's sex?" in the columns of the "List of Residents." That is, the companion instruction book skips the first-page "Start Here" block entirely, and provides no additional residence instructions on who should or should not be included in the resident list. Second, the features described thus far—

the "Start Here" block and basic instructions on page 2—are not solely part of the latest (2005) version of the questionnaire. Rather, these portions of the questionnaire have not changed substantively since 1999. The earliest version of the questionnaire, used in test sites between 1996 and 1998, only made reference to "living and staying here" on page 1 (without mentioning a two-month period), but did provide somewhat more detailed instructions; see the excerpt in Figure 8-3.

A separate item in the column of instructions for "List of Residents" (see Figure 8-2) is the note that the respondent should not list residents "if this place is a **vacation home** or a **temporary residence** where no one in this household stays for 2 months." This is intended to cover cases where no one in the contacted housing unit can be considered a "current resident" under the ACS residence concept. However, the syntax of the statement is awkward because it switches verb tense: "[no one] stays for more than 2 months" rather than "[no one] is staying for more than 2 months." Snowbirds who receive the questionnaire at their seasonal residence in March, shortly before they return "home," could feel compelled to complete the List of Residents and fill out all the person-level information: after all, they do stay at the vacation home for more than 2 months (generally, or in the year). To the extent that this instruction is also meant to account for one of the exceptions to the general 2-month rule—namely, commuter workers who are intended to be counted at their "family home"—it is unclear how effective this instruction will be. An apartment maintained strictly for work would be unlikely to be considered by its resident as either a "vacation home" or a "temporary residence"; moreover, in the aggregate, "usual" residence sense, commuter workers do stay at their work location apartments for more than 2 months (albeit not in a consecutive block), and hence could list themselves as residents of the work location.

Question 25 on the ACS form, shown in Figure 8-4, was first added to the ACS in 2003 to collect more information on seasonal populations. It begins by asking whether any household member "live[s] or stay[s] at this address year round"; follow-up questions are asked only if the answer is no (that is, everyone in the household is only a part-year resident at this location). The question may provide some insight on snowbird and sunbird residences, as well as other situations like groups of college students renting a house or apartment during the academic year. However, by its design, the question falls well short of being able to provide information on a fundamental underlying question: the extent to which individuals' "current residence" coincides with their "usual residence."

The ACS is intended to provide information on social, economic, and demographic characteristics of geographic areas, not population counts for those areas. As a result, it could be argued that there is less of a need for an absolutely complete accounting of "current residents" in the ACS. Moreover, the nature of ACS estimates means that multiple months or years of data are

Step 3 WHOM TO INCLUDE ON THE FORM

Please fill this form for ALL people who are living here, and ALL people who are staying here for more than two months. PRINT the names of these people in the List of Residents column. Begin with the name of the household member in whose name this place is owned or rented. Put this person's name in the PERSON 1 box.

Be sure to list all family members, as well as roommates, foster children, boarders, and live-in employees. Remember to include yourself on the list.

If there are people who live here but are currently staying somewhere else for more than two months, like a college student who is now away at school, DO NOT include him/her on the List of Residents.

If a person is staying here for two months or less and usually lives somewhere else, DO NOT include him/her on the List of Residents.

If EVERYONE staying here is here only temporarily for two months or less and usually lives somewhere else, DO NOT list any names on the List of Residents. However, please answer the questions on pages 4 and 5. Information about short-term visitors is not needed for this survey, but information about the house or apartment is.

If you are not sure whom to include, call 1-800-354-7271.

If there are more than five people who should be listed, use the spaces at the bottom of pages 2 and 3 for their names. We will contact you by telephone to obtain the information for them.

List of Residents

Print the last name, first name, and middle initial for each person who should be included in the list.

PERSON 1

Last name

First name Middle initial

PERSON 2

Last name

First name Middle initial

PERSON 3

Last name

First name Middle initial

PERSON 4

Last name

First name Middle initial

PERSON 5

Last name

Figure 8-3 Excerpt of household roster question and instructions, 1996–1998 American Community Survey

NOTE: Space is provided for full data collection for five household members; additional space to write names of an additional seven household members is provided under the main List of Residents.

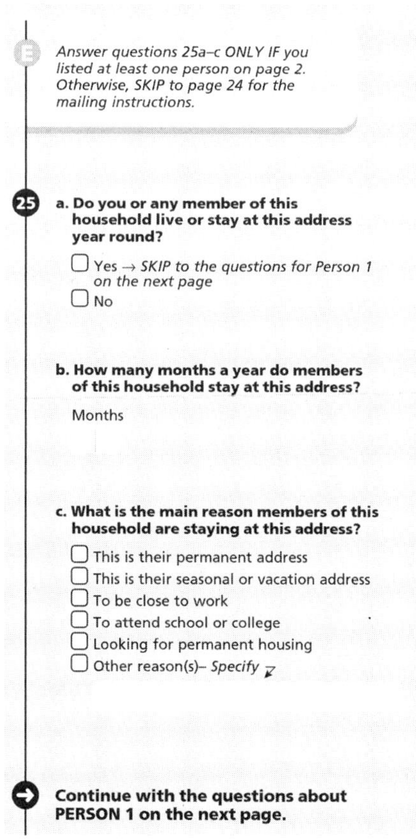
E Answer questions 25a–c ONLY IF you listed at least one person on page 2. Otherwise, SKIP to page 24 for the mailing instructions.

25 **a. Do you or any member of this household live or stay at this address year round?**

- ☐ Yes → *SKIP to the questions for Person 1 on the next page*
- ☐ No

b. How many months a year do members of this household stay at this address?

Months

c. What is the main reason members of this household are staying at this address?

- ☐ This is their permanent address
- ☐ This is their seasonal or vacation address
- ☐ To be close to work
- ☐ To attend school or college
- ☐ Looking for permanent housing
- ☐ Other reason(s)– *Specify*

Continue with the questions about PERSON 1 on the next page.

Figure 8-4 Question 25, 2005 American Community Survey

combined to produce ACS-based proportions and averages; conceptually, it is reasonable to expect some lack of precision in an estimate that is meant to be based on the population that were "current residents" of an area at some point in a 5-year window of time. It is also the case that, despite its oft-stated mandate of replacing the decennial census long form, the ACS should not be held to the exact standards and methodology of past and current censuses. The ACS is more properly thought of as a new and vital data collection system: it must be able to satisfy current uses of long-form census data, but it has unique properties, strengths, and limitations that should be examined but not impeded by complete adherence to census norms.

However, the Census Bureau is now in a position where two flagship products—the decennial census and the ACS—follow two complex residence standards that are conceptually very different. It may be that the aggregation of multiple years of ACS data makes individual residence reporting problems offset each other and produces estimates consistent with what would be found using a "usual residence" standard. Yet it may also be that different applications of residence concepts produce highly discrepant estimates for some areas or population subgroups. Given the newness of full-scale ACS collection and—more fundamentally—the lack of collection of both residence locations on the same questionnaire, it is simply unknown how problematic the disconnect between the two programs will be.

Our recommendation for the ACS is directly analogous to our recommendation for the decennial census. We have suggested a program of research to gain further insight on how individual people's concepts of residence match the decennial census "usual residence" concept, as well as the effectiveness of the census questionnaire in eliciting accurate "usual residence" information. The same line of reasoning holds for the ACS: it is unclear how well the ACS "current residence" concept or "2-month rule" fits with respondents' own notions, and our review above raises considerable uncertainty as to how well the ACS questionnaire items and instructions match the survey's own residence concept.

For the decennial census, we recommend that the Census Bureau collect "any residence elsewhere" information. As a starting point, these data should be collected as a major experiment of the 2010 census so that rigorous evaluation and analysis of those data can inform changes for later censuses. Likewise, we believe that the Census Bureau will ultimately be best served by the inclusion of a usual residence question in the ACS questionnaire, asked of each person and not of a whole household. The collection of both types of residence information is essential to measuring discrepancies between the residence standards and for evaluating the residence concepts of both the census and the ACS. As a first step—a means to gather baseline information for evaluation and refinement of a full-scale implementation of the question—the Bureau should include a usual residence question in its ACS experimentation. Current plans for the ACS include a "methods panel," a subset of the ACS sample that may receive experimental versions of questionnaires or revised wordings of specific items. This methods panel would be an ideal setting for asking respondents whether their current residence is what they consider to be their usual residence and, if not, where their usual residence is.

> *Recommendation 8.3:* **The Census Bureau should plan to ask a question on the usual residence of each household member in the ACS questionnaire, in order to evaluate the extent of incongruity of residence standards between the long-form replacement survey and the decennial census. The usual residence question should first be tested using the survey's experimental**

"methods panel"; the resulting data should be fully evaluated and analyzed to refine final versions of the question.

The frequency of data collection in the ACS makes it a potentially useful vehicle for getting survey coverage of seasonal populations, such as snowbirds, sunbirds, and migrant workers. Yet to make effective sample sizes more consistent with the long-form sample, the Census Bureau is relying on a release schedule where data are aggregated over 1, 3, or 5 years. Adhering to this plan, the Census Bureau has resisted pressures to issue other releases that may capture the seasonal nature of some populations within a year. However, in conjunction with our recommendation to continue studies of trends related to residence and key population subgroups, we encourage the Bureau to consider ways to use the ACS for information on seasonal differences.[2]

8–D TESTING AND RESEARCH IN 2010 AND BEYOND

We argue in this report for a new approach to residence in the census, centered on the derivation of a core set of residence principles. Working with these principles—in particular the precept that usual residence should be an individual-level determination and not an attribute of a specific housing type or population group—suggests the need to ask more residence-related questions on the census form than in the past. We also recommend question-based structure for eliciting resident listings and counts as ultimately more effective than the current instruction-based approach.

Recommendation 8.4: **A major test of census residence concepts, conducted in conjunction with the 2010 census, should be the basis for postcensal development leading to the 2020 census. This test should include both a question-based approach to collecting resident count information and a provision for ARE reporting by all census respondents, including those living in group quarters (nonhousehold) situations.**

In the long term—from the 2010 to the 2020 census—a thorough evaluation of the results of the test and the design of any follow-up work is a major research priority (Recommendation 6.5), as is specific analysis of the returns

[2]Meeting such an objective surely will require obtaining information from respondents on ARE. Furthermore, it will minimally require two additional data releases: (1) estimates of population characteristics for a geographic area that are not controlled to an external census-based population estimate for that area, and (2) estimates based on pooled monthly samples for periods other than a calendar year: for example, estimates from pooled samples covering the months of December through February to contrast with those from pooled samples covering the months June through August.

from incarcerated persons in order to assess the feasibilities of allocating prisoners to a geographic location other than the prison (Recommendation 7.2). It will also be useful to compare and contrast the ARE information with that generated by related operations—the proposed coverage follow-up operation (Box 6-3) and the postenumeration survey used for coverage evaluation—where more detailed banks of coverage and location probes are permissible (Section 6–E.4).

In the nearer term, we suggest throughout this report a number of deficiencies in current research that should be investigated as part of an ongoing research program at the Census Bureau:

- assessing the quality, accessibility, and relevant content of facility and administrative records for group quarters and nonhousehold facilities (Section 7–D.1);
- studies that can provide quantitative information (and validate hypotheses based on qualitative techniques like ethnography) on the magnitude and trends in complex and ambiguous living situations (Section 5.2);
- basic research on living situations as reflected by census operations, such as the tendency for household members to be listed, roughly, in reverse order by age (Section 5–B.3);
- mode effects on response, including both the effects of the mode of administration (e.g., paper versus telephone response) and the general structure of roster types (e.g., whether a question-based or instruction-based approach is easier to follow) (Recommendation 6.4);
- effects of visual layout and wording on census questionnaires (Recommendation 6.7); and
- impact of length of form (number of questions) on survey response (Section 6–G).

It is with problems like those listed as the shorter-term research tasks in mind that we offer additional comment on the shape and direction of the Census Bureau's general research and testing program.

8–E THE CENSUS BUREAU RESEARCH AND TESTING PROGRAM

The most prominent components of the Census Bureau's research program are the suite of formal experiments and tests that accompany each decennial census, the evaluation reports of various census operations that are produced after the decennial count, and the set of large-scale tests scheduled regularly between census years. These tend to be large and complex activities—indeed, one of the usual census tests is a dress rehearsal that tries to mimic every decennial census operation—and the tests attempt to vary several major factors simultaneously. For example:

- The 2003 National Census Test involved approximately 250,000 households using only mailed questionnaires (no field follow-up was performed); it varied both the wording of the race and Hispanic origin questions as well as different cues to respond to the test by mail, Internet, or telephone.
- The 2005 National Census Test included a variety of coverage probe questions related to residence (see Chapter 6). In addition to those changes, though, the test also included experimental panels where Internet response is encouraged, and it included a panel that received a bilingual English/Spanish questionnaire.
- In addition to the Alternative Questionnaire Experiment (AQE) that was conducted alongside the 2000 census, other major experiments conducted at the same time included the C2SS (a prototype for the long-form-replacement ACS), the Response Mode and Incentive Experiment, and the Social Security Number, Privacy Attitudes, and Notification Experiment.
- The 1998 dress rehearsal that preceded the 2000 census was intended to be a true dress rehearsal, but the political circumstances that made it difficult to finalize the basic design of the 2000 census forced the 1998 "dry run" to be a particularly ambitious test. Staged in three sites, the 1998 rehearsal was actually a test of three broad design choices that varied in the degree to which nonresponse follow-up was conducted (either in full or only for a sample) and whether a postenumeration survey was used to adjust the counts for estimated nonresponse.

At the other extreme of testing in the census hierarchy are the small numbers of cognitive tests to which revised questionnaires are routinely submitted; see Box 8-3. Hunter and de la Puente (2005) tested a version of the "worksheet" approach used in the 2005 census test based on 14 cognitive interviews, conducted in the Washington, DC area in early 2005. Other cognitive tests conducted by the Bureau use similar-sized samples. Hunter (2005) reported on a 2003 test intended to see whether a proposed direct question on cohabitation (for possible inclusion on other surveys and not the census) was "direct, gender-neutral, non-offensive, and generally applicable." Conclusions were drawn from a set of interviews with 19 people, all of whom were cohabiting at the time of the interview; the sample included both heterosexual and gay and lesbian respondents. Likewise, Hunter and DeMaio (2005) tested revisions to three separate census questions—housing tenure, age (adding a reminder to report babies as age 0 if they were less than 1 year old), and relationship—based on 18 cognitive interviews, the set of which contained

Box 8-3 Cognitive Testing

When faced with a question during a survey interviewer, a survey respondent must perform several basic tasks in cognitively processing the question and formulating a response. Cognitive testing of survey instruments has emerged as a fairly standard practice for getting a sense of the basic thought processes stimulated by sets of questions. Tourangeau (1984), summarized by Willis (1999), identifies the major focus areas of cognitive testing, grouped by the basic cognitive task being performed by respondents:

1. **Comprehension of the Question**
   (a) *Question intent:* What does the respondent believe the question to be asking?
   (b) *Meaning of terms:* What do specific words and phrases in the question mean to the respondent?
2. **Retrieval from Memory of Relevant Information**
   (a) *Recallability of information:* What types of information does the respondent need to recall in order to answer the question?
   (b) *Recall strategy:* What type of strategies are used to retrieve information? For example, does the respondent tend to count events by recalling each one individually, or does he/she use an estimation strategy?
3. **Decision Processes**
   (a) *Motivation:* Does the respondent devote sufficient mental effort to answer the question accurately and thoughtfully?
   (b) *Sensitivity/social desirability:* Does the respondent want to tell the truth? Does he/she say something that makes him/her look "better"?
4. **Response Processes**—*Mapping the response:* Can the respondent match his or her internally generated answer to the response categories given by the survey question?

Cognitive interviews typically follow one or both of two basic models (Willis, 1999). The first, "think-aloud interviewing," consists of urging and cuing respondents to talk through their thought processes as they answer each question; while this method can generate a rich array of information, it also puts almost the entire burden on the respondent rather than the interviewer. The second technique, verbal probing, is more structured: a question is read and a response is given, and the interviewer then asks a set of probing questions as to how the answer was generated. These probes may include paraphrasing ("Can you repeat the question I just asked in your own words?"), comprehension/interpretation ("What does the term *X* mean to you?"), or more general assessments ("Was that easy or hard to answer?").

Cognitive testing may occur as an iterative process. The first draft of a questionnaire will lead to an initial set of interviews with a small number of subjects (5–10). Based on that initial feedback, results are generated and the questionnaire is designed; other rounds of testing may result as the questionnaire is successively refined. In this manner, when multiple rounds of testing are anticipated, early rounds of cognitive interviews may concentrate on general concepts, while the later rounds focus on specific question wording and structure concerns.

different mixes of people directly affected by the question changes (10 of 18 were renters rather than owners and 6 of 18 were in households with infants).

As we have observed the development of the mid-decade census tests, the panel has grown concerned about the fact that there seems to be very little experimentation and testing by the Census Bureau that operates between these two extremes.

> **Finding 8.2:** The Census Bureau often relies on small numbers (20 or less) of cognitive interviews or very large field tests (tens or hundreds of thousands of households, in omnibus census operational tests) to reach conclusions about the effectiveness of changes in census enumeration procedures. As a consequence many important questions about the effectiveness of residence rules do not get addressed effectively.

To be clear, we do not suggest by this finding that there is anything necessarily wrong with tests that operate at these extremes. In particular, we do not mean in any sense to malign small-sample cognitive testing as a research tool by the Census Bureau; cognitive tests are definitely worth doing, since they are an excellent diagnostic process (and generator of research hypotheses) that can identify major problems with specific questionnaire items and formats and can highlight problems in logic and syntax. What we do argue is that it is possible to put too much weight on cognitive tests, whose sample sizes are too small and unrepresentative to support broad conclusions; filtering possibilities and eliminating potential approaches to practical census problems on the strength of comments from a very small number of interviews is too restrictive.

Likewise, there is benefit to the massive scale census tests (or, more precisely, operational trials) that the Census Bureau regularly conducts. Particularly important is that they allow the Census Bureau to keep its field "machinery" well trained and in good working order; the sheer sample size that is possible in some of these trials also affords a variety and depth of response that is difficult to obtain through different means. However, the omnibus census tests also have conceptual weaknesses, as discussed in this report. By trying to coerce problems into a catch-all test, it is easy to "design" a test for which the great advantage of sample size is offset by the fact that the test reaches relatively few people who are most directly affected. As a previous study (National Research Council, 2004b) concluded, the 2003 census test—a major goal of which was to test the effectiveness of altered wording of the Hispanic origin question—was severely impaired because the test failed to adequately target responses from Hispanic communities. Also, even a relatively simple large-scale test—the 2000 AQE—can suffer from being forced into a large-test framework. Box 6-2 describes how the 2000 AQE questionnaire block varied so many factors simultaneously that the effectiveness of any single change is impossible to determine.

Put simply, the panel's concern is that the Bureau tends to waste test cases because it does not target relevant populations effectively. More significantly, the overall direction of the Bureau's testing efforts is impeded by the lack of a thread of sustained research; test topics seem to arise on essentially an ad hoc basis, rather than following a more iterative series of tests designed to achieve specified goals.

> *Recommendation 8.5:* **The Census Bureau should undertake analytical research on specific problems in order to better evaluate the effectiveness of residence and other questions on the census forms. These studies should be designed to focus on particular populations of interest. Candidates for such research include:**
>
> - **why babies are often omitted from the census form (targeted at households with newborns);**
> - **whether census respondents find a pure *de facto* residence rule easier to follow and interpret than a *de jure* rule (generally, and with specific reference to large households);**
> - **whether additional residence and location probes on questionnaires—increasing the length of the survey—impairs response or other operational activities (e.g., page scanning);**
> - **the difficulty and advantages of including a reference date or time frame;**
> - **multilingual and linguistically isolated households; and**
> - **whether the Census Bureau standard of "live or sleep most of the time" is consistent with respondent notions of "usual residence."**

Sustained research needs to attain a place of prominence in the Bureau's priorities. The Bureau needs to view a steady stream of research as an investment it its own infrastructure that—in due course—will permit more accurate counting, improve the quality of census operations, and otherwise improve its products for the country. Given the scarcity of resources available to it, the Census Bureau needs to explore ways to facilitate additional analysis of its extant data resources by outside researchers. Specific mechanisms by which this may be achieved include public and private partnerships for analysis of census data, renewal and extension of American Statistical Association/National Science Foundation Census fellowships, improved task-order relationships, and enhanced Research Data Centers.

The mechanics of censustaking have changed greatly since marshals were first sent out on horseback in 1790; as times have changed, the "usual residence" concept has endured even though its exact interpretation has shifted. The most recent paradigm shift in defining residence in the census came with the adoption of mail-based enumeration for most of the census population in 1970; that shift included drawing a linkage between census residence and a specific mailing address. Looking ahead, over the long term, the Census Bureau research program needs to consider broader shifts that lie ahead—the impact of the Internet and e-mail and the diminished importance of traditional mailing addresses (and paper mail) in people's lives, more transitory living arrangements, and the changing need for census data as private and public databases grow in completeness.

References

Abramson, F. H. (2003, September 17). *Special Place/Group Quarters Enumeration*. Census 2000 Testing, Experimentation, and Evaluation Program, Topic Report 5. Washington, DC: U.S. Census Bureau.

Adams, T. S. and E. A. Krejsa (2001). *ESCAP II: Results of the Person Followup and Evaluation Followup Forms Review*. Executive Steering Committee for A.C.E. Policy II Supporting Report 24. Washington, DC: U.S. Census Bureau.

Alberti, N. (2003, September 25). *Data Processing in Census 2000*. Census 2000 Testing, Experimentation, and Evaluation Program, Topic Report 7. Washington, DC: U.S. Census Bureau.

Allard, P., M. K. Biklen, and K. D. Levingston (2004). *One Size Does Not Fit All: Why the Census Bureau Should Change the Way It Counts Prisoners*. New York: Brennan Center for Justice at New York University School of Law.

Allard, P. and K. D. Levingston (2004). *Accuracy Counts: Incarcerated People and the Census*. New York: Brennan Center for Justice at New York University School of Law.

Allard, P. and C. Muller (2004). *Incarcerated People and the Census: Painting a Distorted Picture of Virginia*. New York: Brennan Center for Justice at New York University School of Law.

Anderson, M. J. (1988). Planning the future in the context of the past. *Society 25*(March/April), 39–47.

Anderson, M. J. (Ed.) (2000). *Encyclopedia of the U.S. Census*. Washington, DC: CQ Press.

Australian Bureau of Statistics (2004, January 15). *Usual Residence Concepts Sources and Methods Paper*. Number 1389.0. Canberra: Australian Bureau of Statistics.

Baumgardner, S. (2002, November 26). *Analysis of the Primary Selection Algorithm.* Census 2000 Evaluation L.3.a (Executive Summary Only). Washington, DC: U.S. Census Bureau.

Baumgardner, S. (2003, September 10). *Resolution of Multiple Census Returns Using a Re-interview.* Census 2000 Evaluation L.3.b (Executive Summary Only). Washington, DC: U.S. Census Bureau.

Beale, C. (1996). Rural prisons: An update. *Rural Development Perspectives 11*(2), 25–27.

Black, D., G. Gates, S. Sanders, and L. Taylor (2000). Demographics of the gay and lesbian population in the United States: Evidence from available systematic data sources. *Demography 37*(2), 139–154.

Blumstein, A. (1995). Stability of punishment: What happened and what next? In T. Blomberg and S. Cohen (Eds.), *Punishment and Social Control*, pp. 259–274. New York: Aldine de Gruyter.

Bonvillain, N. (1989). *Residence Patterns at the St. Regis Reservation.* Ethnographic Exploratory Research Report #5. Washington, DC: U.S. Census Bureau.

Brownrigg, L. and P. Wobus (1993). The myth of the man under the bed. In *Proceedings of the Survey Research Methods Section, American Statistical Association*, pp. 156–161. Alexandria, VA: American Statistical Association.

Brownrigg, L. A. (2003). *Ethnographic Social Network Tracing of Highly Mobile People.* Census 2000 Evaluation J.2. Washington, DC: U.S. Census Bureau.

Bumpass, L. and H.-H. Lu (2000). Trends in cohabitation and implications for children's family contexts in the United States. *Population Studies 54*(1), 29–41.

Bumpass, L. L. and J. Sweet (1989). National estimates of cohabitation. *Demography 26*, 615–625.

Bureau of Justice Statistics (1998). *State and Federal Corrections Information Systems: An Inventory of Data Elements and an Assessment of Reporting Capabilities.* Report NCJ 170016. Joint project of Association of State Correctional Administrators, Corrections Program Office (Office of Justice Programs), Bureau of Justice Statistics, and National Institute of Justice. Washington, DC: U.S. Department of Justice. Available: http://www.ojp.usdoj.gov/bjs/pub/pdf/sfcis.pdf [12/19/05].

Bureau of the Census (1982). *User's Guide—1980 Census of Population and Housing.* Part A, Text. Washington, DC: U.S. Department of Commerce.

Bureau of the Census (1995, October). *1990 Census of Population and Housing: History—Part C.* 1990 CPH-R-2C. Washington, DC: U.S. Department of Commerce.

Burt, M. R. (2001). Homeless familities, singles, and others: Findings from the 1996 National Survey of Homeless Assistance Providers and Clients. *Housing Policy Debate 12*(4), 737–780.

Burt, M. R. and L. Y. Aron (2000, January 1). *America's Homeless II: Populations and Services*. Washington, DC: Urban Institute. Formatted as presentation; available: http://www.urban.org/url.cfm?ID=900344.

Burt, M. R., L. Y. Aron, T. Douglas, J. Valente, E. Lee, and B. Iwen (1999). *Homelessness: Programs and the People They Serve—Findings of the National Survey of Homeless Assistance Providers and Clients; Technical Report*. Washington, DC: Interagency Council on the Homelessness.

Burt, M. R. and B. S. Cohen (1989). *America's Homeless: Numbers, Characteristics, and Programs That Serve Them*. Washington, DC: Urban Institute.

Cancian, M. and D. R. Meyer (1998). Who gets custody? *Demography 35*(2), 147–157.

Carter, N. (2002, September 25). *Be Counted Campaign for Census 2000*. Census 2000 Evaluation A.3. Washington, DC: U.S. Census Bureau.

Carter, S. B., S. Gartner, M. R. Haines, A. L. Olmstead, R. Sutch, and G. Wright (2006). *Historical Statistics of the United States: Millennial Edition*. New York: Cambridge University Press.

CEC Associates (1987). Decennial census residence rules and enumeration principles. See Council of Professional Associations on Federal Statistics (1987).

Center for Tourism Research (2003). Winter visitor study 2002–2003. PowerPoint presentation. University of Texas–Pan American.

Child Trends (2002). *Charting Parenthood: A Statistical Portrait of Fathers and Mothers in America*. Washington, DC: Child Trends. Available: http://fatherhood.hhs.gov/charting02/ChartingParenthood02.pdf [6/1/06].

Christensen, D., C. Dahl, and K. Rettig (1990). Noncustodial mothers and child support: Examining the larger context. *Family Relations 39*, 388–394.

Christian, L. M. and D. A. Dillman (2004). The influence of graphical and symbolic language manipulations on responses to self-administered questions. *Public Opinion Quarterly 68*(1), 57–80.

Citro, C. F. (2000). Enumeration: Special populations. See Anderson (2000), pp. 34–42.

Citywide Families in SROs Collaborative (2001, October 23). *Report on the Census of Families with Children Living in Single Room Occupancy Hotels in San Francisco*. San Francisco, CA: Citywide Families in SROs Collaborative.

Clarke, S. C. (1995, March 22). Advance report of final divorce statistics, 1989 and 1990. *Monthly Vital Statistics Report 43*(9—Supplement).

Clear, T. and K. Terry (2000). Correction beyond prison walls. See Sheley (2000), pp. 517–537.

Clemence, T. G. (1987). Place of abode. See Council of Professional Associations on Federal Statistics (1987).

Clement, M. and N. Keough (2004, September). *Political Punishment: The Consequences of Felon Disenfranchisement for Rhode Island Communities*. Providence, RI: Rhode Island Family Life Center.

Coates, K. S., R. Healy, and W. R. Morrison (2002). Tracking the snowbirds: Seasonal migration from Canada to the U.S.A. and Mexico. *American Review of Canadian Studies 32*(3), 433–450.

Cohen, P. C. (2000). Decennial census: 1830 census. See Anderson (2000), pp. 120–121.

Cole, J. (2001a, April 21). Town of Fredonia suspects a miscount. *Milwaukee Journal-Sentinel*.

Cole, J. (2001b, July 31). Town's census 'gain' of 800 goes to Mequon. *Milwaukee Journal-Sentinel*.

Conrad, F., M. P. Couper, R. Tourangeau, and A. Peytchev (2006). The use and non-use of clarification features in web surveys. *Journal of Official Statistics*. In press.

Cordray, D. S. and G. M. Pion (1991). What's behind the numbers? Definitional issues in counting the homeless. *Housing Policy Debate 2*(3), 587–616.

Council of Professional Associations on Federal Statistics (1987). *Decennial Census Residence Rules: Proceedings of the December 15–16, 1986 Conference, Washington, DC*. Washington, DC: U.S. Census Bureau.

Courtney, A. (2005a, May). *The 2011 Census: Initial View on Content for England and Wales—Second Residences*. Consultation document. London: Office for National Statistics.

Courtney, A. (2005b, May). *The 2011 Census: Initial View on Content for England and Wales—Socio-Demographic Context*. Consultation document. London: Office for National Statistics.

Cousineau, M. and T. Ward (1992). An evaluation of the S-Night street enumeration of the homeless in Los Angeles. *Evaluation Review 16*(4), 355–364.

Crowley, M. (2003). *Generation X Speaks Out on Civic Engagement and the Decennial Census: An Ethnographic Approach*. Census 2000 Ethnographic Study. Washington, DC: U.S. Census Bureau.

Cullen, F. T. and J. L. Sundt (2000). Imprisonment in the United States. See Sheley (2000), pp. 473–515.

Curran, S. R. and E. Rivero-Fuentes (2003). Engendering migrant networks: The case of Mexican migration. *Demography 40*(2), 289–307.

Curtin, R. T. (2001). *The RV Consumer*. Surveys of Consumers Report, Survey Research Center. Ann Arbor, MI: University of Michigan.

de la Puente, M. (1993). Why are people missed or erroneously included by the census: A summary of findings from ethnographic coverage reports. In *Proceedings of the 1993 Research Conference on Undercounting Ethnic Populations of the U.S. Census Bureau*, pp. 29–66. Washington, DC: U.S. Census Bureau.

de la Puente, M. (2004, February 5). *Census 2000 Ethnographic Studies*. Census 2000 Testing, Experimentation, and Evaluation Program, Topic Report 15. Washington, DC: U.S. Census Bureau.

de la Puente, M. and D. Stemper (2003, September 22). *The Enumeration of Colonias in Census 2000: Perspectives of Ethnographers and Census Enumerators*. Census 2000 Evaluation J.4. Washington, DC: U.S. Census Bureau.

DeBow, J. (1853). *The Seventh Census of the United States: 1850*. Washington, DC: Office of the Public Printer.

Decker, F. H. (2005). *Nursing Homes 1977–99: What Has Changed, What Has Not?* Hyattsville, MD: National Center for Health Statistics.

DeMaio, T. J. and N. A. Bates (1992). *Redesigning the Census Long Form: Results from the 1990 Alternative Questionnaire Experiment*. Research Paper SM92/01. Washington, DC: U.S. Census Bureau.

DeMaio, T. J. and K. A. Hughes (2003, July 7). *Report of Cognitive Research on the Residence Rules and Seasonality Questions on the American Community Survey*. Survey Methodology Study Series #2003–08. Washington, DC: U.S. Census Bureau.

Devine, J. A. and J. D. Wright (1992). Counting the homeless: S-Night in New Orleans. *Evaluation Review 16*(4), 409–417.

Dillman, D. A., J. R. Clark, and M. A. Sinclair (1995). How prenotice letters, stamped return envelopes, and reminder postcards affect mailback response rates for census questionnaires. *Survey Methodology 21*(2), 1–7.

Drabek, J. H. (2005). Unpublished communication to the Panel on Residence Rules in the Decennial Census. Shared with panel staff, March 17.

Duble, A. (2000, May 4). Census confusion: Two forms required for military. Article from Fort Detrick, Maryland, newsletter reprinted at http://www.dcmilitary.com/army/standard/archives/may4/fd_h5400.html [6/1/06].

East Oregonian and Associated Press (2006, February 2). Where's home for prison inmates? *East Oregonian*. Posted to www.eastoregonian.com.

Edin, K. (1992). Counting Chicago's homeless: An assessment of the Census Bureau's 'street and shelter night'. *Evaluation Review 16*(4), 365–375.

Edin, K. and L. Lein (1997). *Making Ends Meet: How Single Mothers Survive Welfare and Low-Wage Work*. New York: Russell Sage Foundation.

Edwards, R. W. (2003, July). Census of population and housing: ABS views on content and procedures. Information paper, Australian Bureau of Statistics, Canberra.

Ericksen, E. P. (2000). Errors in the census. See Anderson (2000), pp. 206–209.

Fay, R. (2004, November 22). Duplicated persons in census 2000: Results by major demographic group. Unpublished report shared with the Panel on Residence Rules in the Decennial Census on December 13, 2004. U.S. Census Bureau.

Fay, R. E. (2002a, March 27). *ESCAP II: Evidence of Additional Erroneous Enumerations from the Person Duplication Study*. Executive Steering Committee for A.C.E. Policy II Supporting Report 9 (revised version). Washington, DC: U.S. Census Bureau.

Fay, R. E. (2002b). *Evidence of Additional Erroneous Enumerations from the Person Duplication Study*. Executive Steering Committee for A.C.E. Policy II Supporting Report 9 (revised). Washington, DC: U.S. Census Bureau.

Federal Committee on Statistical Methodology (1990, April). *Survey Coverage*. Statistical Policy Working Paper 17. Washington, DC: Subcommittee on Survey Coverage, Federal Committee on Statistical Methodology.

Federal Reserve Bank of Dallas (1996). *Texas Colonias: A Thumbnail Sketch of the Conditions, Issues, Challenges, and Opportunities*. Dallas: Federal Reserve Bank of Dallas. Available: http://www.dallasfed.org/ca/pubs/colonias.html [12/2/05].

Fleisher, M. S. (2001). Residential mobility of adult male gang members. Unpublished paper prepared for U.S. Census Bureau.

Fox, G. and R. Kelly (1995). Determinants of child custody arrangements at divorce. *Journal of Marriage and the Family 57*, 693–708.

Frase, R. S. (1998). Jails. See Tonry (1998), pp. 474–508.

Galvez, J. (1997, October). The Florida elusive snowbird. Report, Bureau of Economic and Business Research, University of Florida. Available: http://www.bebr.ufl.edu/Articles/snowbirds.pdf [11/30/05].

Garcia, V. and L. Gonzales (1995). *Finding and Enumerating Migrants in Mexican Enclaves of the U.S. Northeast: The Case of Southern Chester County, Pennsylvania*. Research Paper 95-20. Washington, DC: U.S. Census Bureau.

Gauthier, J. G. (2002). *Measuring America: The Decennial Censuses from 1790 to 2000*. Report POL/02-MA(RV). Washington, DC: U.S. Census Bureau.

Gerber, E. (2003). *Privacy Schemas and Data Collection: An Ethnographic Account.* Census 2000 Ethnographic Study. Washington, DC: U.S. Census Bureau.

Gerber, E., A. Dajani, and M. Scaggs (2002). *An Experiment to Improve Coverage Through Revised Roster Instructions.* Census 2000 Alternative Questionnaire Experiment Report. Washington, DC: U.S. Census Bureau.

Gerber, E. R. (2004). Qualitative research: Residence instructions on the decennial short form. U.S. Census Bureau. Presentation to the Panel on Residence Rules in the Decennial Census, July 15.

Gober, P. and R. C. Mings (1984). A geography of nonpermanent residence in the U.S. *Professional Geographer 35*(2), 154–173.

Grall, T. S. (2003, October). *Custodial Mothers and Fathers and Their Child Support: 2001.* Current Population Reports, Number P60-225. Washington, DC: U.S. Census Bureau.

Greenwood, P. W. (1995). Juvenile crime and juvenile justice. In J. Q. Wilson and J. Petersilia (Eds.), *Crime*, Chapter 5. San Francisco: ICS Press.

Griffin, D. H. (2005). Possible effects of the residence rule discrepancy between the decennial census and the American Community Survey. U.S. Census Bureau. Presentation to the Panel on Residence Rules for the Decennial Census, June 2.

Griffin, D. H. and S. M. Obenski (2001, September 28). *A Demonstration of the Operational Feasibility of the American Community Survey.* Census 2000 Testing, Experimentation, and Evaluation Program Report. Washington, DC: U.S. Census Bureau.

Groves, R. M., D. A. Dillman, J. L. Eltinge, and R. J. Little (Eds.) (2002). *Survey Nonresponse.* New York: Wiley.

Hainer, P. C. (1987). A brief and qualitative anthropological study exploring the reasons for census coverage error among low income black households. Report submitted to U.S. Census Bureau.

Hainer, P. C. (1994). "Strivin,' gettin' over, and keepin' people out of our business:" Pushing household boundaries in survey data collection. In *Proceedings of the Survey Research Methods Section, American Statistical Association*, pp. 337–341. Alexandria, VA: American Statistical Association.

Happel, S. K. and T. D. Hogan (2002). Counting snowbirds: The importance of and the problems with estimating seasonal populations. *Population Research and Policy Review 21*, 227–240.

Harrison, P. M. and A. J. Beck (2005, October). *Prisoners in 2004.* Bureau of Justice Statistics Bulletin NCJ 210677. Washington: U.S. Department of Justice, Office of Justice Programs.

Hernandez, D. J. (2004). Demographic change and the life circumstances of immigrant families. *The Future of Children 14*(2), 17–47.

Heyer, R. and P. Wagner (2004, April). Too big to ignore: How counting people in prisons distorted Census 2000. Available: http://www.prisonersofthecensus.org/toobig/toobig.shtml [12/10/05].

Hill, M. S. (1987). Discussion of 'Dealing with residential ambiguity in the 1990 Census of Population' by Ira S. Lowry. See Council of Professional Associations on Federal Statistics (1987).

Hogan, T. D. and D. N. Steinnes (1996). Arizona sunbirds and Minnesota snowbirds: Two species of the elderly seasonal migrant genus. *Journal of Economic and Social Measurement 22*, 129–139.

Hollmann, W. (1987). An examination of the application of residence rules to the military population. See Council of Professional Associations on Federal Statistics (1987).

Hopper, K. (1992). Counting the homeless: S-Night in New York. *Evaluation Review 16*(4), 376–388.

Hunter, J. (2005, December 21). *Report on Cognitive Testing of Cohabitation Questions*. Survey Methodology Study Series #2005–06. Washington, DC: U.S. Census Bureau.

Hunter, J. and M. de la Puente (2005, March 22). *Report on the Cognitive Testing of the Space Saving Features, Roster Questions, and Revised Instructions for the 2005 National Census Test*. Survey Methodology Study Series #2005–03. Washington, DC: U.S. Census Bureau.

Hunter, J., M. de la Puente, and M. Salo (2003). *Comparative Ethnographic Research on Mobile Populations*. Census 2000 Evaluation J.3. Washington, DC: U.S. Census Bureau.

Hunter, J. and T. DeMaio (2005, December 21). *Report on Cognitive Testing of Tenure, Age, and Relationship Questions for the 2005 National Content Text*. Survey Methodology Study Series #2005–05. Washington, DC: U.S. Census Bureau.

Iversen, R. R., F. F. Furstenberg, Jr., and A. A. Belzer (1999). How much do we count? Interpretation and error-making in the decennial census. *Demography 36*(1), 121–134.

Jackson Citizen Patriot (2006, February 9). Counting convicts: Do it here or there? [editorial]. *Jackson Citizen Patriot*.

Jenkins, C. R. and D. A. Dillman (1995). Towards a theory of self-administered questionnaire design. See Lyberg et al. (1995), Chapter 7, pp. 165–196.

Jonas, K. (2003, August 6). *Group Quarters Enumeration*. Census 2000 Evaluation E.5, Revision 1. Washington, DC: U.S. Census Bureau.

Jones, C. D. (1987, October 15). *Enumeration and Residence Rules for the 1990 Census*. 1990 Decennial Census Policy Memorandum No. 12. Washington, DC: U.S. Census Bureau.

Jones, J. (2000, June 15). *Census Observation of Group Quarters Enumeration: Seattle, Washington*. DSSD Census 2000 Procedures and Operations Memorandum Series O-12. Washington, DC: U.S. Census Bureau.

Kamen, C. S. (2005). The 2008 Israel integrated census of population and housing: Basic conception and procedure. Report, Central Bureau of Statistics, State of Israel. Available: http://www.cbs.gov.il/mifkad/census2008_e.pdf [5/25/05].

Keathley, D. (2003, September 29). *Evaluation of Nonresponse Followup—Mover Probe*. Census 2000 Evaluation I.3. Washington, DC: U.S. Census Bureau.

Kelly, J. B. (1994). The determination of child custody. *The Future of Children 4*(1), 121–142.

King, R. D. (1998). Prisons. See Tonry (1998), pp. 589–625.

Kozak, L. (2002). Hospital transfers to LTC facilities in the 1990s. *Long-Term Care Interface 3*(6), 34–38.

Kreider, R. M. and J. Fields (2005, July). *Living Arrangements of Children: 2001*. Current Population Reports, Number P70-104. Washington, DC: U.S. Census Bureau.

Krosnick, J. and M. Berent (1993). Comparisons of party identification and policy preferences: The impact of survey question format. *American Journal of Political Science 37*, 941–964.

Kusmer, K. L. (2002). *Down and Out, On the Road: The Homeless in American History*. New York: Oxford University Press.

Lestina, Jr., F. A. (2003, September 30). *Analysis of the Linguistically Isolated Population in Census 2000*. Census 2000 Evaluation A.5.a. Washington, DC: U.S. Census Bureau.

Levingston, K. D. and C. Muller (2006, February 16). *"Home" in 2010: A Report on the Feasibility of Enumerating People in Prison at their Home Addresses in the Next Census*. New York: Brennan Center for Justice at New York University School of Law.

Lewis, R. (2005, January 4). IG says US census undercounts, hurts Mass[achusetts]. *Boston Globe*. Posted to http://www.boston.com.

Lin, I.-F., N. C. Schaeffer, J. A. Seltzer, and K. L. Tuschen (2004). Divorced parents' qualitative and quantitative reports of children's living arrangements. *Journal of Marriage and Family 66*, 385–397.

Lobo, S. (2001). American Indian urban mobility in the San Francisco Bay area. Unpublished paper prepared for U.S. Census Bureau.

Lotke, E. and P. Wagner (2004). Prisoners of the census: Electoral and financial consequences of counting prisoners where they go, not where they come from. *Pace Law Review 24*(2), 587–607.

Lowenthal, T. A. (2005). Congressional interest in residence rules: Recent history. Unpublished report commissioned by Panel on Residence Rules in the Decennial Census.

Lowry, I. S. (1987). Discussion of 'Dealing with residential ambiguity in the 1990 Census of Population' by Ira S. Lowry. See Council of Professional Associations on Federal Statistics (1987).

Lujan, C. (1990). *As Simple as One, Two, Three: Census Underenumeration Among the American Indians and Alaska Natives*. Undercount Behavioral Research Group Staff Working Paper #2. Washington, DC: U.S. Census Bureau.

Lyberg, L., P. Biemer, M. Collins, E. deLeeuw, C. Dippo, N. Schwarz, and D. Trewin (Eds.) (1995). *Survey Measurement and Process Quality*. New York: Wiley.

Maccoby, E. and R. Mnookin (1992). *Dividing the Child: Social and Legal Dilemmas of Custody*. Cambridge: Harvard University Press.

Mann, E. S. (1987). Discussion of 'Dealing with residential ambiguity in the 1990 Census of Population' by Ira S. Lowry. See Council of Professional Associations on Federal Statistics (1987).

Martin, E. (1992). Asessment of S-Night street enumeration in the 1990 census. *Evaluation Review 16*(4), 418–438.

Martin, E. (1999). Who knows who lives here? Within-household disagreements as a source of survey coverage error. *Public Opinion Quarterly 63*(2), 220–236.

Martin, E. (2002). *Questionnaire Effects on Reporting of Race and Hispanic Origin: Results of a Replication of the 1990 Mail Short Form in Census 2000*. Census 2000 Alternative Questionnaire Experiment Report. Washington, DC: U.S. Census Bureau.

Martin, E. (2004). The validity of census assumptions and rules about residence (Part 1. Evidence from the Living Situation Survey). Unpublished Census Bureau Memorandum.

Martin, E., E. Gerber, and C. Redline (2003, August 28). *Synthesis Report: Census 2000 Alternative Questionnaire Experiment*. Census 2000 Alternative Questionnaire Experiment Report. Washington, DC: U.S. Census Bureau.

Massey, D. S. (1986). The settlement process among Mexican migrants to the United States. *American Sociological Review 51*(5), 670–684.

Massey, D. S. and C. Capoferro (2004). Measuring undocumented migration. *International Migration Review 38*(3), 1075–1102.

McCormick, J. C. and G. S. Chulis (2003). Growth in residential alternatives to nursing homes: 2001. *Health Care Financing Review 24*(4), 143–150.

McHugh, K. E., T. D. Hogan, and S. K. Happel (1995). Multiple residence and cyclical migration: A life course perspective. *Professional Geographer 47*(3), 251–267.

McMillen, D. (2000a). Americans overseas. See Anderson (2000), pp. 31–33.

McMillen, D. (2000b). Apportionment and redistricting. See Anderson (2000), pp. 34–42.

McNally, T. (2002, November 6). *Service-Based Enumeration*. Census 2000 Evaluation E.6. Washington, DC: U.S. Census Bureau.

Mills, K. M. (1993). Americans overseas in U.S. censuses. Technical Paper 62, U.S. Census Bureau, Population Division. Available: http://www.census.gov/population/www/socdemo/overseas/twps0062.html.

Mings, R. C. (2001). Tracking Arizona snowbirds for census 2000. Unpublished paper prepared for U.S. Census Bureau.

Moul, D. A. (2003, May 9). *Coverage Improvement Followup*. Census 2000 Evaluation I.4. Washington, DC: U.S. Census Bureau.

Mule, T. (2002, December 31). *A.C.E. Revision II Results: Further Study of Person Duplication*. DSSD Revised A.C.E. Estimates Memorandum Series PP-51. Washington, DC: U.S. Census Bureau.

Mumola, C. J. (2000, August). *Incarcerated Parents and Their Children*. Bureau of Justice Statistics Bulletin NCJ 182335. Washington, DC: U.S. Department of Justice, Office of Justice Programs.

National Center for Health Statistics (2005, June 28). Births, marriages, divorces, and deaths: Provisional data for 2004. *National Vital Statistics Reports 53*(21).

National Law Center on Homelessness and Poverty (2004, January 14). *Homelessness in the United States and the Human Right to Housing*. Washington, DC: National Law Center on Homeless and Poverty.

National Research Council (1993). *A Census That Mirrors America*. Panel to Evaluate Alternative Census Methods. Committee on National Statistics, Commission on Behavioral and Social Sciences and Education. Washington, DC: National Academy Press.

National Research Council (1994). *Counting People in the Information Age*. Panel to Evaluate Alternative Census Methods. Duane L. Steffey and Norman M. Bradburn (Eds.), Committee on National Statistics, Commission on Behavioral and Social Sciences and Education. Washington, DC: National Academy Press.

National Research Council (2004a). *Measuring Racial Discrimination*. Panel on Methods for Assessing Discrimination. Rebecca M. Blank, Marilyn Dabady, and Constance F. Citro (Eds.), Committee on National Statistics, Division of Behavioral and Social Sciences and Education. Washington, DC: The National Academies Press.

National Research Council (2004b). *Reengineering the 2010 Census: Risks and Challenges*. Panel on Research on Future Census Methods, Daniel L. Cork, Michael L. Cohen, and Benjamin F. King (Eds.), Committee on National Statistics. Washington, DC: The National Academies Press.

National Research Council (2004c). *The 2000 Census: Counting Under Adversity*. Panel to Review the 2000 Census. Constance F. Citro, Daniel L. Cork, and Janet L. Norwood (Eds.), Committee on National Statistics, Division of Behavioral and Social Sciences and Education. Washington, DC: The National Academies Press.

National Research Council (2005). *Principles and Practices for a Federal Statistical Agency* (Third ed.). Margaret E. Martin, Miron L. Straf, and Constance F. Citro (Eds.), Committee on National Statistics, Division of Behavioral and Social Sciences and Education. Washington, DC: The National Academies Press.

National Research Council and Institute of Medicine (2001). *Juvenile Crime, Juvenile Justice*. Joan McCord, Cathy Spatz Wisdom, and Nancy A. Crowell (Eds.), Committee on Law and Justice and Board on Children, Youth, and Families. Washington, DC: National Academy Press.

New York State Office of Children and Family Services (2004). *2004 Monitoring and Analysis Profiles with Selected Trend Data: 2000–2004*. Albany: State of New York.

New York Times (2005, September 26). Phantom constituents in the census [editorial]. *New York Times*.

New York Times (2006, April 18). Counting noses in prison [editorial]. *New York Times*.

Nguyen, N. T. and M. F. E. Zelenak (2003, June 24). *Coverage Gains from Coverage Questions C1 and C2 on Enumerator-Completed Questionnaires for Census 2000*. Census 2000 Evaluation I.5. Washington, DC: U.S. Census Bureau.

Office for National Statistics (2003, October). *Census Strategic Development Review—Alternatives to a Census: Review of International Approaches*. Information Paper. London: United Kingdom Office for National Statistics.

Office for National Statistics (2004, March). *The 2011 Census: A Design for England and Wales*. Information Paper 1. London: Office for National Statistics.

Office for National Statistics (2005). Discussion paper: Population base for 2011 census enumeration. Unpublished discussion paper; undated, but assumed to be 2005 to match others in series, Office for National Statistics, London.

Office for National Statistics, General Register Office for Scotland, and Northern Ireland Statistics and Research Agency (2004). *Census 2001: Definitions.* London: United Kingdom Office for National Statistics.

O'Hare, W. P. (1999). The overlooked undercount: Children missed in the decennial census. KIDS COUNT working paper, Annie E. Casey Foundation, Washington, DC.

Orasi, A. and A. Ferruzza (2001, July). Population and housing censuses in Italy. Paper presented at the Symposium on Global Review of 2000 Round of Population and Housing Censuses: Mid-Decade Assessment and Future Prospects, United Nations Statistics Division.

Palerm, J. V. (1994). *Immigrant and Migrant Farm Workers in the Santa Maria Valley, California.* Research Paper EX95-21. Washington, DC: U.S. Census Bureau.

Parent, D. G., V. Lieter, S. Kennedy, L. Livens, D. Wentworth, and S. Wilcox (1994). *Conditions of Confinement: Juvenile Detention and Corrections Facilities.* NRJ 145793. Washington, DC: Office of Juvenile Justice and Delinquency Prevention.

Pendall, R. (2003, August). *Upstate New York's Population Plateau.* Washington, DC: The Brookings Institution. Available: http://www.brookings.edu/es/urban/publications/200308_Pendall.htm [1/25/06].

Pérez, A., K. O'Neil, and S. Gesiriech (2003). *Demographics of Children in Foster Care.* Research Report. Washington, DC: Pew Commission on Children in Foster Care.

Petersilia, J. (1998). Probation and parole. See Tonry (1998), pp. 563–588.

Petersilia, J. (2001). When prisoners return to communities: Political, economic and social consequences. *Correctional Management Quarterly* 5(1), 1–10.

Pew Commission on Children in Foster Care (2004). *Fostering the Future: Safety, Permanence and Well-Being for Children in Foster Care.* Washington, DC: Pew Commission on Children in Foster Care.

Price, A. (2005, November 30). Census to study how prisoners are counted: Change could have important political changes in Texas. *Austin American-Statesman.* Posted to newspaper Web site and available: http://www.statesman.com/metrostate/content/metro/stories/11/30CENSUS.html [11/30/05].

Redline, C., D. Dillman, A. Dajani, and M. A. Scaggs (2002). *The Effects of Altering the Design of Branching Instructions on Navigational Performance in Census 2000.* Census 2000 Alternative Questionnaire Experiment Report. Washington, DC: U.S. Census Bureau.

Redline, C. D. and D. A. Dillman (2002). The influence of alternative visual designs on respondents' performance with branching instructions in self-administered questionnaires. See Groves et al. (2002), pp. 179–193.

Rolark, S. J. (1995, October 20). Residence rules for the 2000 census. Memorandum to Arthur J. Norton, Chief, Population Division, U.S. Census Bureau.

Rollinson, P. A. (1990). The everyday geography of poor elderly hotel tenants in Chicago. *Geografiska Annaler. Series B, Human Geography 72*(2/3), 47–57.

Roman, C. G. and J. Travis (2004, March 8). *Taking Stock: Housing, Homelessness, and Prisoner Reentry*. Washington, DC: Justice Policy Center, Urban Institute.

Ruggles, S. (2003). Measurement of household and family composition in census 2000: An update. *Population and Development Review 29*(3), 480–482.

Ruggles, S. and S. Brower (2003). Measurement of household and family composition in the United States, 1850–2000. *Population and Development Review 29*(1), 73–101.

Schachter, J. P. and J. J. Kuenzi (2002). Seasonality of moves and the duration and tenure of residence: 1996. Working Paper 69, U.S. Census Bureau, Population Division.

Schaeffer, N. C., J. A. Seltzer, and J. Dykema (1998). Methodological and theoretical issues in studying nonresident fathers: A selective review. Working Paper 98–02, National Center on Fathers and Families, University of Pennsylvania, Philadelphia.

Schwarz, N., C. E. Grayson, and B. Knäuper (1998). Formal features of rating scales and the interpretation of question meaning. *International Journal of Public Opinion Research 19*(2), 177–183.

Schwede, L. (2003). *Complex Households and Relationships in the Decennial Census and in Ethnographic Studies of Six Race/Ethnic Groups*. Census 2000 Testing, Experimentation, and Evaluation Program Report. Washington, DC: U.S. Census Bureau.

Schwede, L., R. L. Blumberg, and A. Y. Chan (Eds.) (2005). *Complex Ethnic Households in America*. Lanham, MD: Rowman and Littlefield.

Seltzer, J. A. (2004). Cohabitation and family change. In M. Coleman and L. H. Ganong (Eds.), *Handbook of Contemporary Families: Considering the Past, Contemplating the Future*, Chapter 4, pp. 57–78. Thousand Oaks, CA: Sage Publications.

Sheley, J. F. (Ed.) (2000). *Criminology: A Contemporary Handbook* (Third ed.). Belmont, CA: Wadsworth.

Sheppard, D. (2003, July 29). *Coverage Edit Followup*. Census 2000 Evaluation I.1. Washington, DC: U.S. Census Bureau.

Shiono, P. H. and L. S. Quinn (1994). Epidemiology of divorce. *The Future of Children 4*(1), 15–28.

Shryock, H. S. (1950). Plans for the 1950 census. *Population Index 16*(1), 3–13.

Sickmund, M. (2002, December). *Juveniles Residential Facility Census, 2000: Selected Findings*. Juvenile Offenders and Victims National Report Series Bulletin. Pittsburgh: Office of Juvenile Justice and Delinquency Prevention, Office of Justice Programs, U.S. Department of Justice.

Sickmund, M. (2004, June). *Juveniles in Corrections*. Juvenile Offenders and Victims National Report Series Bulletin. Pittsburgh: Office of Juvenile Justice and Delinquency Prevention, Office of Justice Programs, U.S. Department of Justice.

Simmons, T. and G. O'Neill (2001). *Households and Families: 2000*. Census 2000 Brief C2KBR/01-8. Washington, DC: U.S. Census Bureau.

Singer, A. and D. S. Massey (1998). The social process of undocumented border crossing among Mexican migrants. *International Migration Review 32*(3), 561–592.

Smith, A. C. and D. I. Smith (2001). *Emergency and Transitional Shelter Population: 2000*. Census 2000 Special Reports CENSR/01-2. Washington, DC: U.S. Census Bureau.

Smith, C. W. (2005, May). Assessment of the implications of the choice of population base for the continuity of population measurement. Unpublished discussion paper, Office for National Statistics, London.

Smith, D. M. and G. J. Gates (2001). *Gay and Lesbian Families in the United States: Same-Sex Unmarried Partner Households: A Preliminary Analysis of 2000 United States Census Data*. Washington, DC: Human Rights Campaign.

Smith, D. S. (1992). The meanings of family and household: Change and continuity in the mirror of the American census. *Population and Development Review 18*(3), 421–456.

Smith, S. and M. House (2004a). Temporary migration flows in Florida. Unpublished paper presented at the annual meeting of the Southern Demographic Association, October 14–16, 2004.

Smith, S. K. (1989). Toward a methodology for estimating temporary residents. *Journal of the American Statistical Association 84*(406), 430–436.

Smith, S. K. and M. House (2004b, October). Snowbirds and other temporary residents: Florida, 2004. Report, Bureau of Economic and Business Research, University of Florida. Available: http://www.bebr.ufl.edu/Articles/Temp_Residents_2004.pdf [11/30/05].

Smith, S. K. and M. House (2005). Snowbirds, sunbirds, and stayers: Seasonal migration of the elderly in Florida. Paper presented at the annual meeting of the Population Association of America, Philadelphia. Available: http://www.bebr.ufl.edu/Articles/PAA2005_ElderlyMigr.pdf [7/26/05].

Stark, L. (1992). Counting the homeless: An assessment of S-Night in Phoenix. *Evaluation Review 16*(4), 400–408.

Statistics New Zealand (2001, December). *2001 Census of Population and Dwellings: Definitions and Questionnaires*. Wellington: Statistics New Zealand.

Statistics New Zealand (2003, October). *2006 Census of Population and Dwellings: Final Report on Content*. Wellington: Statistics New Zealand.

Steuart, W. M. (1921). The conduct of the fourteenth census. *Quarterly Publications of the American Statistical Association 17*(133), 569–583.

Stokes, P. (2005a, May). *The 2011 Census: Initial View on Content for England and Wales—Population Base*. Consultation document. London: Office for National Statistics.

Stokes, P. (2005b, January). Comparison of composite questionnaire options for 2011 census. Unpublished discussion paper, Office for National Statistics, London.

Strayhorn, C. K. (2004). *Forgotten Children: A Special Report on the Texas Foster Care System*. Austin, TX: State of Texas, Office of the Comptroller.

Sweet, E. M. and N. S. Alberti (1994). Coverage implications of the term 'usual residence'. In *Proceedings of the Survey Research Methods Section, American Statistical Association*, pp. 319–324. Alexandria, VA: American Statistical Association.

Sweet, J. A. (1987). Enumeration of persons with ambiguous residence [discussion of 'Dealing with residential ambiguity in the 1990 Census of Population' by Ira S. Lowry]. See Council of Professional Associations on Federal Statistics (1987).

Sweet, J. A. and L. L. Bumpass (1996). The National Survey of Families and Households—waves 1 and 2: Data description and documentation. Technical report, Center for Demography and Ecology, University of Wisconsin–Madison.

Sweet, J. A. and L. L. Bumpass (2002). The National Survey of Families and Households—waves 1, 2, and 3: Data description and documentation. Technical report, Center for Demography and Ecology, University of Wisconsin–Madison.

Sweet, J. A., L. L. Bumpass, and V. Call (1988). The design and content of the National Survey of Families and Households. Technical report, Center for Demography and Ecology, University of Wisconsin–Madison.

Taeuber, C. M. and P. M. Siegel (1991). Counting the nation's homeless population in the 1990 census. In C. M. Taeuber (Ed.), *Conference Proceedings for Enumerating Homeless Persons: Methods and Data Needs*. Washington, DC: Bureau of the Census.

Thornburgh, R. (2001). *State of Kansas: Adjustment to the 2000 U.S. Decennial Census*. Topeka, KS: Secretary of State, State of Kansas.

Tonry, M. (Ed.) (1998). *The Handbook of Crime and Punishment*. New York: Oxford University Press.

Tourangeau, R. (1984). Cognitive sciences and survey methods. In *Cognitive Aspects of Survey Methodology: Building a Bridge Between Disciplines*, pp. 73–100. Report of the Advanced Research Seminar on Cognitive Aspects of Survey Methodology. Thomas B. Jabine, Miron L. Straf, Judith M. Tanur, and Roger Tourangeau (Eds.), Committee on National Statistics, Commission on Behavioral and Social Sciences and Education. Washington, DC: National Academy Press.

Tourangeau, R., F. Conrad, Z. Arens, S. Fricker, S. Lee, and E. Smith (2006). Everyday concepts and classification errors: Judgments of disability and residence. *Journal of Official Statistics*. In press.

Tourangeau, R., M. P. Couper, and F. G. Conrad (2004). Spacing, position, and order: Interpretive heuristics for visual features of survey questions. *Public Opinion Quarterly 68*, 368–393.

Tourangeau, R., G. Shapiro, A. Kearney, and L. Ernst (1997). Who lives here? Survey undercoverage and household roster questions. *Journal of Official Statistics 13*, 1–18.

Travis, J., A. L. Solomon, and M. Waul (2001, June). *From Prison to Home: The Dimensions and Consequences of Prisoner Reentry*. Justice Policy Center. Washington, DC: Urban Institute.

Trewin, D. (2005). *Census of Population and Housing: Nature and Content, Australia, 2006*. Information Paper 2008.0. Canberra: Australian Bureau of Statistics.

United Nations Economic Commission for Europe and Statistical Office of the European Communities (1998). *Recommendations for the 2000 Censuses of Population and Housing in the ECE Region*. Statistical Standards and Studies Report Number 49. New York and Geneva: United Nations.

United Nations Statistics Division (1997). *Principles and Recommendations for Population and Housing Censuses: Revision 1*. Statistical Paper, ST/ESA/STAT/SER.M/67/Rev.1. New York: United Nations.

United Nations Statistics Division (2004, August). Topics of emerging policy relevance and suggested minimum and essential census topics: Implications for updating the *Principles and Recommendations for Population and Housing Censuses*. Paper prepared for the United Nations Expert Group Meeting to Review Critical Issues Relevant to the Planning of the 2010 Round of Population and Housing Censuses.

U.S. Bureau of the Census (1955). *The 1950 Censuses—How They Were Taken*. Procedural Studies of the 1950 Censuses, No. 2: Population, Housing, Agriculture, Irrigation, Drainage. Washington, DC: U.S. Government Printing Office.

U.S. Bureau of the Census (1970). *1970 Census of Population and Housing—Procedural History, PHC(R)-1*. Washington, DC: U.S. Government Printing Office.

U.S. Census Bureau (1998). *Current Population Survey, March/April 1998—Match File: Child Support*. Washington, DC: U.S. Census Bureau. Data documentation file. Available: http://www.census.gov/apsd/techdoc/cps/cpsapr98.pdf [7/1/06].

U.S. Census Bureau (1999). *Navigating the Census: A Guide for Merchant Mariners and Their Onshore Advocates*. Publication D-3278 (8-99). Washington, DC: U.S. Census Bureau.

U.S. Census Bureau (2001). *Technical Documentation, Summary File 1, 2000 Census of Population and Housing*. Washington, DC: U.S. Census Bureau.

U.S. Census Bureau (2002). *Design and Methodology*. Current Population Survey Technical Paper 63RV. Washington, DC: U.S. Census Bureau.

U.S. Census Bureau (2006a, May). *Design and Methodology: American Community Survey*. Washington, DC: U.S. Census Bureau. Technical Paper 67, Unedited Version.

U.S. Census Bureau (2006b, February 21). *Tabulating Prisoners at their "Permanent Home of Record" Address*. Report to Congressional Appropriations Committees. Washington, DC: U.S. Census Bureau.

U.S. Census Bureau, Population Division (2004). Residence situations to inform the 2010 census residence rules. Background material presented at meeting of the Panel on Residence Rules in the Decennial Census, December 13.

U.S. Census Office (1902). *Twelfth Census of the United States, Population, Part II*. Washington, DC: U.S. Government Printing Office. Available: http://www.census.gov/prod/www/abs/decennial/1900.htm [3/6/06].

U.S. Department of Commerce (2000). Nearly 100 million census questionnaires in the mail, census workers delivering the rest. Press release, March 13. Available: http://www.census.gov/Press-Release/www/2000/cb00cn24.html [6/1/06].

U.S. Department of Labor (2005). *Findings from the National Agricultural Workers Survey (NAWS) 1997–1998: A Demographic and Employment Profile of United States Farmworkers*. Research Report 9. Office of the Assistant Secretary for Policy, Office of Program Economics: U.S. Department of Labor.

U.S. General Accounting Office (1999). *Formula Grants: Effects of the Adjusted Population Counts on Federal Funding to States*. GAO/HEHS-99-69. Washington, DC: U.S. Government Printing Office.

U.S. General Accounting Office (2003). *Decennial Census: Methods for Collecting and Reporting Data on the Homeless and Others Without Conventional Housing Need Refinement*. GAO-03-227. Washington, DC: U.S. Government Printing Office.

U.S. General Accounting Office (2004). *2010 Census: Overseas Enumeration Test Raises Need for Clear Policy Direction*. Report GAO-04-470. Washington, DC: U.S. Government Printing Office.

U.S. Government Accountability Office (2004a). *2010 Census: Counting Americans Overseas as Part of the Decennial Census Would Not Be Cost-Effective*. Report GAO-04-898. Washington, DC: U.S. Government Printing Office.

U.S. Government Accountability Office (2004b). *American Community Survey: Key Unresolved Issues*. Report GAO-05-82. Washington, DC: U.S. Government Printing Office.

U.S. Government Accountability Office (2005). *Data Quality: Improvements to Count Correction Efforts Could Produce More Accurate Census Data*. Report GAO-05-463. Washington, DC: U.S. Government Printing Office.

U.S. House of Representatives, Committee on Government Reform (2000). *Oversight of the 2000 Census; Examining the Bureau's Policy to Count Prisoners, Military Personnel, and Americans Residing Overseas*. Number 106–39. Washington, DC: U.S. Government Printing Office. Hearing before the Subcommittee on the Census, June 9, 1999.

Usher, C. L., K. A. Randolph, and H. C. Gogan (1999). Placement patterns in foster care. *Social Service Review 73*(1), 22–36.

Valentine, C. A. and B. L. Valentine (1971a). *A Comparative Methodological Study of Underenumeration and Related Problems*. Report prepared under Joint Statistical Agreement with the Brookdale Hospital Center. Washington, DC: U.S. Census Bureau.

Valentine, C. A. and B. L. Valentine (1971b). *Missing Men: A Comparative Methodological Study of Underenumeration and Related Problems*. Report prepared under Joint Statistical Agreement with the Brookdale Hospital Center. Washington, DC: U.S. Census Bureau.

Valley Markets and Tourism Research Center (2005, October). The valley winter visitors and local market report 2004–2005: Fast facts. PowerPoint presentation. University of Texas–Pan American.

Vincent, V., W. Thompson, J. Pagan, and E. Galy (2000, August). South Padre Island spring break 2000. PowerPoint presentation. Center for Tourism Research, University of Texas–Pan American.

Wagner, P. (2002, April 22). Importing constituents: Prisoners and political clout in New York. Available: http://prisonpolicy.org/importing/index.shtml [1/25/06].

Wagner, P. and E. Lotke (2004, December 6). Feasibility of using administrative records or personal surveys to enumerate people in correctional facilities. Memorandum to Paul Voss, Chair, Panel on Residence Rules in the Decennial Census.

Wagner, P., E. Lotke, and A. Beveridge (2006). Why the Census Bureau can and must start collecting the home addresses of incarcerated people. Submitted to the U.S. Census Bureau on February 10, 2006.

Waite, L. (1995). Does marriage matter? *Demography 32*, 483–507.

West, K. K. and J. G. Robinson (1999). What do we know about the undercount of children? Working Paper 39, U.S. Census Bureau, Population Division.

Willis, G. B. (1999). Cognitive interviewing: A "how to" guide. Documentation for short course presented at 1999 Joint Statistical Meetings. Available: http://appliedresearch.cancer.gov/areas/cognitive/interview.pdf [7/26/05].

Wright, J. D. and J. A. Devine (1992a). Counting the homeless: The Census Bureau's 'S-Night' in five U.S. cities. *Evaluation Review 16*(4), 355–364.

Wright, J. D. and J. A. Devine (1992b). The United States census effort to count the homeless: An assessment of street enumeration procedures during 'S-Night' in New Orleans. *Journal of Official Statistics 8*(2), 211–222.

Young, A. A. (1901). The enumeration of children. *Publications of the American Statistical Association 7*(53), 21–48.

Zimring, F. E. and G. Hawkins (1991). *The Scale of Imprisonment*. Chicago: University of Chicago Press.

Zitter, M. (1987). Enumeration of Americans living abroad. See Council of Professional Associations on Federal Statistics (1987).

Appendixes

– A –

Residence Rules of the 2000 Census

HOUSEHOLD POPULATION

Rule 1 Person lives in this household but is temporarily absent on Census Day on a visit, business trip, vacation, or in connection with a job (e.g., bus driver, traveling salesperson, boat operator). This includes foreign nationals whose usual place of residence is in the U.S. and American citizens traveling overseas.

Count person at: This household

Rule 2 Person has multiple residences and, as of Census Day, travels between one residence and another on a "weekly cycle," a "monthly cycle," a "yearly cycle," or some other cycle (e.g., commuter workers, "snowbirds," and children in joint custody situations).

Count person at: The residence where they spend **most** of time during the week, month, or year, etc. If an individual cannot identify such a place for himself/herself, count him/her at the residence where he/she was on Census Day. (See "Guiding Principles" attached for more information [Box A-1].)

Rule 3 Person lives in this household, but is in a general or Veterans Affairs hospital on Census Day. Including newborn babies who have not yet been brought home.

Count person at: This household, unless in a psychiatric or chronic disease hospital ward, or a hospital or ward for the mentally retarded,

Box A-1 Guiding Principles for the Residence Rules as They Apply to Individual(s) with Multiple Residences

The following provides guidance for determining "usual residence" for an individual with more than one residence.

Weekly Cycle

If a person is on a "weekly cycle," he/she should be counted at the residence where he/she spends most of their time during the week. For example: Some individuals live part of the week at a residence near where they work, and live at their "family home" the rest of the week. We consider these people to be on a "weekly cycle," and they should be counted at the residence where they spend most of their time during the week (e.g., commuter workers).

Monthly Cycle

If a person is on a "monthly cycle," he/she should be counted at the residence where he/she spends most of his/her time during the month. For example: Some children live with one parent for one week out of the month and the other parent the remaining three weeks during the month. We consider these individuals to be on a "monthly cycle" and they should be counted at the residence where they spend most of their time during the month (e.g., children in joint custody situations).

Yearly Cycle

If a person is on a "yearly cycle," he/she should be counted at the residence where he/she spends most of his/her time during the year. For example:

a) Some individuals live in one state during the spring, summer, and fall, but move to a state in a warmer climate during the winter months ("snowbirds"). We consider these people to be on a "yearly cycle," and they should be counted at the residence where they spend most of their time during the year.

b) Some college students live at the college during the school year and at the "family home" during holidays or the summer. We consider these people to be on a "yearly cycle," and they should be counted at the residence where they spend most of their time during the year.

No Clearly Defined Cycle

If a person is on no clearly defined "cycle," he/she should be counted at the residence where he/she was on Census Day. For example: Temporary workers may establish another residence for an undefined period of time for work. We consider these people to be on an "undefined cycle," and they should be counted at the residence where they were on Census Day.

Time Split Equally Among Two or More Residences

No matter what the cycle, if time is split equally among multiple residences, a person should be counted at the place where he/she was on Census Day.

the physically handicapped, or drug/alcohol abuse patients. If so, the person should be counted in the hospital.

Rule 4 Person is a member of the U.S. Armed Forces stationed on a nearby military installation or ship but on Census Day is living in this off-base household.

Count person at: The off-base household

Rule 5 Person is a college student not living in this household during the school year and is only here during break or vacation (see Rules 6 and 25).

Count person at: The residence where the person lives while attending college (Usual Home Elsewhere [UHE] not allowed).

Rule 6 Person is a college student living in this household during the school year (see Rules 5 and 25).

Count person at: This household

Rule 7 Person is a student attending school away from home below the college level, such as a boarding school or a Bureau of Indian Affairs boarding school.

Count person at: This household

Rule 8 Person is an officer or crew member of a merchant vessel and on Census Day is engaged in inland waterway transportation.

Count person at: This household

Rule 9 Person works for and lives in this household and has no other home (e.g., a domestic worker or nanny who "lives in").

Count person at: This household

Rule 10 Person is staying temporarily in this household on Census Day and has another home.

Count person at: DO NOT LIST. (This person will be counted at the other household.)

Rule 11 On Census Day, person is a citizen of a foreign country who has established a household (or is part of an established household) in the U.S. while working or studying. This includes any family member living with the person.

Count person at: This household

Rule 12 Person is a citizen of a foreign country and on Census Day is living on the premises of an Embassy, Ministry, Legation, Chancellery, or Consulate in the U.S.

Count person at: This household, that is, the Embassy, etc. (The person has the right to refuse to provide any or all information.)

GROUP QUARTERS POPULATION, UHE ALLOWED

Rule 13 Person is a member of the U.S. Armed Forces and on Census Day is living on a military installation in the United States, or is living on a military vessel which is assigned to a home port in the United States.

Count person at: The residence where the person spends **most** of his/her time (UHE allowed) [GQ code 601 for military barracks on base; GQ code 602 for transient quarters for temporary residents; GQ code 603 for military ships]. If the person does not claim a UHE, count him/her at the military installation or at the home port of the vessel.

Rule 14 On Census Day, person is at a camp for temporary workers, such as agricultural or migrant workers; or logging, pipeline, or construction workers.

Count person at: The residence where the person spends **most** of his/her time (UHE allowed) [GQ code 901 for agriculture workers' dormitories on farms; GQ code 902 for other workers' dormitories]. If the person does not claim a UHE, count him/her at the camp.

Rule 15 On Census Day, person is at a hostel, YMCA/YWCA, or transient location, such as a commercial or public campground, racetrack, park, or carnival (see also Rule 16).

Count person at: The location where they spend **most** of their time (UHE allowed) [GQ code 908 for hostels or YMCAs/YWCAs; GQ code 910 for commercial or public campgrounds, racetracks, fairs, or carnivals]. If the person does not claim a UHE, count them at the special place.

Rule 16 On Census Day, person is at a recreational camp (i.e., a commercial or public campground). This rule is targeted to persons known as "full-timers" or "good-sams" who live and travel in a recreational vehicle, and the recreational vehicle is their only or usual residence.

Count person at: The location where the person spends **most** of his/her time (UHE allowed). If the person does not claim a UHE, count them at the camp. (Note that if the recreational vehicle is their only or usual residence, it is considered a housing unit [HU] and tabulated as an HU. It is part of GQ enumeration but not part of the GQ population.)

Rule 17 On Census Day, person is at a soup kitchen or outreach program (e.g., mobile food van).

Count person at: The location where these individuals spend **most** of their time (UHE allowed) [GQ code 704 for soup kitchens; GQ code 705 for outreach program]. If the person does not claim a UHE, count them at the special place.

Rule 18 Person is an officer or crew member of a U.S. flag merchant vessel and on Census Day is docked in a U.S. port or is sailing from one U.S. port to another U.S. port.

Count person at: These persons are allowed to claim a UHE [GQ code 900]. If they do not claim a UHE, count them at the merchant vessel.

Rule 19 Person is a resident staff member or a member of a special place. For example, a staff member living in a hospital or nursing home, or a member of a religious order living in a monastery or convent.

Count person at: These persons are allowed to claim a UHE [GQ code 904 for staff members living in military hospitals; GQ code 905 for staff members living in civilian group quarters; GQ code 906 for religious group quarters]. If they do not claim a UHE, they are counted at the special place.

GROUP QUARTERS POPULATION, UHE NOT ALLOWED

Rule 20 On Census Day, person is under formally authorized, supervised care or custody, in a correctional institution, such as a federal or state prison, local jail or workhouse, federal detention center, or halfway house.

Count person at: The special place (UHE not allowed)

Rule 21 On Census Day, person is under formally authorized, supervised care or custody, in a nursing, convalescent, or rest home for the aged and dependent.

Count person at: The special place (UHE not allowed)

Rule 22 On Census Day, person is under formally authorized, supervised care or custody, in a juvenile institution such as a residential school for delinquents.

Count person at: The special place (UHE not allowed)

Rule 23 On Census Day, person is under formally authorized, supervised care or custody, in a home, school, hospital, or ward for the physically handicapped, mentally retarded, or mentally ill.

Count person at: The special place (UHE not allowed)

Rule 24 On Census Day, person is at an emergency shelter, including shelters with sleeping facilities for individuals without a usual residence; shelters for abused women; shelters for runaway, neglected, or homeless children; or shelters for other homeless persons.

Count person at: The shelter (UHE not allowed)

Rule 25 Person is a college student living in a group quarters (e.g., a dormitory, or sorority or fraternity house) (see Rules 5 and 6).

Count person at: The group quarters (UHE not allowed)

OVERSEAS POPULATION

Rule 26 Person is a member of the U.S. Armed Forces and on Census Day is stationed on a military vessel which is assigned to a home port in a foreign country.

Count person at: DO NOT LIST. (This person will be counted as part of the overseas population.)

Rule 27 Person is a member of the U.S. Armed Forces and on Census Day is assigned to a military installation outside the United States. This rule includes family members living with him/her.

Count person at: DO NOT LIST. (This person will be counted as part of the overseas population.)

Rule 28 Person is an American citizen overseas employed by the U.S. government and on Census Day has a place of duty abroad. This rule includes family members living with him/her.

Count person at: DO NOT LIST. (This person will be counted as part of the overseas population.)

DO NOT LIST POPULATION

Rule 29 Person is an American citizen and on Census Day is working, studying, or living abroad, but not employed by the U.S. government.

Count person at: DO NOT LIST

Rule 30 Person is a citizen of a foreign country who on Census Day is temporarily traveling or visiting in the U.S.

Count person at: DO NOT LIST

Rule 31 Person is an officer or crew member of a U.S. flag merchant vessel which on Census Day is docked in a foreign port, is sailing from one foreign port to another foreign port, is sailing from a U.S. port to a foreign port, or is sailing from a foreign port to a U.S. port.

Count person at: DO NOT LIST

– B –

Residence Concepts and Questions in Selected Foreign Censuses

In this appendix, we summarize general approaches to residence concepts used in foreign censuses, as well as specific questions and instructions used on their census forms. We begin with a description of two sets of guidelines promulgated by cross-national agencies before describing practices in individual nations. This analysis only covers select countries that continue to conduct traditional head-count censuses (as opposed to records-based censuses or rolling censuses that infer population counts from a sample), and is also limited to nations for which English-language versions or translations of the census instrument are available from the Internet.

B.1 UNITED NATIONS/ECONOMIC COMMISSION OF EUROPE GUIDELINES

In 1998, the United Nations and the Statistical Office of the European Communities jointly issued a set of suggested guidelines for population censuses (U.N. Economic Commission for Europe and Statistical Office of the European Communities, 1998; hereafter, UNECE, 1998). Those guidelines explicitly linked the definition of "usual residence" to sleeping: "A person's usual residence should be that at which he/she spends most of his/her daily night-rest" (UNECE, 1998:10–11). However, the guidelines immediately list several caveats to that definition:

> . . .problems [in stating a usual residence] may arise in dealing with the following groups of persons:
>
> (a) persons who maintain more than one residence, e.g., a town house and a country home;
> (b) students who live in a school or university residence, as boarders in a household or as a one-person household for part of the year and elsewhere during vacations;
> (c) persons who live away from their homes during the working week and return at weekends;
> (d) persons in compulsory military service;
> (e) members of the regular armed forces who live in a military barrack or camp but maintain a private residence elsewhere;
> (f) persons who have been an inmate of a hospital, welfare institution, prison, etc., for a sufficiently long time to weaken their ties with their previous residence to which they may return eventually;
> (g) persons who have been at the place where they are enumerated for some time but do not consider themselves to be residents of this place because they intend to return to their previous place of residence at some future time;
> (h) persons who have recently moved into an area and may not feel that they have lived there long enough to claim it as their place of usual residence—this may apply in particular to immigrants from abroad;
> (i) persons who have left the country temporarily but are expected to return after some time; . . .and
> (j) nomads, homeless and roofless persons, vagrants and persons with no concept of a usual address.

"The treatment of all these cases should be set out clearly in the census instructions," note the guidelines, and, "if possible, objective rules should be formulated for dealing with them" (UNECE, 1998:11).

The guidelines further suggest that "people in groups (a) to (i) should treat the address at which they spend the majority of their daily night-rest to be their usual residence. For persons with a spouse/partner and/or children, the usual residence should be that at which they spend the majority of the time with their family. . . . People in group (j) should be treated as usually resident where they are enumerated."

Of particular note is point (f) on the treatment of persons in "institutional households," which the guidelines later define as "persons whose need for shelter or subsistence are being provided by an institution" (UNECE, 1998:42). The language of (f) evokes the "enduring ties" concept of *Franklin v. Massachusetts* (Box 2-5), but suggests that the tie can decay after a "sufficiently long time." In the discussion of "usual residence," the guidelines provide no specific guidance on what constitutes such a time period, but in discussing "institutional household" the guidelines note that "people who are normally members of private households but who are living in institutions . . .are only considered members of institutional households if their absence from the private households exceeds the one-year time limit specified for the place of usual residence topic" (UNECE, 1998:42).

B.2 UNITED NATIONS STATISTICS DIVISION

The joint recommendations of UNECE (1998) with regard to usual residence are consistent with—albeit more detailed than—general guidance provided by the United Nations Statistics Division (1997) to member nations. Indeed, the Statistics Division guidance merely provides the literal definition of "place of usual residence" as "the geographical place where the enumerated person usually resides." Acknowledging that some groups may have difficulty specifying a usual residence, the 1997 guidelines state only that "the treatment of all such cases should be clearly set forth in the census instructions."

A planning paper suggesting revisions to the 1997 principles acknowledges their deficiencies, noting in particular that "defining the place of usual residence as the geographical place where the enumerated person usually resides implies a time element," yet "the recommendations do not offer any time limit for considering oneself a usual resident of a place" (United Nations Statistics Division, 2004:3). Citing companion recommendations on the measurement of international migration statistics, the planning paper notes a trend toward defining "country of usual residence" as "the country in which he or she has a place to live where he or she normally spends the daily period of rest." The planning paper offers no specific recommendation, but suggests that a revision of census principles for the 2010 round of censuses "could recommend time periods to be used for defining a 'usual' resident of a place" and offer "more guidance [to help determine] the treatment of persons who appear to have more than one residence" (United Nations Statistics Division, 2004:4).

B.3 AUSTRALIA

The Australian Bureau of Statistics (ABS) takes a multifaceted approach to residence concepts; the core quinquennial census is operated as a *de facto* census, enumerating people where they are on a designated Census Night.[1] However, the census asks questions on usual residence and usual residence 5 years ago. Moreover, ABS formally defines "usual residence" in two ways, one for use in the census and the other for its regular demographic surveys.

In 2001, as in previous censuses, the Australian census was conducted by field visits; enumerators dropped off forms at households and later returned

[1] In 2001, that night was August 7; in 2006, it will be August 8.

to collect them. For the 2006 census, ABS plans to increase its reliance on mailback census responses (particularly for areas with high concentrations of apartment buildings), and it will permit Internet responses (Office for National Statistics, 2003). Though the census objective is a *de facto* count, information on usual residence is also gathered in order to compute the estimated resident population (ERP), ABS' official population estimate series, which is used for electoral and fund allocation purposes. To convert from the census figures to ERP, three statistical adjustments are made: one each for estimated census coverage errors (undercount and overcount), Australian residents temporarily overseas, and backdating from Census Night to an official reference date of June 30.

An ABS review of the "usual residence" concept concluded that it is "impractical . . .to devise a single standard concept of Usual Residence that satisfies the full range of analytical requirements of users and operational exigencies" (Australian Bureau of Statistics, 2004). In simplest terms, ABS considers a usual residence "the dwelling a person calls home and resides in on a permanent basis" and it is defined "based on the fact that each person has a basic attachment to a particular dwelling." That attachment does not lend itself to a single definition, and so ABS defines two separate usual residence concepts:

- Attachment to the dwelling *in which a person lives the majority of the time*, which is the concept used in the quinquennial census. The formal definition allows for prospective definition of residence; the concept refers to the address at which "a person has lived or intends to live" for 6 months or more in a year, "even if they do not regard it as their home and do not have a strong social, economic or familial attachment to it."
- Attachment to the dwelling *which a person considers to be their "home," or family home*, which is used in the Australian Monthly Population Survey and other household surveys. This concept, based on self-perception, "embodies social, economic and familial attachment to a dwelling because it contains their household or family home." Under this concept, a person may "be considered as a usual resident of the dwelling in which their family home is located even if they do not live there the majority of the time."

"As the majority of people live in their family home the majority of the time the same dwelling constitutes their usual residence in both cases."

Figure B-1 illustrates the form of the usual residence question as it is planned to be asked in the 2006 Australian census; as noted in the figure, the 2006 format is similar to the presentation used in 2001. ABS household surveys like the Monthly Population Survey depend greatly on enumerator interviews, including those by telephone. The ABS usual residence concept paper (Australian Bureau of Statistics, 2004) notes that the usual residence

8 Where does the person usually live?

- For persons who usually live in another country and who are visiting Australia for less than one year, mark 'Other country'.
- For other persons, 'usually live' means that address at which the person has lived or intends to live for a total of six months or more in 2006.
- For persons who now have no usual address, write 'NONE' in the 'Suburb/Locality' box.
- For boarders at boarding school, write the address of the boarding school or college.
- Remember to mark box like this: ▬

○ The address shown on the front of this form

○ Elsewhere in Australia – please specify address

Apartment/Flat/Unit number (if any)

Street number

Street name

Suburb/Locality

State/Territory Postcode

○ **Other country**

NOTE: The 2006 version of the question is very similar to the one used in 2001; "Apartment/Flat/Unit number (if any)" has been added, and the 2001 instructions instructed "persons who now have no usual address" to write "no usual address" but did not specify where to write that text. The bulleted point on "persons who usually live in another country" was a new addition to the question in 2001.

SOURCE: Trewin (2005).

Figure B-1 Proposed form of basic usual residence questionnaire item (UR1), 2006 Census of Population and Housing, Australia

question on the household survey forms first asks "What are the names of all the people who usually live here?" and then asks "Will any of these people be staying away tonight?" If a respondent answers yes to the latter question, he or she is then asked, "Will [name of person] be away for more than six weeks altogether?" If the answer to the third question is yes, that person's record is retained for household (family) coding but they are excluded from being counted in the survey.

The usual residence question has been on the Australian census form for every census since 1961, except for 1966. In 1961, only the person's state or territory of usual residence was requested; in all other instances full address information was collected. A question asking respondents for their full address of usual residence 5 years ago has also appeared in all Australian censuses since 1971, and a question on usual residence 1 year ago became a fixture in 1976 (although there has been variation in whether a full address was requested or merely a state/territory). A planning document indicates that the three usual residence questions will be reviewed prior to the 2006 census and that "the standard for usual residence is under review" (Edwards, 2003:25), and similar language is included in the first report detailing content for the 2006 census (Trewin, 2005).

B.4 CANADA

The Canadian *de jure* census is conducted every 5 years. Since 1971, questionnaires have been dropped off by enumerators and are expected to be returned by mail. Broader mailout of questionnaires to two-thirds of census addresses was implemented in 2006, after completion of an address register (Office for National Statistics, 2003).

The 2001 Canadian census form devotes a full page to a set of include/exclude instructions; this instruction block is reproduced in Figure B-2. The 32-page "Census Guide" prepared by Statistics Canada to provide additional information for respondents lists a basic rationale for the usual residence questions in Steps B and C: "These steps help you to decide who **should** be included and who **should not** be included in the questionnaire. They tell us that we have counted everyone we need to count and that we have not counted anyone twice."

For the 2006 Canadian census, the residence instructions are much the same as in 2001 but are edited for syntax, so that all the bulleted points read as complete phrases rather than colon-separated definitions. Some proposed changes are more substantial:

- The potentially confusing point on "**ABSENT SPOUSES**" is modified to read: "**SPOUSES OR COMMON-LAW PARTNERS WHO LIVE ELSEWHERE** while working or studying, but who return here periodically."
- The point on persons in the country with some form of work permit is simplified to: "**PERSONS FROM ANOTHER COUNTRY WITH A WORK OR STUDY PERMIT** and family members living here with them."
- Emphasis is added to parts of the last two bulleted points, on institutional residents and persons with no other home. They are now slated to read "**PERSONS** who usually live here but are now **IN AN INSTITUTION** (such as a home for the aged, a hospital or a prison), **IF THEY HAVE BEEN THERE LESS THAN SIX MONTHS**" and "**PERSONS** staying here on May 16, 2006, **WHO HAVE NO USUAL HOME ELSEWHERE.**"

WHOM TO INCLUDE IN STEP B

- **EVERYONE WHO USUALLY LIVES HERE, AT THIS ADDRESS:** including newborn babies and room-mates;
- **STUDENTS:** students who return to live with their parents during the year should be included at their parents' address, even if they live elsewhere while attending school or working at a summer job;
- **CHILDREN IN JOINT CUSTODY:** children in joint custody who live here most of the time. Children who spend equal time with each parent should be included in the home of the parent where they are staying on May 15, 2001;
- **ABSENT SPOUSES:** spouses or common-law partners who live elsewhere while working or studying but who return here periodically;
- **LANDED IMMIGRANTS:** landed immigrants who usually live at this address;
- **REFUGEES:** persons claiming refugee status and family members living here with them;
- **PERSONS FROM ANOTHER COUNTRY WITH A WORK, STUDENT OR MINISTER'S PERMIT:** persons from another country who have an employment authorization, a student authorization or a Minister's permit, and family members living here with them;
- **PERSONS IN INSTITUTIONS:** persons who usually live here but are now in an institution (such as a home for the aged, a hospital or a prison), if they have been there less than six months;
- **PERSONS WITH NO OTHER HOME:** persons staying here on May 15, 2001, who have no usual home elsewhere.

DO NOT INCLUDE IN STEP B

- Persons who have their usual home at another address in Canada and who are staying here temporarily (for example, persons visiting or persons who have their secondary residence here, at this address);
- Residents of another country visiting Canada (for example, on a business trip or on vacation);
- Government representatives of another country or members of the Armed Forces of another country and family members.

IF ALL PERSONS STAYING AT THIS ADDRESS on May 15, 2001, have their usual home elsewhere in Canada **OR** are visitors, government or military representatives of another country, mark this circle . . . . 90. 01 ○ → Do not complete this questionnaire. Mail it in the enclosed postage-paid envelope.

Figure B-2 Residence instructions, 2001 Census of Population, Canada

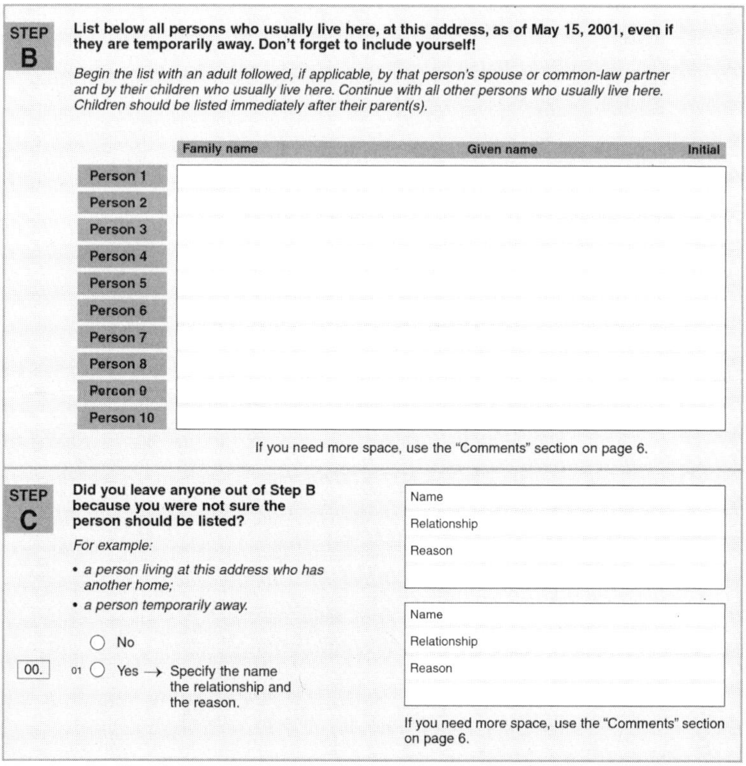

STEP B

List below all persons who usually live here, at this address, as of May 15, 2001, even if they are temporarily away. Don't forget to include yourself!

Begin the list with an adult followed, if applicable, by that person's spouse or common-law partner and by their children who usually live here. Continue with all other persons who usually live here. Children should be listed immediately after their parent(s).

	Family name	Given name	Initial
Person 1			
Person 2			
Person 3			
Person 4			
Person 5			
Person 6			
Person 7			
Person 8			
Person 9			
Person 10			

If you need more space, use the "Comments" section on page 6.

STEP C

Did you leave anyone out of Step B because you were not sure the person should be listed?

For example:

- *a person living at this address who has another home;*
- *a person temporarily away.*

○ No

00. 01 ○ Yes → Specify the name the relationship and the reason.

Name
Relationship
Reason

Name
Relationship
Reason

If you need more space, use the "Comments" section on page 6.

Figure B-3 Basic residence questions, 2001 Census of Population, Canada

- If all the (similar to 2001) "do not include" instructions apply, respondents are still asked to fill in a circle but are prompted to give their name and telephone number before mailing in the (not completed) questionnaire.

In 2001 the actual residence questions were presented as shown in Figure B-3; a coverage probe question asking the respondent to consider persons who might be missing is included in Step C. For 2006, Statistics Canada plans to keep the same listing of up to 10 household members, but the roster question will be preceded by a basic count query: "Including yourself, how many persons usually live here, at this address, as of May 16, 2006? *Include all persons who usually live here, even if they are temporarily away.*" [Because the new question begins with "including yourself," the reminder in 2001's Step B—"Don't forget to include yourself!"—is omitted in 2006. The Step C coverage probe is identical to the 2001 version.

Though not a formal list of residence rules, Statistics Canada posted an overview of "2006 Census Collection" on the Internet that does illustrate some underlying residence concepts; see http://www222.statcan.ca/ccr01/ccr01_001_e.htm [8/1/06]. In particular, the document identifies some "special enumeration" categories:

- Students away from home are enumerated in both places but counted at the parental home. "Students attending school out of town but who return home when school is not in session should be included on their parents' questionnaire, as part of the regular household." However, because the "school residence is considered a collective dwelling," they are required to complete the first two pages of the census form at the school.
- People living in "non-institutional collective units"—"inns, hotels, motels, campgrounds, YMCA/YWCAs, and military bases"—are enumerated there; if they are only "staying temporarily," they are counted at their usual place of residence.
- Statistics Canada makes explicit that "residents at institutions such as detention facilities, hospitals, residences for senior citizens, orphanages or prisons are enumerated using the institution's administrative records." However, exception is made for "seniors who reside in institutions or residences with distinct, separate living quarters that do not blend with units such as those for chronic care;" if those persons are able to complete the census questionnaire, they are allowed to do so. As mentioned in the questionnaire instructions, they are counted at the facility if they have been there at least six months.

The document also specifies that the objective of the census is to include "persons alive at midnight between May 15 and 16, 2006," so that babies born on May 16 are not included. Though the document says that the census "will take place on Tuesday, May 16," and that "on [that date], the majority of households will complete a census questionnaire and return it online or mail," the document also includes a contradictory instruction: "householders are asked to complete the questionnaire. . .and return it either online or [by mail] by May 16th, Census Day."

The Canadian census includes a long-form sample with additional questions. A long-form question on usual residence 5 years ago has appeared in the Canadian quinquennial census since 1961; the question on usual residence 1 year ago first appeared in 1991. In 2001, respondents were asked for the city, province, and postal code (but not full street address) for the 1-year and 5-years-ago residences if they were located in Canada; if they were out of the country, only the name of the other country was requested. The 2006 versions of these questions appear identical to their 2001 counterparts.

B.5 ESTONIA

Estonia's most recent population census—and the first since reacquiring its independence—was conducted in 2000. The objective of the census was to characterize the population and housing characteristics of the nation as of the "census moment," midnight (00:00) on March 31, 2000. The census was designed in accord with the UNECE (1998) guidelines.

Section 7 of the Population and Housing Census Act of 1998, which authorized conduct of the census, mandated that the Estonian census include "persons who are in the Republic of Estonia at the moment of the Census [with exceptions noted below]," "persons who reside in the Republic of Estonia but who are in foreign states temporarily for a term of up to one year," and "diplomatic staff of diplomatic missions and consular posts of the Republic of Estonia and their family members, who are in a foreign state at the moment of the Census." The census explicitly excludes "diplomatic staff of foreign diplomatic missions and consular posts and their family members" as well as "persons in active service in a foreign army."

Though the focus is on a *de facto* count at the census moment, the Estonian census also collected *de jure* residence information, as well as usual residence in the previous (1989) census. Section 4 of the Census Rules for 2000 Population and Housing Census holds that "every enumerated person covered by the Census has to determine his/her permanent place of residence (permanent, main, usual place of residence), from the temporary absence of which he/she can elsewhere have a temporary place of residence. The permanent place of residence is the place of residence where he/she lives permanently, regardless of whether he/she is in-registered at that place or has the right to use the dwelling." Consistent with the UNECE suggestions, the census rules hold that a person's permanent or usual residence "is the place where he/she spends the majority of his/her daily night-rest." If there is doubt as to which of two places should be judged the permanent residence, the rules say that "the place should be preferred where he/she intends to live for most of the time of the year."

According to the census rules, Estonia's handling of certain special populations is quite similar to the current U.S. model, with some slight deviations. College students and "pupils of professional secondary or other educational institutions" are deemed to be usual residents of the place where they study and "not the place of residence of parents." With regard to prisons and other institutions, "the permanent place of residence of persons who have lived in an institution for more than a year or who will stay there for more than a year is the institution." However, persons in military service are counted at "the place where they departed from for military service."

B.6 IRELAND

Since 1951 the Central Statistical Office of Ireland has conducted a quinquennial census, with only slight deviations; the 1976 census was canceled due to budget constraints and replaced with an abbreviated census in 1979, and the most recent census in 2002 reflected a postponement from 2001 due to an outbreak of foot and mouth disease.[2] The Irish census is oriented as a *de facto* count as of census night and instructs its respondents to fill out the census form on the designated night.

The 2002 census of Ireland was conducted on Sunday, April 28, 2002. Page 1 of the census instrument includes the following instructions on "How to complete your Census form:"

1. The form should be completed on the night of Sunday 28 April.
2. Please answer questions about the household on page 2.
3. Identify on page 3:
    - all persons (including visitors) who spent the night of Sunday 28 April in the household;
    - any household members who are normally resident in the household but who are temporarily away on the night of Sunday 28 April.
4. Answer the questions beginning on page 4 for all persons present on the night of Sunday 28 April.
5. Answer the questions on pages 22–23 in respect of any household members temporarily away on the night of Sunday 28 April.
6. Sign the declaration on the back page.

The draft form for the 2006 census, constructed following a 2004 pilot test, only modifies the date of the census.

Page 3 of the 2002 questionnaire asks the respondent to "**List every person who spent the night of Sunday 28 April in the household or who arrived the following morning not having been enumerated elsewhere.**" A bulleted list provides further inclusion and exclusion instructions:

INCLUDE

- all persons alive at midnight on Sunday 28 April.
- persons staying temporarily in the household.

DO NOT INCLUDE

- babies born after midnight on Sunday 28 April.
- anyone who is temporarily away from home on the night of Sunday 28 April. However, these persons should be listed as being absent in List 2 below.

---

[2]See http://www.cso.ie/census/When_is_Census.htm [5/30/05].

- students who are away from home on the night of Sunday 28 April; they should be listed as being absent in List 2 below.

List 1 allows 12 names to be listed as persons present in the household. List 2 allows 6 names to be listed as persons absent from the household, and is preceded by the reminder to "**Include in particular students who are living away from home during term time who were not present on the night of Sunday 28 April.**" The corresponding page of the 2006 draft questionnaire is identical to the 2002 version, with one difference: the ordering of exclusion instructions is rearranged so that students are listed first, babies second, and temporarily absentees third.

Both the 2002 form and 2006 draft form contain the same stark question 7: "Where do you usually live?" No further instructions are provided. Responses are "HERE at this address," "Elsewhere in IRELAND (including Northern Ireland), write in the COUNTY," and "Elsewhere ABROAD, write in the COUNTRY." Question 8 asks, "Where did you usually live one year ago?" and adds the reminder "*Answer if aged 1 year or over.*" the responses are the same as to question 7.

Later in the 2002 form, the respondent is asked eight questions about "any household members who usually live here at this address but who are NOT present on the night of Sunday 28 April. Include in particular students who are living away from home during term time who are NOT present at this address on the night of Sunday 28 April." The eight questions include name, relationship to reference person, date of birth, and marital status. A question asks, "How long altogether is this person away for?" Responses are "less than 3 months" and "3 months or more." The final questions are "Was this person in the Republic of Ireland on Sunday 28 April?" and "Is this person a student away at school or college?" These questions are identical on the 2006 draft form.

B.7 ISRAEL

Israel has held censuses on an irregular schedule since the creation of the nation in 1948. The most recent Israeli census was held in 1995; a census originally planned for 2006 was postponed until 2008 due to a change in methodology.

Summarizing initial plans for the Israeli census of 2008, Kamen (2005) indicates that the plan is to switch from a "conventional" short-form/long-form census to an "integrated" census, combining the results of a smaller survey with data available from administrative records and registers. Specifically, Kamen (2005:3) argues that there remains a need for a survey measure because "Israel conducts a *de jure* census" and that "Israel's Population Registry cannot now substitute for a census." Israel defines "usual residence" as, "for persons listed in the Population Registry, [those] present in Israel or absent for less than one year, and for those not listed in the Population Registry, as persons who have lived in Israel for one year or longer." The population coverage in the registry is acknowledged to not be exactly equal to the *de jure* census population; moreover, geographic information in the registry is known to be of "poor quality," and the registry does not include housing information or other demographic data items of interest.

B.8 ITALY

A major purpose of the Italian decennial census is to update the administrative registers (*anagrafi*) that are maintained by municipalities: these registers are known to be incorrect due to delays in enrolling and removing residents. Accordingly, the objective of the census is a *de jure* enumeration. However, cases such as students living in a different municipality during their education and other dual-residence matters caused complications in the 1991 population and housing counts. Practice in 1991 was to classify a housing unit as "not occupied" if no person from the municipal register lived there; hence, any dwellings occupied by "temporary residents" was considered vacant (Orasi and Ferruzza, 2001). Accordingly, the 2001 census had as a major focus the designation of "persons temporarily living in a dwelling" as a unit of analysis. Specifically, "a person who is registered as a usual resident in a certain municipality, where he will be enumerated as a resident, but who lives, lodges or works in a another municipality, where he will be enumerated as 'temporarily resident' " (Orasi and Ferruzza, 2001).

B.9 JAPAN

Japan conducts a census every 5 years, alternating between large-scale censuses (that include demographic and economic questions) in years ending in 0 and smaller-scale simplified censuses in years ending in 5. The 2000 census of Japan was taken with respect to midnight (0:00 am) on October 1, 2000, and is intended to get a *de jure* count at that time.

According to an English-language description of the methodology of the 2000 census (http://www.stat.go.jp/english/data/kokusei/2000/outline.htm [April 2005]) by the Japanese Statistics Bureau, Japan uses a 3-month rule to define usual residence: "persons usually living" was defined as "those persons who had lived or were going to live for three months or more at their respective households at the census date." Persons with no usual residence were counted at their current location. The methodological description outlines the following exceptions to the usual residence rule:

(1) Students and pupils of regular schools as well as those attending special training schools or miscellaneous schools who were living in school dormitories, boarding houses or the like were enumerated at the places where they lived regardless of their period of stay.

(2) In-patients of hospitals were enumerated in the hospitals only when they had been hospitalized for 3 months or more at the census date. Otherwise, they were enumerated at their homes even if they were expected to be in the hospital for 3 months or more.

(3) Crews aboard ships except ships of the self-defence forces were enumerated at their residential places on land, if any. Otherwise, they were enumerated on the ships if the ships were of the Japanese flag and were at anchor at a port of Japan at the census date or if the ships left a port of Japan before the census date and entered a port of Japan within 5 days after the census date without calling at any foreign ports.

(4) Residents in the camps of the self-defence forces were enumerated in the camps. Crews aboard ships of the self-defence forces were enumerated at the places of the local general headquarters to which their ships belonged.

(5) Persons in prisons or detention houses whose penalties had been fixed and inmates of reformatories or women's guidance homes were enumerated at those institutions.

"In accordance with the rules described above, all persons living in Japan were enumerated whether they were foreigners or not." However, foreign diplomats and military personnel and their families were excluded.

The Japanese census asks for information on the length of stay at the usual residence as well as the usual residence 5 years ago.

B.10 NEW ZEALAND

New Zealand's quinquennial census has traditionally been conducted through enumerator drop-off and pick-up of questionnaires; due to its *de facto* goal of recording the population present on census night, census forms were even distributed on overnight trains on the 2001 census night of March 6 (Office for National Statistics, 2003). The main product of the census is the "census night population count," which includes all people who were "on New Zealand soil, on a vessel in New Zealand waters, or on a passage between New Zealand ports" on census night. This count includes "overseas residents and other people in diplomatic residences in New Zealand" as well as "foreign military personnel and their families located in New Zealand on census night" (Statistics New Zealand, 2001:11).

Questionnaire probes also allow construction of the "census usually resident population count," so that "if a person usually lives in Christchurch but was in Wellington on census night, they will be included in the census usually resident population count for Christchurch and the census night population count for Wellington" (Statistics New Zealand, 2001:12).

Like the Australian census, the New Zealand census is intended as a *de facto* count, but usual residence information is collected for purposes of generating estimates for fund allocation and other purposes. New Zealand's official definition of "usual residence" is interesting because it is explicitly defined to the "meshblock" (equivalent to census blocks in the United States) level. The official definition used in the 2001 census is lengthy (Statistics New Zealand, 2001:16):

> Usual residence is the [census block] of the dwelling where a person considers himself or herself to usually reside, except in the following cases:
>
> - people who board at another residence to attend primary or secondary school, and return to their parent's(s') or guardian's(s') home for the holidays, usually reside at the address of their parent(s) or guardian(s). Post-secondary students usually reside at the address where they live while studying
> - children in joint custody usually reside at the place where they spend more nights, or if they spend equal amounts of time at each residence, they usually reside at the place where they are at the time of the census
> - people who are in rest homes, hospitals, prisons or other institutions, usually reside where they consider themselves to live, and this may include the institution
> - a person whose home is on any ship, boat or vessel permanently located in any harbour shall be deemed to usually reside at the wharf or landing place (or main wharf or landing place) of the harbour
> - a person from another country who has lived, or intends to live, in New Zealand for 12 months or more usually resides at his or her address in New Zealand (as in external migration)
> - people who spend equal amounts of time residing at different addresses, and can not decide which address is their usual residence, usually reside at the address they were at on census night, or
> - if none of the above guidelines apply, the person usually resides at the address he or she was surveyed at.
>
> The definition of usual residence does not include a time-criterion and instead uses the approach of self definition. This is because a time criterion can lead to households and families being classified on an arbitrary basis. Furthermore, most people know where they usually live (reside) and as such this involves feelings of belonging, association and participation in and with a household.

Address of usual residence has been a question in New Zealand censuses since 1921. A question asking how long the respondent had lived at that address was added in 1976, dropped in 1981, and has been asked in the 1986 and subsequent censuses. A question on usual residence 5 years ago has been asked since 1971; a question on usual residence 1 year ago appeared only in the 1971 and 1981 censuses (Statistics New Zealand, 2001:33).

Despite the length of the formal Statistics New Zealand definition of usual residence, the census instruments for 2001 are relatively free of residence instructions or cues. However, the questionnaire includes repeated cues to the respondent to include babies in the enumeration.

The New Zealand census relies on completion of two forms: a dwelling form, filled out by one member of the household that includes housing questions, and an individual form, one of which is completed for each person in the house. The principal residence instruction appears in a small column beside the first question on the dwelling form and reads:

> These people fill in an Individual Form here in this dwelling:
>
> - everyone, including babies, who is spending the night of 6 March here; and
> - anyone who arrives on 7 March, who has not filled in an Individual Form anywhere else. This includes babies.

On the second page of the dwelling form, a household roster (up to 10 people) is constructed in question 4: "List all the people who are filling in a blue Individual Form here in this dwelling (and people having one filled in for them), starting with yourself as Person 1." Two special instructions are highlighted in circles under person 1 (where relationship to household questions would normally appear): those read "**Remember to list any babies who live here!**" and "If a baby is aged under one year, print [graphic showing a mark of zero]."

Several pages later in the dwelling form, question 18 asks, "Will everyone who usually lives in this dwelling fill in a blue Individual Form (or have one filled in for them here)?" If the answer is no, the respondent is prompted to go on to question 19, which asks, "How many people who usually live here WON'T fill in a blue Individual form here (and WON'T have one filled in for them here)?" Beside these two questions is a text box:

- Count as usually living here
  - children away at boarding school
  - people who are away on holiday, away for work, in hospital for a short time, etc.
- DON'T count
  - university or other tertiary students who live somewhere else for most of the year

For up to five such absentees, the dwelling form respondent is asked to provide the missing person's name, age, gender, and relationship to respondent;

they are also asked whether the absentee is in New Zealand on census night and, if not, "how long altogether is s/he away from NZ?" (responses are "less than 12 months" or "12 months or more").

The blue-colored Individual Form is bilingual and contains only the instruction, "One of these forms must be filled in for every person in New Zealand on the night of 6 March 2001." Question 5 asks, "Where do you usually live?" and a blank box is provided for respondents to write the full address. Question 6 asks how long the respondent has lived at that address (either "less than one year," or a write-in for the number of years). Question 7 asks, "Where did you usually live 5 years ago on 6 March 1995?" and allows the responses "not born 5 years ago," "at the address you gave in question 5," "in New Zealand at another address" [blank for full address], and "NOT living in New Zealand" [blank to write in country name].

Question 8 asks for census night residence information: "On the night of Tuesday 6 March, what address are you at?" Responses are "at the address you gave in question 5," "at another address" [blank for full address].

Based on feedback to initial content plans for the 2006 census, Statistics New Zealand reported that it was considering enhancing the geographic detail coded for absentees from their usual residence. In 2001, absentees were only coded to the meshblock (census block) of usual residence, not explicitly linked to specific residences. Final determination as to whether absentees would be repatriated back to their usual home of residence (for reconstruction of family statistics) was said to depend on final census tests (Statistics New Zealand, 2003). The standard of usual residence "was raised repeatedly during consultation with stakeholders," with particular regard to "overseas students studying in New Zealand" and "New Zealand students who are away from home on census night." The Statistics New Zealand planning document also acknowledges "known difficulties with this topic, such as a respondent's interpretation of the word 'usual' " (Statistics New Zealand, 2003:10).

Though some analysts suggested reinstating a question on usual residence 1 year ago, Statistics New Zealand declined to include it in the 2006 census, judging that "usual residence five years ago gives the best indication of intercensal migration." The bureau also declined to add questions on the number of moves made in the last year and the reasons for moving (Statistics New Zealand, 2003:11).

B.11 SOUTH AFRICA

The South African census of 2001 was intended to provide a *de facto* count of persons present in South Africa on the night of October 9–10, 2001. Limited information was collected on usual residence, which was based on where a person spends 4 or more nights a week. Question 11 asked "Does (the per-

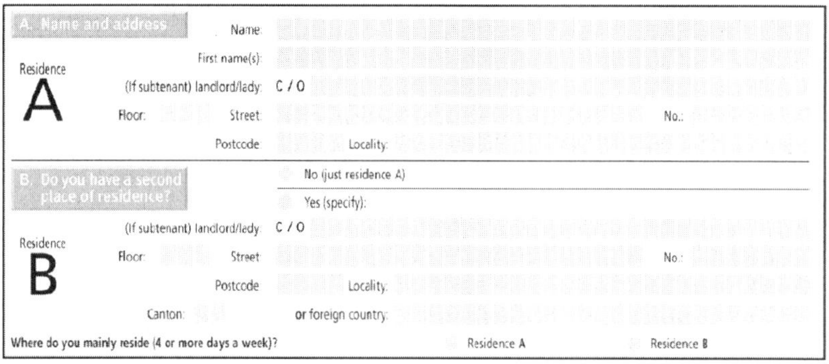
A. Name and address
Residence A
Name:
First name(s):
(If subtenant) landlord/lady: C / O
Floor: Street: No.:
Postcode: Locality:
B. Do you have a second place of residence?
No (just residence A)
Yes (specify):
Residence B
(If subtenant) landlord/lady: C / O
Floor: Street: No.:
Postcode: Locality:
Canton: or foreign country:
Where do you mainly reside (4 or more days a week)? Residence A Residence B

Figure B-4 Questionnaire items to collect primary and secondary address information, 2000 Census of Population, Switzerland

son) usually live in this household for at least four nights a week?" If no, respondent was asked a follow-up as question 11a: "Where does (this person) usually live?" If the usual residence is also the place of data collection (current residence), that could be indicated by checking a box; if different, respondents were supposed to answer with the province of residence and the name of the place (for South Africa) or country name (for residences in another country).

The census form also asks if their place of usual residence is the same as it was 5 years ago, during the preceding census; if the answer is no, a follow-up prompts for the year in which the person moved (if more than one move was involved, the year of the most recent move was to be indicated).

B.12 SWITZERLAND

The Swiss *de jure* decennial census defines the "resident population" as "all persons who officially reside in Switzerland over a given period of time regardless of their citizenship, duration of residence and type of permit. Persons who do not officially reside in Switzerland (e.g. cross-border commuters working in Switzerland, tourists, visitors or business travellers) do not fall into this category" (http://www.bfs.admin.ch/bfs/portal/en/index/themen/volkszaehlung/definitionen.html [April 2005]).

The 2000 Swiss census form collected address information on both a primary and a secondary residence with a very limited amount of space on the physical page of the personal questionnaire (see Figure B-4). Arguably, it may be too tight a space: the key follow-up question as to which of the two addresses is the place where "you mainly reside" is not well distinguished and could be missed or skipped.

Question 5 on the questionnaire is "Place of residence 5 years ago; where were you living on 5 December 1995?" (Because Switzerland's censuses are decennial, the 5-year-ago request does not correspond with the immediately previous census.) Four responses are permitted: "At the same address as now (residence A)," "In the same commune (as residence A) but at another address," "In another commune (specify)," or "Abroad" (respondent is asked for the country).

B.13 UNITED KINGDOM

The British decennial census was a *de facto* enumeration for most of its history, from 1801 to 1971. In 1981 and 1991, hybrid approaches collecting both *de facto* and *de jure* information were tried before the 2001 census switched solely to *de jure* counting (Smith, 2005). Early in the planning cycle leading to the 2011 United Kingdom census, the Office for National Statistics considered switching back from a *de jure* to a *de facto* count. They have since determined that the population base for the 2011 census will continue to be usual residents, augmented by visitors to the nation present on census night.

The 2001 British decennial census was executed pursuant to the "Census Order 2000" passed by Parliament and enacted on March 15, 2000; the target census day for the 2001 count was April 29. Somewhat akin to the 31 formal residence rules for the 2000 U.S. census, the order formally defined "usual residence" with respect to a set of eight groups (see Table B-1). Specifically, the legislation defines "usually resident" as including:

> persons who
>
> (a) have a usual address in England and Wales;
>
> (b) are present at an address in England and Wales on census night and have no other usual address in England and Wales or elsewhere;
>
> (c) in the case of Groups I and IV, are in full-time education and are residing at their term-time address;
>
> (d) in the case of Groups II, III, VI and VII, have resided or intend to reside in the premises or vessel for a total period of 6 months or more beginning on or before census day; and
>
> (e) in the case of Group V, are spending a period of 6 months or more in custody whether at the premises or elsewhere.

A formal set of definitions published after the census provided an easier-to-parse definition of usual residence (Office for National Statistics, General Register Office for Scotland, and Northern Ireland Statistics and Research Agency, 2004:17):

> A usual resident is generally defined as someone who spends the majority of their time residing at that address. It includes

Table B-1 Usual Residence Categories as Delineated by the Census Order 2000, United Kingdom

Group	Premises, vessel or other place [of enumerations]	Persons [to be counted]
I	Any dwelling or part of a dwelling (including a caravan, houseboat or other temporary building or structure used as living accommodation) occupied by a household (whether consisting of one or more persons).	Every person who—(a) is usually resident in the dwelling or the part of the dwelling whether as a member of the household, paying guest or boarder, or as an employee of any such person; (b) not being a person to whom paragraph (a) above refers, is in full-time education and has a home address in the dwelling or the part of the dwelling.
II	Any hotel or guest house.	Every person who is usually resident in the premises and who has not been included in any other return in the United Kingdom.
III	Any hospital, nursing home, religious or charitable community or other residential establishment whatsoever, not being an establishment mentioned elsewhere in this Schedule.	Every person who is usually resident in the premises and who has not been included in any other return in the United Kingdom.
IV	Any residential school, college or other educational establishment.	Every person who is usually resident in the premises.
V	Any civil prison or other place of detention.	Every person who is usually resident in the premises and who has not been included in any other return in the United Kingdom.
VI	Any vessel which is at a port in England and Wales at midnight ending census day, barracks, station or other premises under naval, military or air force discipline.	Every person who is usually resident on the vessel or in the premises and who has not been included in any other return in the United Kingdom.
VII	Any vessel which is at a port in England and Wales at midnight ending census day not included in Group VI.	Every person who is usually resident on the vessel and who has not been included in any other return in the United Kingdom.
VIII	Any other place not included in the above Groups.	Every person who has no usual address.

- people who usually live at that address but who are temporarily away from home (on holiday, visiting friends or relatives, or temporarily in a hospital or similar establishment) on Census Day;
- people who work away from home for part of the time, or who are members of the Armed Forces;
- a baby born before 30th April 2001 even if it was still in hospital; and
- people present on Census Day, even if temporarily, who have no other usual address. . . .

The usual resident population did not include:

- people present at an address on Census Day whose usual address was elsewhere; or
- people away from their home address who had been living, or intended to live, in a special establishment such as a residential home, nursing home or hospital for six months or more (they were enumerated as usually resident at the special establishment).

College and boarding school students were to be counted at their school address. British military personnel permanently stationed in Northern Ireland were to be recorded at their "actual address of residence unless they were married and unaccompanied by the spouse," in which case they were to be counted at the address shared with the spouse.

The 2001 count represented a shift toward a more pure *de jure* census; while the 1991 census sought information from both usual residents and visitors present on census night, the 2001 instrument collected data only on usual residents.

Space to complete rosters of household members and visitors took up a full second page of the individual census form used in the United Kingdom census of 2001. First, respondents are asked to complete Table 1 pursuant to the following instructions:

List all members of your household who usually live at this address, including yourself.

- Start with the Householder or Joint Householders.
- Include anyone who is temporarily away from home on the night of 29 April 2001 who usually lives at this address.
- Include schoolchildren and students if they live at this address during the school, college or university term.
- Also include schoolchildren and students who are away from home during the school, college or university term and for whom only basic information is required.
- Include any baby born before 30 April 2001, even if still in hospital.
- Include people with more than one address if they live at this address for the *majority of time*.

- Include anyone who is staying with you who has no other usual address.
- Remember to include a spouse or partner who works away from home, or is a member of the armed forces, and usually lives *at this address*.

Space is then provided for the names of 10 household members.

Directly below Table 1, Table 2 asks the respondent to list up to 5 names and addresses for visitors to the household: "To help you complete the form you may use Table 2 to list any visitors at this address, on the night of 29 April 2001, who usually live elsewhere." The instructions further direct that "if there are only visitors at this address," only the first five questions on the nature of the housing unit should be completed; "no further questions need to be answered."

Later in the questionnaire, at the beginning of collection of personal characteristics, full-time students living elsewhere are immediately screened from further questioning. If a person answers "yes" to question 5, "Are you a schoolchild or student in full-time education?," they are asked "Do you live at the address shown on the front of this form during the school, college or university term?" If the answer is yes, the student is guided through the complete set of person-level questions (race, employment, health, etc.). If the answer is "no," they are routed to the end of questions for that person.

The possible switch back to a *de facto* or "persons present" count in 2011 was described in the initial design document (Office for National Statistics, 2004). The rationale for the switch included the argument that the usual residence definition in 2001 "introduced ambiguity about who should be included and may have resulted in higher non-response amongst certain population groups" (Office for National Statistics, 2004:3). The design document also noted strong arguments from the census user base on the need for counts of the daytime or working population of areas, for better planning of services.

Further consideration reported by Smith (2005) raised serious issues of discontinuity if the *de facto* model is adopted. Specifically, 2011 results would not be directly comparable with the 2001 census returns, but they could also differ from post-2011 counts, for which the Office for National Statistics would likely rely heavily on administrative records data for which usual residence is coded. Maintaining a count of the usual resident base but with additional emphasis on collecting some information from visitors (particularly "workers who are not usually resident but contributing to the national economy" and other short-term visitors) was judged to be the best compromise and recommended by Smith (2005). This recommendation was corroborated by Office for National Statistics (2005) and Stokes (2005b), and issued as a formal Office for National Statistics recommendation in Stokes (2005a).

Stokes (2005a) further notes that exact definitions for both "usual resident" and "visitor" remain to be decided.

The Office for National Statistics is still considering the addition of a series of questions related to second residence: "whether a person lives at another address for part of the year, and if so, what this address is, what the address is used for, and the amount of time spent at this address" (Courtney, 2005a:3). The switch is due to perceived increases in the complexity of living situations in the United Kingdom; analysis of auxiliary data sources including the stand-alone Survey of English Housing suggests a fluctuating but generally increasing trend in ownership of second residences (Courtney, 2005a,b).

– C –

Americans Residing Overseas

As described at the outset of this report, our panel's charge pointedly excludes consideration of who should or should not be counted in the census (see Section 1–A). The issue of whether noncitizens (and particularly illegal immigrants) should be included in census counts (as they currently are) is the source of great political debate and has periodically drawn attention from Congress and other policy makers, but it is one that is out of our panel's scope. Likewise, the current practice of excluding American citizens living overseas, who are not military personnel or employees of the federal government, is one on which we do not offer advice.

However, while we do not comment on the appropriateness of counting Americans living outside the country in the census, some comment on the way this group has been handled in previous censuses and in census tests is in order. Several of the residence rules in the 2000 census deal directly with this group, and the handling of a subset of Americans living abroad was the focus of the first legal challenge to the 2000 census count. Americans living outside the United States are a natural and important part of any listing of situations for which residence standards may be difficult, and so the issues are important to consider in residence rules. We describe procedures for counting the military population in Section 3–F; in this appendix, we briefly outline the issues associated with counting American citizens living abroad in the census.

Following the same procedures used in the 1970 and 1990 censuses (as described below), the 2000 census included military and federal government personnel stationed overseas in the census apportionment counts but *not* in other products, including redistricting totals. These people were allocated to their home states based on "home of record" information from the Department of Defense or the employees' parent agency (e.g., the Department of State).[1] This overseas population count tallied 576,367 military and federal personnel, just over 59,000 of whom were allocated to California and 52,000 to Texas. Other states with large military installations gained from the inclusion of these residents, such as the 22,187 added to Virginia's total. As would be significant shortly after the 2000 census, North Carolina—with its large naval facilities—was credited with 18,360 overseas residents while Utah received 3,545.

Estimates suggest that the overseas military and federal employees (and their dependents) represent a small share of the total American citizenry living in other countries. As of July 1999, the U.S. State Department's Bureau of Consular Affairs estimated that 4,163,810 private American citizens lived in foreign nations (not counting military and government personnel and their dependents).[2] Private American citizens living in Mexico account for 25 percent of that total, and Canada another 17 percent, by far the largest single-country contingents. Regionally, 28 percent of private American citizens overseas reside in Europe, 12 percent in Asia and Oceania, 7 percent in the Middle East, and 2 percent in Africa. Other estimates put the number of Americans outside the country at up to 10 million.

C.1 TREATMENT IN PAST CENSUSES

The 1830 and 1840 censuses were the first to include the counts of some segment of the American overseas population when they included U.S. naval crews in the counts (Mills, 1993:3); otherwise, consideration of the overseas population is absent from both census tabulations and enumerator instructions until the beginning of the 20th century.[3] The 1900 census counted 91,219 Americans living abroad, consisting of military and federal civilian personnel at military stations and their dependents, along with the crews of naval vessels. Census authorities noted that these were counted and printed in the census volume because, "except for conditions arising out of the Spanish-American War, they would have been found residing largely within the limits of the United States proper, and they are, therefore, included properly in the

[1]This included 2,037 persons allocated to the District of Columbia; the District's population is excluded from the subsequent calculation of seats in the House of Representatives.

[2]The 4.1 million total is shown on versions of the State Department data such as that published by American Citizens Abroad (http://www.aca.ch/amabroad.pdf [10/5/05]) and cited by U.S. Government Accountability Office (2004a). An archived version of the Bureau of Consular Affairs tabulation at http://www.pueblo.gsa.gov/cic_text/state/amcit_numbers.html [10/5/05] records the total as 3,784,693.

[3]The 1870 and 1880 census enumerator instructions did include the somewhat vague rule that "seafaring men are to be reported at their land homes, no matter how long they may have been absent, if they are supposed to be still alive," without specifying naval vessels (Mills, 1993:10).

total for 1900 for purposes of comparison with the population returned at preceding censuses" (Mills, 1993:11). The enumerator instructions for the 1900 census include the more generic directive that "persons temporarily residing in foreign lands"—seemingly casting eligiblity wider than military and federal employees—"should be enumerated as part of their family at their home or usual place of abode" (Gauthier, 2002:35); however, the published statistics include only those on naval vessels and in territories associated with the Spanish-American War, principally the Philippines (Mills, 1993:11).

The 1910 census reversed that determination and included federal civilian employees in a separate overseas population count, an approach that continued through 2000 (McMillen, 2000a:32). Persons stationed abroad continued to be counted with the cooperation of the War and Navy Departments. Curiously, the 1910 enumerator instructions for the domestic census count included a clear directive to permit the reporting of American civilians who were living abroad "temporarily" (Mills, 1993:18):

> **Citizens abroad**—Any citizen of the United States who is a member of a family living in your district, but abroad temporarily at the time of the enumeration, should be enumerated as of your district. It does not matter how long the absence abroad is continued, provided the person intends to return to the United States.

No inference was made on possible duplication between people included in the overseas count and reported in the stateside enumeration.

The exact means by which data were collected, and the groups covered, varied slightly over the next few decades. Both the 1900 and 1910 censuses used enumerator schedules identical to the general population schedule to gather information on military and naval personnel, federal civilian employees, and their dependents. The 1920 census saw the switch to a different basis schedule for these groups along with a modest expansion of scope—"persons in the service of the American Red Cross or U.S. consular service" were added to the target population (Mills, 1993:8). The 1920–1940 enumerator instructions repeated the "citizens abroad" advisory from 1910.

The 1950 census was the first in which the Census Bureau entered into cooperative agreements with the U.S. Departments of Defense and State, as well as the Maritime Administration, to provide information on personnel and dependents under their authority. A Bureau summary of innovations for 1950 suggested that, prior to 1950, "no systematic effort [was] made to enumerate American citizens abroad"; "since their numbers were small, this omission made little practical difference." The situation changed after World War II, though: "now that we have hundreds of thousands of our people overseas in occupation and other military duty and on the staffs of federal civilian agencies, it has become important to have an inventory of them" (Shryock, 1950:8). The same Bureau summary judged that "it is still not feasible to enumerate all those overseas who claim American citizenship," and McMillen (2000a) found that the 1950 census included a limited effort to collect data on "other citizens" living abroad. Mills (1993:32) comments that "scattered" forms "were received from private U.S. citizens living abroad" who picked up a form from embassies or consulates, but the response was said to be low.

The Bureau also sought and received an opinion from the U.S. Attorney General in 1949 on the legality of its plans for enumerating Americans overseas; the Attorney General favorably approved the notion of enumerating them, but not including them in the total population figures for any state or in the total population of the continental United States. Thus, the overseas population counts were intended for use as information, but not for apportionment.

The 1960 census repeated the same procedures for the military and federal employee enumeration, including an emphasis on direct distribution of individual census report forms. The 1960 census did mark "a special effort . . . to obtain voluntary reports from private U.S. citizens not affiliated with the Federal Government who were abroad for an extended period" (Mills, 1993:3). The Department of State agreed to play a larger role in disseminating questionnaires from its embassies and consulates, and some of those foreign stations used local newspapers to generate word of the census operations. "Religious groups with missionaries abroad, as well as large corporations with overseas employees," were also advised of the procedures and included notices in their periodicals and newsletters (Mills, 1993:37). Again, though, "the reporting of these private U.S. citizens was not complete" and "understated the true number of these persons abroad" (Mills, 1993:3). The effort did result in a separate census report on the social and economic characteristics of the population abroad, the first of its kind (Mills, 1993:36).

The 1970 census marked a major change in the use of counts of the overseas population. Specifically, military personnel and federal civilian employees stationed overseas were included in the state-level census totals for purposes of apportionment. Two major reasons were cited for this change: the larger number of persons expected to be overseas at the time of the census because of the Vietnam War, and the then-recent court decisions on "one person, one vote." Subcommittees in both houses of Congress held hearings on the issue in 1969, and they favored the inclusion of the military and federal civilian personnel in apportionment counts; advice was again sought from the U.S. Justice Department, and then-Assistant Attorney General William Rehnquist concluded that the addition was within the powers of the census director (Mills, 1993:4). In terms of operations, the State Department handled the enumeration of civilian employees of the federal government, their dependents, and other Americans residing abroad through contacts at embassies and consulates. Much of the overseas military count was obtained from administrative records rather than direct questionnaire enumeration. The enumeration of "other citizens" was again conducted on a voluntary basis and the results—as in previous tentative tests—showed low reporting. Mills (1993:46) suggests that "a post-1970 census comparison of data from the 1970 overseas census with country of birth/citizenship data from individual foreign censuses indicated that the census counts for private U.S. citizens represented a substantial undercount, particularly in Canada and Mexico, where the underenumeration probably exceeded 90 percent."

The Census Bureau reversed itself in the 1980 census, electing to exclude all overseas population from the apportionment counts. The then substantially smaller deployment of military personnel overseas was the major reason for the reversal; the Census Bureau also expressed concerns on the reliability of data on a person's "home state." The 1980 census was also distinctive in that no attempt was made to directly enumerate any part of the overseas population; instead, administrative record counts (but no data on characteristics) for armed forces, civilian employees, and their overseas dependents were obtained from the Department of Defense and the Office of Personnel Management and tallied as a separate "overseas population" count. Citing low participation in the 1960 and 1970 efforts, no data or counts of "other citizens" not affiliated with the federal government were collected. Zitter (1987) argued that the Census Bureau's attempt to count the overseas population in 1980 was minimal in comparison with earlier censuses; anything that distracted from the main census event was deemed an unnecessary risk and was not to be undertaken without a compelling reason, such as a congressional mandate.

C.2 THE 1990 CENSUS

For the 1990 census, overseas military and federal employees were again counted by administrative records. The overseas military population and their dependents living with them were counted by using Department of Defense records; federal employees and their dependents were similarly addressed using Office of Personnel Management records. No effort was made to count "other citizens" living abroad.

However, the Bureau reversed itself on including military and federal civilian employees in apportionment totals, returning to the 1970 policy of including them in their home states' counts. The change owed a great deal to increased congressional interest in the issue, which grew as both the overseas Americans issue and the inclusion of undocumented immigrants emerged as major issues and potential sources of litigation (McMillen, 2000a).

In a background paper to inform the panel's deliberations, Lowenthal (2005) summarized the congressional interest, despite the absence of large numbers of American troops abroad:

In April 1988, the House Subcommittee on Census and Population held a hearing to review Census Bureau policy with respect to members of the armed forces and federal employees, and their dependents, stationed overseas during the census. The Census Bureau and the Department of Defense (DoD) announced that they had agreed to administer a census on overseas military bases, but Census Bureau Director John Keane expressed doubt that his agency could allocate this population back to their states of usual residence within the legal timeframe for reporting apportionment counts to the President.

[Subcommittee chair Mervyn Dymally (D-CA)], along with the census subcommittee's senior Republican member, Rep. Constance Morella (R-MD), subsequently sponsored legislation to include military personnel and federal civilian employees stationed overseas during the census in the state population totals used for apportionment (H.R. 4720). While the Census Bureau expressed reservations about the bill, the Committee on Post Office and Civil Service approved the measure by voice vote, and House leaders scheduled the bill for consideration under Suspension of the Rules, a procedure used almost exclusively for non-controversial legislation because it requires a two-thirds vote for passage. Surprisingly, though, in September 1988, H.R. 4720 was defeated by a recorded vote of 93–317, with debate focused on the criteria for assigning overseas military and federal personnel to a home state [*Congressional Record,* September 28, 1988, Roll Call No. 361.]. Many members, concerned that states with large military facilities would benefit unfairly, objected to counting overseas personnel in the state where they last resided for six months before being deployed or assigned to a foreign post. (Some Members also suggested that Congress should, at the same time, exclude undocumented residents from the apportionment counts.)

Despite the defeat of H.R. 4720, congressional support for including military and federal civilian personnel in the apportionment base remained high in the 100th and 101st Congresses. Additional bills were introduced in both the House and Senate in the 101st Congress, with various proposals for determining the state of residence of this population for apportionment purposes. Rep. Dymally introduced another bill (H.R. 3016) that set the place of residence for this population as "the individual's last place of general abode within the United States for a period of 6 consecutive months or longer." Sen. John Heinz (R-PA) sponsored a bill (S. 290) that would have left the decision on where to count overseas military and federal civilian personnel to the discretion of the Secretary of Commerce. Rep. [William Goodling (R-PA)] suggested a combination of criteria to establish usual residence, proposing to count overseas military personnel at their "home of record" and overseas government employees in the state where they had last lived for 12 consecutive months or longer (H.R. 1468).

In August 1989, with the start of the decennial census fast approaching, the Department of Commerce announced that it had decided to in-

> clude military personnel and federal civilian employees assigned to overseas posts, and their dependents living with them, in the state population totals used for apportionment. At the time, the Defense and Commerce Departments were still preparing to administer a modified census on overseas military bases, to collect the necessary information on members of the armed forces (who comprised the vast majority of federal personnel stationed abroad). In December, however, DoD concluded that it could not administer an overseas census, which would have provided an opportunity for self-enumeration. Instead, it agreed to provide state-by-state counts of military personnel stationed abroad using information from administrative records. According to the Census Bureau, DoD identified three possible criteria, obtainable from administrative files, for assigning overseas military personnel to a home state: home of record; legal residence (state declared for income tax purposes); and last duty station (facility where armed forces member was assigned before deploying overseas) (Mills, 1993:5).
>
> Concerned about the consequences of the various alternative methods, senior members from the House census authorizing committee wrote to the Census Bureau, urging that overseas military personnel be counted at their home of record. Nevertheless, with the 1990 census well underway, the Secretary of Commerce determined that these individuals would be allocated to the state where they last resided for six consecutive months or longer before deployment overseas, even though the House of Representatives had rejected just such a proposal (H.R. 4720) in the previous Congress.
>
> Clearly displeased that the Commerce Department did not consult Congress before making this decision, Rep. [Thomas Ridge (R-PA)] and Rep. Thomas C. Sawyer, now the ranking minority member and chairman, respectively, of the Subcommittee on Census and Population, introduced legislation in May 1990, requiring the inclusion of overseas military and federal personnel, and their dependents, at their home of record (H.R. 4903) [Statement of Rep. Thomas C. Sawyer, *Congressional Record*, May 24, 1990, Extension of Remarks]. The House of Representatives passed the bill by voice vote in June, but the Senate did not take any further action. The clear signal sent by the House, however, with its primary interest in the outcome of apportionment, was finally heard by the Census Bureau, which concluded in July that "home of record" for military personnel "was the most consistent with the concept of 'usual residence' in the census (Mills, 1993:5). The Commerce Department subsequently adopted "home of record" as the basis for assigning overseas military personnel and their dependents to a state for purposes of apportionment.

When the apportionment counts and home-of-record overseas allocations were released in December 1990, the overseas figures appeared to sway the placement of a congressional seat. If the overseas counts were not included, the state of Massachusetts would have retained its eleventh seat in the House of Representatives; instead, the state of Washington gained its ninth seat. This occasioned Massachusetts' May 1991 legal challenge to the policy of including overseas personnel in the apportionment totals; we summarize the resulting case—*Massachusetts v. Franklin*—and the "enduring tie" standard it suggested in Box 2-5.

C.3 THE 2000 CENSUS

The 2000 rules for overseas populations followed those from 1990. As in 1990, the 435th and final seat in the House of Representatives was won by a small population total, and the overseas count became a focus of litigation.

Prior to the 2000 census, organizations representing Americans living abroad pressed for congressional interest in the issue. The House Subcommittee on the Census held a 1999 hearing to review various legislative proposals to include nongovernmental overseas Americans in the count that had been introduced in both chambers of Congress.[4] At the hearing, which also considered proposals to change the Bureau's policy on counting prisoners at the prison location (see Chapter 3), Census Bureau director Kenneth Prewitt argued that it was far too late to consider enumerating private American citizens in the 2000 census and reiterated basic concerns about conducting the count (see below).

Lowenthal (2005) writes that "while they did not seriously consider last minute proposals to extend Census 2000 coverage abroad, legislators did pursue the idea with an eye toward the future":

> In 1999, in their reports on the Fiscal Year 2000 Commerce Department funding bill, both House and Senate appropriators directed the Census Bureau to develop a plan for counting overseas Americans in the census and to inform Congress of its progress.[5] The following year, appropriators issued a more specific directive to the Census Bureau for a report on methodological, operational, and policy issues associated with including all American citizens living abroad in the census, as well as an estimate of the number of Americans living or working overseas, both for the federal

[4]H.R. 2444 was introduced in the 106th Congress by Rep. Carolyn Maloney (D-NY), the ranking minority member of the House census subcommittee; a companion bill was introduced in the Senate by Sen. Jay Rockefeller (D-WV) as S. 1715. Rep. Benjamin Gilman (R-NY) and Sen. Spencer Abraham (R-MI) sponsored less-binding "sense of Congress" resolutions opining that all Americans residing abroad should be counted in the 2000 census as H. Con. Res. 129 and S. Con. Res. 38, respectively (Lowenthal, 2005).

[5]House Report 106-283, p. 68; Senate Report 106-76, p. 77. The Senate report actually urges the Census Bureau to work with the State Department to include Americans living overseas in the 2000 census, but it is worth bearing in mind that the Senate had conducted no oversight of the issue and little oversight of census preparations generally. Committee report language is not legally binding.

government and for private organizations.[6] Rep. Carolyn Maloney continued to push the issue forward, as well, sponsoring bills to authorize and pay for an interim census of Americans abroad in 2003 and to begin planning for the inclusion of all Americans living overseas in the 2010 census (H.R. 3649 and H.R. 4568).

When the apportionment counts were released in December 2000, North Carolina edged Utah for the 435th House seat by fewer than 1,000 people. Utah brought suit, charging that the Census Bureau's treatment of Americans living abroad was unfair in that military personnel and other federal employees stationed overseas are included in the apportionment counts but other civilians are not. Specifically, Utah challenged the exclusion of approximately 11,000 missionaries of the Church of Jesus Christ of Latter-Day Saints: these missionaries may live abroad temporarily, and many of them would ordinarily be counted in Utah. For uniformity, Utah held that federal employees stationed overseas should be dropped from the apportionment counts. The U.S. District Court for Utah ruled against the state in April 2001, a ruling affirmed without comment by the U.S. Supreme Court later that year.[7]

C.4 THE 2004 OVERSEAS CENSUS TEST

Following the Utah challenge, as well as the Bureau's required report to congressional appropriators on issues of counting Americans overseas, plans developed for an overseas enumeration test in July 2004. For the 2004 test, the Bureau chose France, Kuwait, and Mexico as test sites. Following procedures like those used in the 1960 and 1970 censuses, the test relied on citizens who lived abroad to obtain a form from a U.S. embassy or consular office. The Bureau mounted some publicity efforts in English-language newspapers and media in the test countries. In all, the Bureau estimated that the planning, execution, and analysis of the test cost $7.8 million.

The Government Accountability Office issued two reports on the overseas test, one noting general concern about the direction and utility of the test (U.S. General Accounting Office, 2004) and the other analyzing the test's results (U.S. Government Accountability Office, 2004a). The reports concluded that the test suffered from design shortcomings that fundamentally

[6]House Report 106-1005 (conference report covering a range of federal departments and agencies, including the Department of Commerce), pp. 256–257.

[7]Even as their overseas challenge was in process, Utah filed a second legal challenge—like the first, known as *Utah v. Evans*—claiming that certain types of imputation for nonresponse used by the Census Bureau constituted statistical sampling and were thus prohibited by a 1999 Supreme Court ruling. The case was appealed to the U.S. Supreme Court and decided in favor of the Census Bureau on June 20, 2002 (*Utah v. Evans*, 536 U.S. 452). See National Research Council (2004c:Box 2.2 and generally) for additional detail on the case and imputation methods in the 2000 census.

limited its capacity to answer key research questions about the feasibility and cost of gathering data on Americans living abroad; they also questioned the resulting data quality. The response rates on the test were extremely poor: the Bureau had printed 520,000 questionnaires for the test, yet only 5,390 questionnaires (1,783 paper and 3,607 on the Internet) were returned from all three sites. By comparison, the July 1999 State Department estimates suggested that 1,036,300 American citizens reside in Mexico, 101,750 in France, and 7,710 in Kuwait.

In carrying out the test, the Bureau experienced some country-specific problems that would likely be more significant if the enumeration were conducted in additional countries. Perhaps most significant was the problem experienced with collecting even short-form census information in France; under French privacy laws, collection of data on race and ethnicity is generally prohibited. The Bureau also had difficulty overseeing the contractor responsible for raising public awareness of the test. The U.S. Government Accountability Office (2004a) commented that the approach used to test the viability of counting this group by means of a voluntary survey that relied on marketing to ensure a complete count would be very costly and yield poor results.

At the outset of planning efforts for the 2004 overseas test, a follow-up test in 2006 had been scheduled. However, on the basis of the 2004 test results, funds for the 2006 test were not provided in the Bureau's appropriation.

C.5 CONCEPTS IN COUNTING AMERICAN CIVILIANS OVERSEAS

The U.S. General Accounting Office (2004:8) usefully summarizes the basic "logistical, conceptual, policy, and other questions that surround the counting of overseas Americans." Similar issues were also raised by then-Census director Kenneth Prewitt in his testimony on the matter prior to the 2000 census (U.S. House of Representatives, Committee on Government Reform, 2000:44–46):

- *Who should be counted?* U.S. citizens only? Foreign-born spouses? Children born overseas? Dual citizens? American citizens who have no intention of ever returning to the United States? Naturalized citizens?
- *How should overseas Americans be assigned to individual states?* For certain purposes, such as apportioning Congress, the Bureau would need to assign overseas Americans to a particular state. Should one's state be determined by the state claimed for income tax purposes? Where one is registered to vote? Last state of residence before going overseas? These and other options all have limitations that would need to be addressed.

- *How should the population data be used?* To apportion Congress? To redistrict Congress? To allocate federal funds? To provide a count of overseas Americans only for general informational purposes? The answers to these questions have significant implications for the level of precision needed for the data and ultimately, the enumeration methodology.
- *How can the Bureau verify U.S. citizenship?* Administrative records such as passports and Social Security data have limitations. For example, Americans can reside in Mexico and Canada without a passport and many Americans overseas do not have Social Security numbers, especially dependents.
- *How can the Bureau ensure a complete count without a master address list?* The foundation of the stateside decennial census is a master address list. Because the list is essentially the universe of all known living quarters in the United States, the Bureau uses it to deliver questionnaires, follow up with nonrespondents, determine vacancies, and determine individuals the Bureau may have missed or counted more than once. The Bureau lacks a complete and accurate address list of overseas Americans. Consequently, these operations would be impossible and the quality of the data would suffer as a result.
- *Can administrative records be used to help locate and count overseas Americans?* Administrative records such as passport and visa files, voter registration forms, as well as records held by private companies and organizations have the potential to help the Bureau enumerate Americans abroad. However, the accuracy of these records, the Bureau's ability to access them, confidentiality issues, and the possibility of duplication all remain open questions.
- *Do certain countries have requirements that could restrict the Bureau's ability to conduct a count?* According to the Bureau, in planning the overseas test, the Bureau was informed that French privacy laws prohibit asking about race and ethnicity, two questions that are included on the U.S. census questionnaire. Although the Bureau worked with French officials to address this problem, the extent to which the Bureau will encounter restrictions in other countries, or whether other countries will cooperate with the Bureau at all, is unknown.

Still, the issue of counting all Americans overseas has been contentious in the past two decennial censuses, and the directions of modern business—and the degree to which advances in transportation and technology continue to make the world seem a smaller place—suggest that the issue will endure. Students continue to take advantage of opportunities to study abroad; jobs in foreign countries remain attractive possibilities, and the nature of modern businesses can lead to employee's living in foreign sites for weeks or more. Generally, questions may be raised as to why federal government employees should be counted but not American employees of multinational companies.

Mills (1993:1) soundly notes a "major observation" that arises from reviewing the history of overseas Americans in the census—namely, "the lack of a single conceptual thread running through the censuses concerning how Americans abroad fit into the overall decennial enumeration. It was partly this absence that led to the inconsistencies—evident in this report—in census treatment of Americans overseas." The voluntary survey approach tried by the Bureau in 1960 and 1970—and retested in 2004—seems unlikely to produce satisfactory results. Much as is the case for prisoners and other domestic institutional populations, the quality of enumeration of overseas Americans will likely depend on the use and reliability of administrative records. In addition to the records maintained by the Defense and State Departments, one might also need to consider the employee rosters maintained by companies and religious organizations, as well as other government sources (e.g., Social Security Administration rosters of American retirees receiving checks in other nations). A useful test would involve assembling these data sources (under appropriate confidentiality agreements) and studying the resulting picture of overseas Americans, possibly comparing the results with survey measures.

– D –

Biographical Sketches of Panel Members and Staff

Paul R. Voss is emeritus professor of rural sociology at the University of Wisconsin–Madison and has served as director of the university's Applied Population Laboratory. He previously served as assistant director of the Roper Public Opinion Research Center. His research interests are in applied demography, including small-area demographic models of population estimation and projection, as well as human migration and environmental demography and the analysis of spatial statistical data. He served as a member of the Committee on National Statistics' Panel on Estimates of Poverty for Small Geographic Areas and has written extensively on the use and applicability of census and American Community Survey data in small communities. He has served on the U.S. Department of Commerce's Decennial Census Advisory Committee as representative from the Population Association of America as well as the Census Bureau's advisory committee of professional associations. He received masters' and Ph.D. degrees in sociology (demography) from the University of Michigan.

Jorge Chapa is director of the Center on Democracy in a Multiracial Society at the University of Illinois at Urbana-Champaign. From 1999 to 2006, he was professor and founding director of the Latino Studies Program at Indiana University. Previously, he held research and faculty appointments at the Julian Samora Research Institute at Michigan State University, the Lyndon B. Johnson School of Public Affairs at the University of Texas, and the Tomas

Rivera Center in San Antonio, Texas. His research has focused on Latino educational, occupational, and economic mobility, and has included extensive analysis of census data. He has written extensively on Latino immigration patterns and has reported on housing patterns and dynamics of *colonia* communities along the U.S.-Mexico border. He served on the Census Bureau's Advisory Committee on the Hispanic Population for many years and served on the editorial board of the *Encyclopedia of the U.S. Census* (Anderson, 2000). As a member of committees established by the Texas Higher Education Coordinating Board, he helped develop the Texas "Ten Percent Plan" on university admissions and contributed to an amicus curiae brief to the U.S. Supreme Court in *Gratz v. Bollinger* (2003). He received masters' degrees in sociology and demography and a Ph.D. in sociology from the University of California, Berkeley.

Daniel L. Cork is a senior program officer for the Committee on National Statistics, currently serving as study director of the Panel on Residence Rules in the Decennial Census and the Panel on the Feasibility, Accuracy, and Technical Capability of a National Ballistics Database. He previously served as co-study director of the Panel on Research on Future Census Methods and program officer for the Panel to Review the 2000 Census. His research interests include quantitative criminology, particularly space-time dynamics in homicide; Bayesian statistics; and statistics in sports. He holds a B.S. degree in statistics from George Washington University and an M.S. in statistics and a joint Ph.D. in statistics and public policy from Carnegie Mellon University.

Don A. Dillman is regents professor and the Thomas S. Foley distinguished professor of government and public policy in the Departments of Sociology and Community and Rural Sociology at Washington State University. He also serves as deputy director for Research and Development in the Social and Economic Sciences Research Center (SESRC), and he founded the SESRC's Public Opinion Laboratory, one of the first university-based telephone survey laboratories in the United States. His current research emphasizes how visual design and layout influences respondent answers to self-administered surveys. He served as the senior survey methodologist in the Office of the Director, U.S. Bureau of the Census, leading the development of new questionnaire designs and procedures for the 2000 decennial census and other government surveys. He is a past president of the American Association of Public Opinion Research, a fellow of the American Association for the Advancement of Science, the American Statistical Association, and 2002 recipient of the Lester F. Ward Award for Distinguished Contributions to Applied Sociology of the Society for Applied Sociology. He holds a master's degree in rural sociology and a Ph.D. in sociology from Iowa State University.

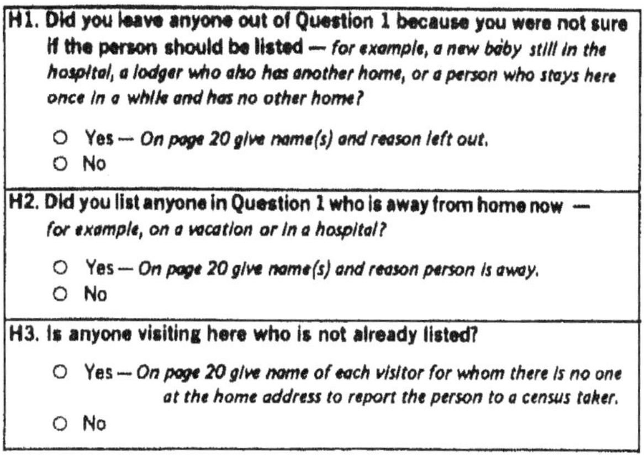

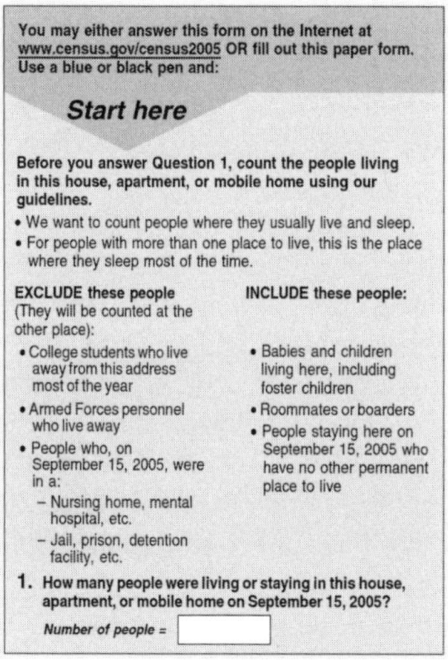

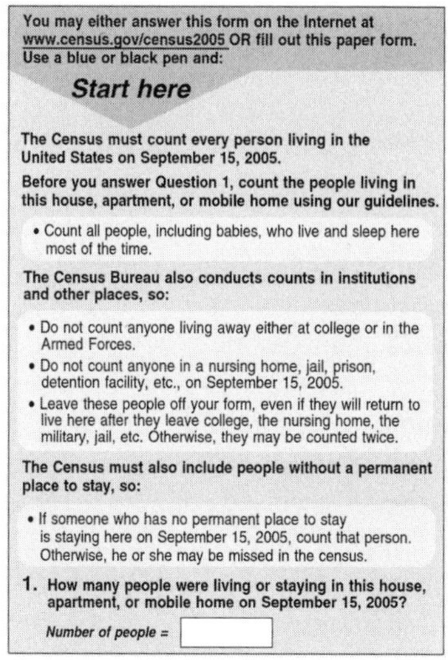

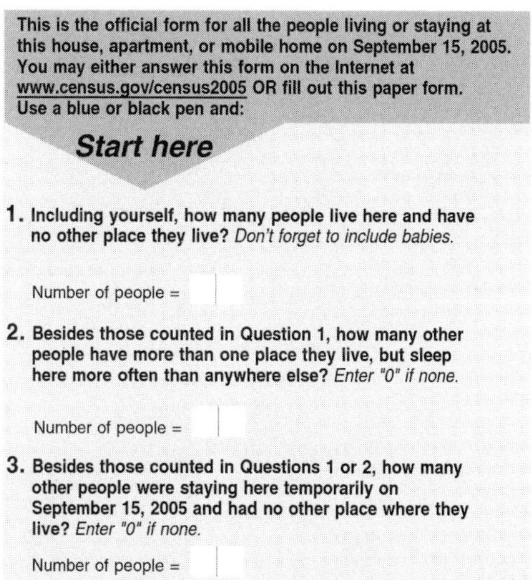

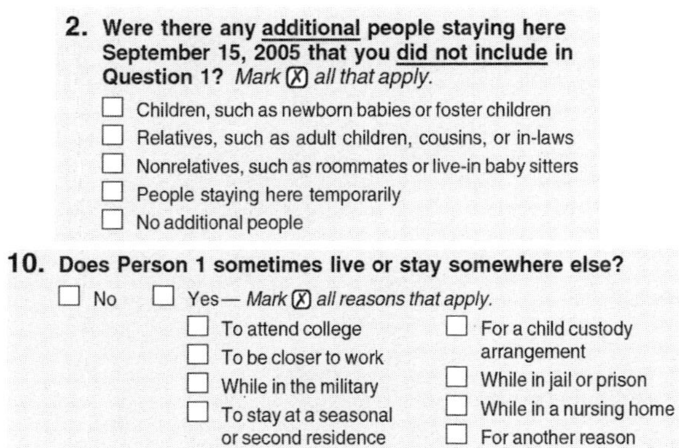

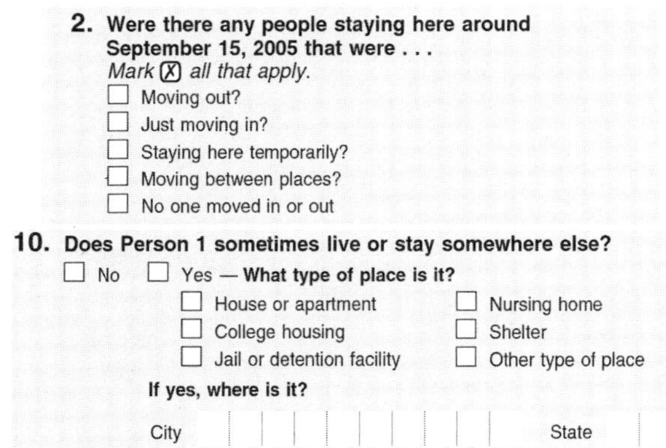

Kathryn Edin is associate professor of sociology at the University of Pennsylvania and an associate fellow of the Institute for Research on Poverty at the University of Wisconsin–Madison and of the Joint Center for Poverty Research of Northwestern University and the University of Chicago. She previously held positions at the Institute for Policy Research at Northwestern University, the Population Studies Center at the University of Pennsylvania, the Center for Urban Policy Research at Rutgers University, and the Russell Sage Foundation. Her principal research interests include poverty and social inequality, urban and community sociology, and family and gender. She is currently principal investigator or co-principal investigator on research projects on the social role of children and marriage among low-income adults, the role of local labor markets and child support programs in affecting fathers' economic and emotional involvement with their children, and couple dynamics among low-income married and unmarried couples with young children. She served as a member of the Committee on National Statistics Panel on Data and Methods for Measuring the Effects of Changes in Social Welfare Programs. She received her masters' and Ph.D. degrees in sociology from Northwestern University.

Colm A. O'Muircheartaigh is professor in the Irving B. Harris Graduate School of Public Policy Studies and vice president for statistics and methodology in the National Opinion Research Center at the University of Chicago. His research encompasses survey sample design, measurement errors in surveys, cognitive aspects of question wording, and latent variable models for nonresponse. He has served as a consultant to a wide range of public and commercial organizations in the United States, the United Kingdom, Ireland, Italy, the Netherlands and elsewhere. Through his work with the United Nations and other international organizations, he has worked in China, Myan Mar (Burma), Kenya, Lesotho, and Peru. Prior to moving to Chicago, he was founding director of the Methodology Institute of the London School of Economics and Political Science. He has served as president of the International Association of Survey Statisticians and is chair-elect of the Social Statistics Section of the American Statistical Association. He is a fellow of the Royal Statistical Society and the American Statistical Association and is an elected member of the International Statistical Institute.

Judith A. Seltzer is professor of sociology at the University of California, Los Angeles. Previously, she was on the faculty of the University of Wisconsin, where she contributed to the development and implementation of the National Survey of Families and Households. Her research interests include kinship patterns, intergenerational obligations, relationships between nonresident fathers and children, and how legal institutions and other policies affect family change. As part of a cross-university consortium of researchers, she is developing new models for explaining family change and variation, in which family dynamics and residence patterns will be important components. She has also participated in the Los Angeles Family and Neighborhood Survey. She has collaborated on research to improve the quality of data on children's living arrangements, transfers, and contact with nonresident parents using information from surveys and administrative data. She received her master's and Ph.D. degress in sociology from the University of Michigan.

C. Matthew Snipp is a professor in the Department of Sociology at Stanford University. He has written extensively on American Indians and has written specifically on the interaction of American Indians and the U.S. census. He was a member of the CNSTAT Panel on Research on Future Census Methods, which reviewed early plans for the 2010 decennial census. He has served on the Census Bureau's Technical Advisory Committee on Racial and Ethnic Statistics and the Native American Population Advisory Committee, both of the U.S. Census Bureau. He received his M.S. and Ph.D. in sociology from the University of Wisconsin.

Roger Tourangeau is director of the Joint Program in Survey Methodology at the University of Maryland and a senior research scientist at the University of Michigan. Previously, he was a senior methodologist at the Gallup Organization, where he designed and selected samples and carried out methodological studies, and at the National Opinion Research Center, where he founded and directed the Statistics and Methodology Center. His research focuses on attitude and opinion measurement and on differences across methods of data collection; he also has extensive experience as an applied sampler, and is well known for his work on the cognitive aspects of survey methodology. He is a fellow of the American Statistical Association and has served on the editorial board of *Public Opinion Quarterly* and on the Census Joint Advisory Panel. He has a Ph.D. in psychology from Yale University.

Index

A

F

G

H

V

W

Y